Pigments in Fruits

FOOD SCIENCE AND TECHNOLOGY

A SERIES OF MONOGRAPHS

Series Editor

Bernard S. Schweigert
University of California, Davis

Advisory Board

S. Arai
University of Tokyo, Japan

C. O. Chichester
Nutrition Foundation, Washington, D.C.

J. H. B. Christian
CSIRO, Australia

Larry Merson
University of California, Davis

Emil Mrak
University of California, Davis

Harry Nursten
University of Reading, England

Louis B. Rockland
Chapman College, Orange, California

Kent K. Stewart
Virginia Polytechnic Institute
and State University, Blacksburg

A complete list of the books in this series appears at the end of the volume.

Pigments in Fruits

Jeana Gross
Hebrew University of Jerusalem, Israel

1987

ACADEMIC PRESS

Harcourt Brace Jovanovich, Publishers

London   Orlando   San Diego   New York   Austin
Boston   Sydney   Tokyo   Toronto

ACADEMIC PRESS INC. (LONDON) LTD.
24–28 Oval Road
London NW1 7DX

U.S. Edition published by
ACADEMIC PRESS INC.
Orlando, Florida 32887

Copyright © 1987 by
ACADEMIC PRESS INC. (LONDON) LTD.

All rights Reserved

No part of this book may be reproduced in any form by photostat, microfilm, or any other means, without written permission from the publishers

British Library Cataloguing in Publication Data

Gross, Jeana
  Pigments in fruits—(Food science and technology)
  1. Plant pigments   2. Fruit—Physiology
  I. Title   II. Series
  664'8   QK898.P7

ISBN 0–12–304200–3

Printed in Great Britain at the Alden Press, Oxford

Preface

One of nature's splendours is its display of colours. When fruits ripen, their wonderful colours in rainbow shades appear and stand out against the contrasting green of the surrounding leaves. The role of these yellows, oranges, reds, blues and purples is to attract insects and other animals, thereby ensuring seed dispersion. Thus, attractive coloration is supposed to be the mechanism used by plants for the purpose of propagation of the species.

Equally attracted by the brilliant colours, mankind responded to their appeal and gradually chemists and biologists discovered that these natural colours are due to plant pigments.

A pigment is a chemical substance capable of absorbing visible light due to the action of a special part of the molecule called the chromophore (Greek for "carrying colour"), its colour being complementary to that of the absorbed light.

Fruit colours are produced by three main classes of pigments, the green chlorophylls, the yellow to red carotenoids and the red, blue and violet anthocyanins. The chlorophylls are porphyrins, containing tetrapyrrole rings. The carotenoids are isoprenoid polyenes and the anthocyanins are a sub-class of flavonoids, the structure of which is a pyran ring bonding together two benzene rings.

This book deals separately with each pigment category. Structure, physical and chemical properties and analytical methods are discussed first. This is followed by a description of the location, biosynthesis of pigments and biochemical changes which occur during early fruit development, maturation and senescence and, finally, factors affecting these changes.

The emphasis is on the qualitative and quantitative distribution of pigments in the most widespread edible fruits, which are given in tabulated form in the Appendix.

To Harry, Ruth, Asaf and Amit

Acknowledgements

My greatest debt is to Professor F. Lenz of the Institut für Obstbau und Gemüsebau der Universität Bonn, West Germany, who, aware of the absence of a comprehensive book on the subject of pigments in fruits, encouraged me to try to fill this gap. Professor Lenz also kindly read the completed manuscript and made useful comments on it.

I am grateful to the Deutsche Forschungsgemeinschaft for the financial help which allowed me to devote my time to the writing of this book.

Many thanks are also due to Professor I. Ohad—Department of Biological Chemistry, Hebrew University of Jerusalem, Israel—who commented on the chapter on "Chlorophylls" and to Professor R. Ikan—Department of Organic Chemistry, Laboratory of Natural Products, Hebrew University of Jerusalem, Israel—who went over the chapter on "Anthocyanins".

Contents

Preface v

Acknowledgements vii

1 Chlorophylls
- I Introduction 1
- II Structure 1
- III Properties 2
- IV The chloroplast 9
- V Spectroscopic properties of chlorophyll *in vivo* 14
- VI Chlorophyll biosynthesis 14
- VII Chlorophyll degradation 19
- VIII Chlorophyll in fruits 21
- IX Chlorophyll changes during fruit ripening 22
- X Fruits containing chlorophyll in the pulp when ripe 27
- XI Photosynthesis in developing fruits 37
- XII The role of phytohormones in fruit ripening 41
- XIII Other factors influencing fruit chlorophyll content 55

2 Anthocyanins
- I Introduction 59
- II Structure 59
- III Properties 64
- IV Biosyntheis of anthocyanins 71
- V Anthocyanins in fruits 74
- VI Characterization of fruit anthocyanins 76
- VII Pigment changes during ripening 79
- VIII Factors influencing anthocyanin levels in fruits 82

3 Carotenoids
- I Introduction 87
- II Definition and nomenclature 88
- III Structure and classification 90
- IV Physical properties 92

V	Stereochemistry 98	
VI	Chemical reactions 101	
VII	Analytical methods 104	
VIII	Biosynthesis of carotenoids 111	
IX	Function of carotenoids 116	
X	Use of carotenoids 121	
XI	Localization of carotenoids 122	
XII	Carotenoids in fruits 126	
XIII	Carotenoids of various fruits 136	
XIV	General considerations on carotenoid distribution	162
XV	Importance of carotenoids in taxonomy 163	
XVI	Carotenoid changes during fruit ripening 166	
XVII	Changes of the carotenoid pattern during ripening	169
XVIII	Plastid changes during ripening 175	
XIX	Xanthophyll esterification during ripening 177	
XX	Carotenoids as an index of maturity 178	
XXI	Other carotenoid changes during ripening 180	
XXII	Factors affecting carotenoid biosynthesis in fruits 182	

Appendix: Pigment Distribution in Fruits   187

References   259

Index   289

1
Chlorophylls

I. Introduction

The green plants owe their colour to the most important plant pigments, the chlorophylls, which participate in the fundamental life process of photosynthesis, the transformation of light energy into chemical energy.

All green plants contain chlorophylls a and b. These two pigments were separated for the first time in 1906 by Tswett, who achieved this by passing a leaf extract through a column of precipitated chalk (Tswett, 1906). This method is the basis of chromatographic adsorption analysis. A vertical column is packed with an adsorptive material such as alumina or powdered sugar, an extract of the plant pigments is added to the top of the column and the pigments are washed through the column with an organic solvent. The pigments separate into a series of discrete coloured bands on the column, divided by regions free of pigments. Thus a "chromatogram" is formed, always consisting of two green and four yellow pigments, universally found in every photosynthetic tissue. The two green pigments are chlorophylls a and b. Chlorophyll a is the less polar and of blue-green colour while the more polar chlorophyll b is yellow-green and is retained more strongly at the top of the column. The pioneering work on the chlorophylls' structure elucidation was initiated by Willstätter and Stoll (1913) and continued by Fischer and co-workers (Fischer and Wenderoth, 1940).

II. Structure

Chlorophylls are porphyrins. In porphyrins the basic skeleton is a tetrapyrrole ring obtained by the joining together of four pyrrole residues by methine groups (—C≡) (Fig. 1.1). In chlorophyll a the four nitrogen atoms are coordinated with the Mg^{2+} ion. The chlorophyll contains an additional ring (E) between C_6 and the γ bridge. At C_7 a long hydrophobic side-chain deriving

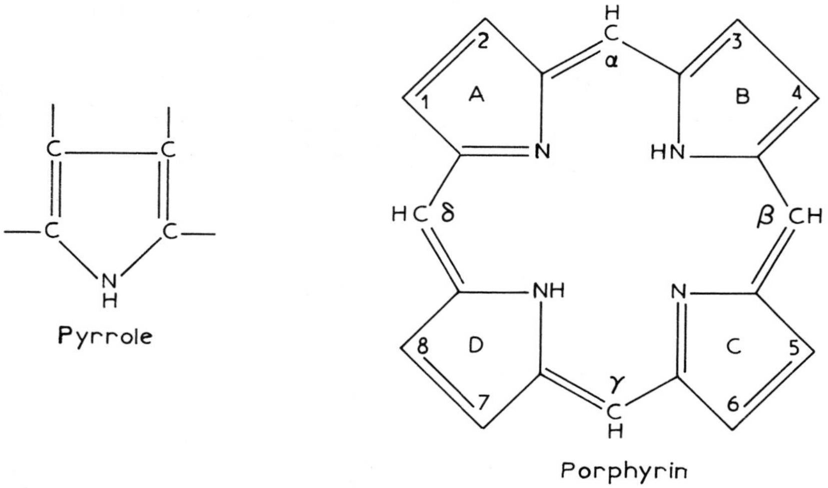

Fig. 1.1 Pyrrole and the porphyrin skeleton.

from phytol ($C_{20}H_{39}OH$) is esterified to a propionic acid. The rest of the molecule is hydrophilic (Fig. 1.2). Chlorophyll *b* differs from chlorophyll *a* only in having an aldehyde group (—CHO) in place of the methyl group at position 3 (Fig. 1.2).

III. Properties

A. Absorption spectra

The chlorophylls are of green colour because they absorb strongly in the blue and red regions of the spectrum. They also exhibit weaker bands at lower wavelengths around 600 and 500 nm. The intense band at around 400 nm is called the Soret band. The small difference in the structures of the two chlorophylls produce differences in the absorption spectra, the absorption maxima of chlorophyll *b* being shifted towards the green region of the spectrum. This produces the different green hues of the two pigments—blue-green for chlorophyll *a* and yellow-green for chlorophyll *b*, as observed when they are separated chromatographically. The chromophore of the chlorophylls, as in most organic compounds, is a system of conjugated double bonds that is a regular succession of single and double bonds between the carbon atoms and nitrogen atoms (Figs 1.1, 1.2).

The absorption maxima are not absolute but vary with the nature of the

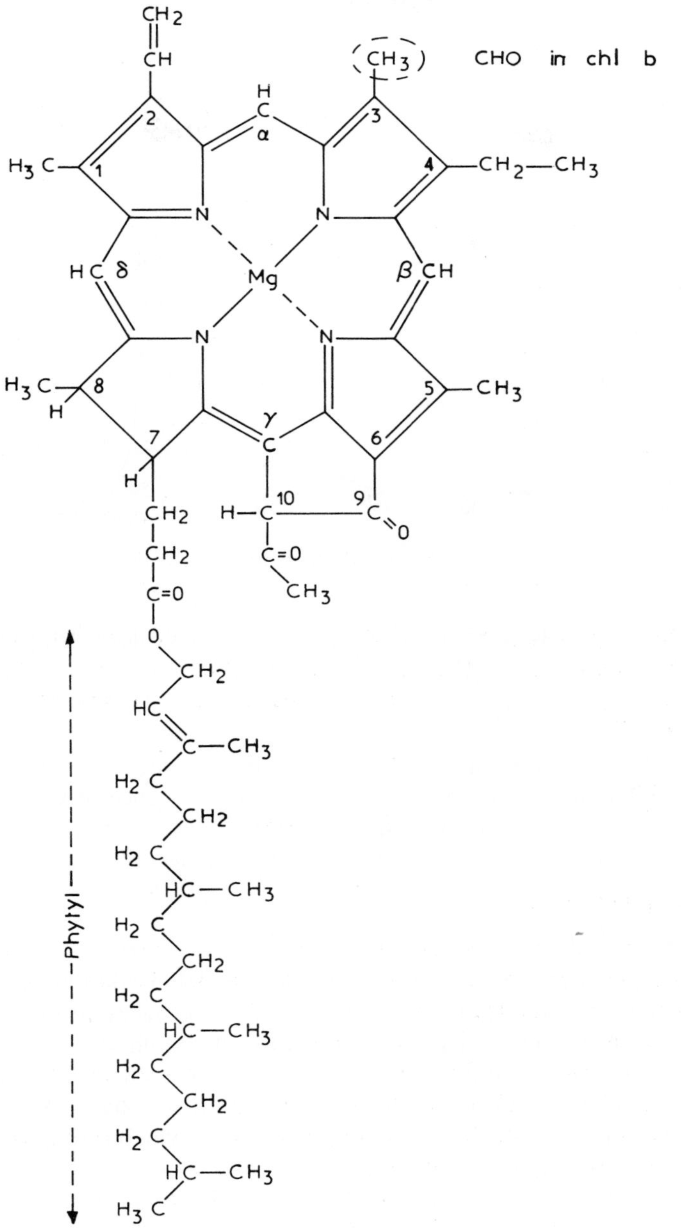

Chlorophyll *a*

Fig. 1.2 Structure of chlorophylls *a* and *b*.

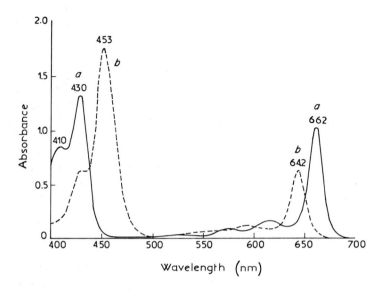

Fig. 1.3 Absorption spectra of chlorophylls *a* and *b* in diethyl ether. Chlorophyll *a* (———); chlorophyll *b* (– – –).

solvents. In solution both chlorophylls are fluorescent; however *in vivo* only chlorophyll *a* fluoresces. The absorption and fluorescence spectra *in vivo* are different as chlorophylls are found as complexes with proteins in the photosynthetic tissue. These will be discussed further (p. 14).

B. Chemical Properties

The most important characteristic is the extreme lability of the chlorophylls. They are sensitive to light, heat, oxygen and chemical degradation. The phytyl ester can easily be hydrolysed by dilute alkali or by the enzyme chlorophyllase commonly found in green plant tissue. The results of the hydrolysis are the free acid derivative chlorophyllide and the C_{20}-alcohol, phytol. Magnesium is lost by the action of dilute acids giving pheophytin, which through hydrolysis gives pheophorbide.

C. Analytical Methods

Owing to the lability of the chlorophylls a series of precautions must be taken to avoid their alteration. All the manipulations must be carried out rapidly in dim light to avoid photobleaching, avoiding heating, oxidation and preventing chlorophyllase action.

1. *Extraction*

The material is extracted in a homogenizer where grinding and extraction occur simultaneously. Solvents like acetone and methanol are used which break the chlorophyll–protein complex non-covalent linkage and extract the pigments quantitatively. Undiluted solvents also block the chlorophyllase activity. To prevent pheophytin formation the extract is generally neutralized with alkaline carbonates. For fruits tris-buffer ((tris-hydroxymethyl) aminomethane) gives better results. To obtain a clear solution the extracts are filtered by suction through a sintered glass filter or a layer of filter aid (Kieselguhr). Another possibility is to transfer the pigments into diethyl ether, a water immiscible solvent, by adding sodium chloride (Smith and Benitez, 1955).

(a) *Chlorophyll determination.* In these extracts chlorophyll can be determined using various methods based on spectrophotometry or spectrofluorometry. Colorimetric methods are based on the Lambert–Beer law, which is a relationship between the concentration of the substance and its colour intensity:

$$E = \alpha l c$$

The extinction E is proportional to α, the specific extinction coefficient of the substance (which is known), l, the length of the light path in cm and c the concentration in g l^{-1}. The extinction is usually measured at the absorption maximum of the substance.

The absorption of the total chlorophyll extract in the region between 600 and 700 nm where carotenoids do not absorb may be considered as that of a two-substance mixture. The measurement is based on the fact that the specific extinctions of the two chlorophylls are additive (Smith and Benitez, 1955). The estimation is not precise when pheophytin is present in higher amounts as its absorption maximum (abs. max.) is at 665 nm in ethyl ether.

The calculation is based on two equations in which the extinction of each chlorophyll is corrected by subtracting the contributed absorption of the second pigment. Simultaneous equations for the various solvents used are given using the respective specific extinction coefficients determined by Mackinney (1941).

For 80% acetone extracts the following equations given by Arnon (1949) are used:

chlorophyll *a* (mg l^{-1}) = 12.7 E_{663} − 2.69 E_{645}
chlorophyll *b* (mg l^{-1}) = 22.9 E_{645} − 4.68 E_{663}

For a diethyl ether extract the equations given are (Smith and Benitez, 1955):

chlorophyll a (mg l^{-1}) = 10.1 E_{662} − 1.01 E_{644}
chlorophyll b (mg l^{-1}) = 16·4 E_{644} − 2.57 E_{662}

When the absorption spectrum of the extract is recorded it is difficult to read the E value for chlorophyll b as its absorption maximum is located on the steep part of the curve. To avoid calculations nomograms which give a rapid solution of the above equations can be used.

There are methods in which the chlorophylls are determined after their conversion in pheophytins (Vernon, 1960). Other methods determine chlorophylls and their breakdown products in the same mixture by measuring the extinction before and after the conversion. For calculation a rather complicated system of suitable equations has been developed (White et al., 1963).

2. *Spectrofluorometric Methods*

A sensitive fluorescence method has been developed for the particular case when chlorophyll a (chl a) is accompanied by only small amounts of chlorophyll b (chl b) and the ratio chl a/b is high. The usual spectrophotometric methods are suitable when the ratio chl a/b is about 3, as found in normal plants (Boardman and Thorne, 1971).

The fluorescence technique, emission and excitation spectrophotometry, has recently been improved by making measurements at low temperatures (77 K) where high resolution can be obtained (Rebeiz and Lascelles, 1982).

3. *Non-destructive Methods*

In the previously described methods the pigments were extracted after the destruction of the plant tissue. Chlorophyll and other pigments may also be estimated directly in the tissue, leaving the sample intact.

For fruits the external colour is considered an important quality parameter. Consumer acceptance is first determined by colour and then by other attributes like texture and flavour. Maturity classifications are based on the amount and quality of certain pigments, as colour changes occur during ripening. This is important for establishing the optimal harvest time, and the optimal storage time for subsequent post-harvest changes.

External fruit colour and colour quality of pulp and juice may be assessed visually or by physical methods. Highly intricate instruments are available for measuring the light reflected from or transmitted through the sample. These are more reliable than the human eye because they are not afflicted by subjective factors.

A colour is characterized by three attributes, hue, lightness and saturation, which translated into a generally accepted system become hue, value and chroma. The relationship of these attributes can be explained by what is known as a colour solid which is a graphical representation of the three attributes of colour. In the Munsell solid, hue extends clockwise around a vertical axis which stands for lightness increasing upwards, while saturation extends horizontally. Each colour is characterized by three coordinates.

The visual colour rating is made by comparing with standardized colour atlases, charts, fans or special colour comparators which are adapted to match specific fruit. They all evolved from studies of the basic properties of colour.

The retina of the human eye has three types of colour receptor molecules, contained in the cone cells. Each pigment corresponds to a primary hue red, blue and green. A given coloured stimulus would evoke response in all three receptors and the pattern of these responses would determine the quality of the sensation (Francis, 1970). The tristimulus colorimeters simulate the human eye by replacing the receptor molecules with filters—one for each primary hue. The colour solid used by both Gardner and Hunter laboratory instruments is the Hunter solid. Its coordinates are $+a$ (red) to $-a$ (green), $+b$ (yellow) and $-b$ (blue) (Francis, 1970, 1980). For fruits the a/b ratio is usually computed. This ratio is negative for green fruits, approximately 0 for yellow fruits and positive for orange fruits.

For determination of reflectance and transmittance, the wavelengths at which the highest correlations exists between the measurements and the fruit colour are chosen. Reflectance measurements determined with a spectrophotometer in citrus leaves showed a maximum at 540 nm which was inversely proportional to the total chlorophyll concentration (Turrell et al. 1961). For apples the wavelength at 678 nm is the most reliable, as here the strictest correlation exists between percentage reflectance measurements and fruit colour (Schulz, 1976).

Reflectance measurements at the wavelengths 648 and 674 nm allowed the changes in proportions of chlorophyll a and b during colour development in citrus fruit to be followed (Jahn and Young, 1976). For ripening tomatoes light reflectance and light transmittance measurements at 500, 540 and 595 nm were tested. It was found that light transmittance systems were more sensitive to small changes in colour than the light reflectance system. Measurements at the longest wavelength were highly correlated with visual colour ratings during ripening (Jahn, 1975).

A special reflectance spectrometer designed to measure green (546 nm) and red (640 nm) reflectance of a rotating object and to display the mathematical ratios of the two colour modes in percentages is especially suitable for changes in the developing fruits in which green and red are the dominant colours, as in the tomato (Kader and Morris, 1978).

New methods developed in Japan use the delayed light emission (DLE) as a means of automatic sorting of different fruits. The relationship between the chlorophyll content and the DLE is almost linear. The procedure involves keeping the sample in the dark for a certain period of time followed by exciting illuminance. A special apparatus has been devised by Chuma *et al.* (1982). A method using a portable colorimeter method for measurements of fruit peel and pulp in the field has been developed recently by Hürter (1982).

These methods represent a small sample of those in current use.

4. *Separation of Plant Pigments*

The spectrophotometric determination of chlorophylls permits the direct determination of both chlorophylls and does not require separation into individual pigments. The pigments of the green parts of plants however are always a mixture of two chlorophylls and at least four carotenoid pigments. A ready separation of these pigments was based on their different solubility in various solvents. By partition between two immiscible solvents methanol and carbon disulphide the two chlorophylls were separated as early as 1873 (Humphrey, 1980). This method is still used as well as the counter current separation, a complicated method based on partition between solvents (Holden, 1976).

(a) *Chromatography.* Chromatography is the most important technique for separating chemical substances on the basis of their relative adsorption rate on a stationary phase from a mobile phase, liquid or gas.

The technique invented by Tswett, described earlier, was intensively elaborated, stimulated by the general interest in plant pigments, resulting in the invention of several modern methods which enabled the discovery of hundreds of new natural substances. The history and modifications of Tswett's method has been amply reviewed by Strain and Sherman (1972).

As the lability of the pigments was gradually discovered, techniques were developed to avoid their alteration. For column chromatography, which may also be used preparatively, multiple adsorbents like alumina, magnesia and polyethylene are used. Organic adsorbents, particularly sucrose, are more widely used as they do not alter the chlorophylls. Column chromatography was first replaced by paper chromatography and subsequently by thin-layer chromatography, which is much more sensitive and therefore requires very small amounts of sample.

The separation on column and thin-layer chromatography is mainly based on adsorption. Partition chromatography, which involves liquid–liquid phases, is also used, the stationary phase being the water bound to the granules of the adsorbents. The solute molecules are partitioned between the stationary liquid phase and the mobile phase, which is an immiscible solvent. Methods of thin-layer chromatography of chlorophylls were described by Stahl (1967), and Bacon (1965) and others. A method based on partition chromatography worked out by Hager and Meyer-Bertenrath is still widely in use. The chromatographic methods (1962, 1966) have been reviewed by Šesták (1975).

Gradually the thin-layer chromatographic methods are being replaced by high performance liquid chromatography (HPLC), which is more rapid, more sensitive and highly reproducible. It can be used for preparative work and analysis. During the last few years reversed-phase liquid chromatography (rp-HPLC) has been introduced, providing a solution for the multiple separation problem. The technique relies on the hydrophobic interactions between the solute mobile phase and the stationary phase. The silica gel particles are chemically bonded so that the elution is similar to that in conventional partition system, the more polar compounds being eluted first (Eskins *et al.*, 1977; Prenzel and Lichtenthaler, 1982; Braumann and Grimme, 1981).

IV. The Chloroplast

The chlorophylls are located in small sub-cellular organelles, plastids, conferring to them green colour and their name, chloroplasts. They are the site of photosynthesis. Plastids are dealt with in the comprehensive book by Kirk and Tilney-Bassett (1978a). Chloroplasts are found mostly in leaves and also in other green tissues such as immature fruits.

The chloroplasts are lens-shaped bodies with a diameter about 7 μm and a thickness of 3 μm. The structure of the chloroplast has been intensively investigated by electron microscopy. It is surrounded by an envelope which is a double membrane. The envelope acts as a selective barrier to the transport of various metabolites into or out of the chloroplast. Additionally it plays a role in formation of new internal lamellae.

The chloroplast is filled with a matrix called stroma. Embedded within the stroma is a complex system of membranes or lamellae. The basic sub-units of the lamellae are double membranes closed in on themselves to form flattened sacks called "thylakoids", from the Greek *thylakos*, sack-like. The thylakoids occur in regular packed stacks called grana. There are 5–20 thylakoids in a granum. In the chloroplasts of shade plants there are 100 thylakoids per granum (Meier and Lichtenthaler, 1981). Similar high stacked grana occur in the chloroplasts of some fruits (Cran and Possingham, 1973; Flügel and Gross, 1982). About every other thylakoid of the granum extends on one or other side, interlinking the grana by the so-called stroma thylakoids. It follows from the existence of all these inter-thylakoid and inter-granal connections

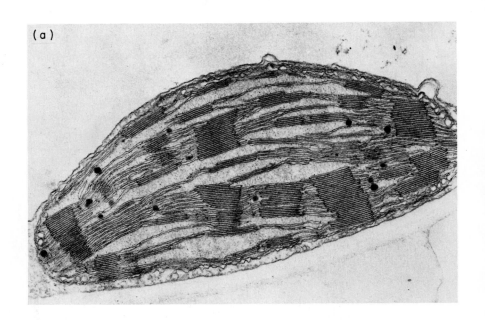

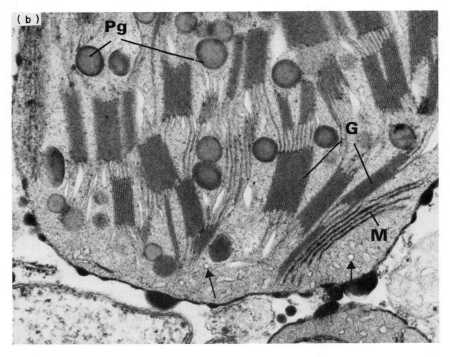

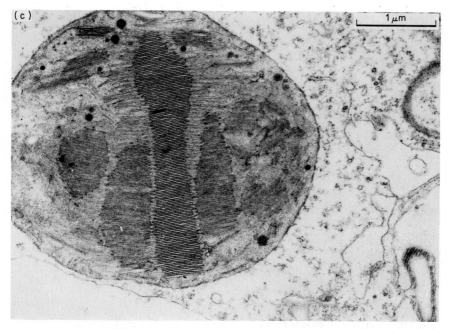

Fig. 1.4 Different chloroplasts. (a) Leaf chloroplast (mesophyll of maize leaf); (b) chloroplast of unripe fruit (flavedo of kumquat); (c) uncommon chloroplast of unripe fruit (green-fleshed melon peel). From (a) Vrhovec and Wrischer (1970), (b) Huyskens *et al.*, (1985) and (c) Flügel and Gross (1982).

that in fact the thylakoid system constitutes a single enormously complex membrane-enclosed cavity. The concept that the thylakoid is a separate membrane-bound space is now abandoned but is still used for interpreting chloroplast structure (Kirk and Tilney-Bassett, 1978b).

The stroma contains a number of particulate structures like the ribosomes and strands of DNA. Starch grains are also usually present. Stroma components which are present in all plastids are the osmiophilic globules or plastoglobuli which are apparent in sections fixed with osmium. They contain mainly the excess plastid quinones but no chlorophylls. Different types of chloroplasts are shown in Fig. 1.4.

A. The Chemical Composition of Chloroplasts

The water content of the chloroplast is about 70–80%, protein representing about 60% of the dry matter. There are two kinds of proteins: structural proteins which are bound in the lipoprotein complex of the lamellae and water-soluble proteins, the enzymes, which are found both in the complex and in the stroma. The lipids bound in the lamellae amount to c. 30–40% and the pigments, chlorophylls and carotenoids, to 5–10% of the total dry matter.

In higher plants chlorophylls *a* and *b* are present, the ratio *a/b* usually being between 2.5 and 3.5. The ratio of chl *a* to chl *b* is a common parameter with respect to the characterization of the physiological status of green plants and the differentiation of partial structures of the photosynthetic apparatus. Species from sunny habitats tend to have higher *a/b* ratio (3.5–4.5) than species from shady habitats (2.6–3.2) (Lichtenthaler, 1971). The increased proportion of chlorophyll *b* is due to its absorption properties, which differ from those of chlorophyll *a*. Absorbing strongly in the 450–480 nm range it partially fills the gaps in the chlorophyll *a* spectrum and thus captures effective light at low intensity.

The main carotenoids of the chloroplasts in higher plants are β-carotene, lutein, violaxanthin and neoxanthin. Both chorophylls and carotenoids exist in chloroplasts as chromoprotein complexes associated with the membrane structure.

B. Function of the Chloroplast

The function of the chloroplast is to capture light energy and use it for the conversion of carbon dioxide to carbohydrate. The photosynthetic process occurring in all oxygen evolving plants can be represented by the equation:

$$6\,CO_2 + 6\,H_2O \xrightarrow{light} C_6H_{12}O_6 + 6\,O_2$$

carbon dioxide   water   glucose   oxygen

In the process besides carbohydrates amino acids, proteins and other organic compounds are also synthesized and molecular oxygen is liberated. Photosynthesis is the source of all forms of food and oxygen in the Earth's atmosphere.

The complex photosynthesis process includes a photochemical or light stage and a dark stage that involves chemical reactions catalysed by enzymes. In the light reactions which take place in the thylakoid membrane system the energy of the sun is captured in two relatively unstable molecules: reduced nicotinamide dinucleotide (NADPH) and adenosine triphosphate (ATP). In the dark reactions which take place in the stroma of the chloroplast ATP and NADPH supply the energy and reducing power needed to form glucose from carbon dioxide.

The photosynthetic pigments all in the form of membrane–protein complexes are organized into two functional assemblies called photosystem I

(PS I) and photosystem II (PS II) which carry out two different light-dependent processes.

Most of the pigment complexes, about 99% of the total chlorophyll, function as antennae, light-harvesting complexes. They include both chlorophylls *a* and *b*, other accessory pigments and carotenoids, so that nearly the whole region of the spectrum is exploited. The incident photons absorbed produce an electronic excitation with is transferred ultimately to the pigment complex of the reaction centre where the actual conversion of light to chemical energy occurs.

In PS I the reaction centre contains chlorophyll *a* associated in a specific way with protein, having its absorption peak at 700 nm (P_{700}), while the reaction centre of PS II contains another form of chlorophyll *a* (P_{680}), absorbing at a shorter wavelength.

Photosystem II is concerned with the removal of hydrogen from water which is split into oxygen, protons and electrons, while photosystem I is concerned with the reduction of NADP. The electrons flow from PS II to PS I through a series of intermediate electron carriers consisting of proteins and lipids, of which most are known. During this electron flow, protons are moved from outside the thylakoid membrane to within it. ATP formation is linked to proton movement. The mechanism of the process of photophosphorylation in which ATP is formed from ADP is not yet totally understood.

C. The Photosynthetic Unit

The photosynthetic membranes are formed of repeated assembled photosynthetic units, each consisting of antenna chlorophyll–proteins and reaction centre complexes linked to one complete set of electron carriers. Even if the exact model is not well characterized, generally each reaction centre of each photosystem is associated with an intermediate small antenna and a second larger antenna which forms a separate entity.

There are at least four chlorophyll–protein complexes found in a photosynthetic unit. Two of them have been well characterized; namely the P_{700}-chl *a*–protein complex and the light-harvesting chl *a/b*–protein complex (Thornber *et al.*, 1976). The former represents the pigment molecules associated with the photosystem I reaction centre; the latter, associated with photosystem II, represents the major light-harvesting chl *a/b*–protein complex or antenna. Two additional complexes containing chlorophyll *a* are less well known, one of which is supposed to form the reaction centre of PS II (Anderson *et al.*, 1978). Several other minor light-harvesting and reaction centre chlorophyll proteins have been described recently (Lichtenthaler *et al.*, 1981; Braumann *et al.*, 1982).

The P_{700}-chl *a*–protein complex is ubiquitous in the photosynthetic plant kingdom. The light-harvesting chl a/b–protein complex transfers most of the absorbed light energy to the reaction centre of PS II. It contains all the chlorophyll b of the chloroplast together with an equimolecular quantity of chlorophyll a, accounting for more than 50% of the total chlorophyll.

Variations in its content, due to genetic or to environmental factors, are fully explained by changes in both chlorophyll contents and chlorophyll a/b ratio.

V. Spectroscopic Properties of Chlorophyll *In Vivo*

In vivo the chlorophylls are bound into lipoprotein membranes and occur there in several distinct spectroscopic forms. Chlorophyll a in living cells has a much broader absorption peak than in extracts, at about 676 nm. At low temperature (77 K) there is a superior resolution that permits the detection of five spectral forms of chl a with maxima at 662, 669, 677, 686 and 690 nm in the P_{700}-chl a–protein. In the light-harvesting chl a/b–protein four spectral forms with maxima at 649, 650, 670 and 677 nm, the shorter peak belonging to chlorophyll b, are found. The different forms are probably different aggregates formed in the thylakoid membrane.

The fluorescence emission spectra also give useful information on the state of organization of chlorophyll within the thylakoids. As with the absorption spectra, the resolution is superior at low temperatures.

The chlorophyll–protein complexes can be identified by their specific fluorescence emission peaks at 77 K. These are the chl a/b–protein complex or light-harvesting complex of photosystem II (λ_F 686 nm), possibly the reaction centre–antenna complex of photosystem II (λ_F 696 nm) and the reaction centre–antenna complex of photosystem I (λ_F 730–740 nm) (Papageorgiou, 1975).

In most photosynthetic tissue frozen in the light the fluorescence emission of PS I at 730 is higher than that of PS II. This is seen in Fig. 1.5, in which the fluorescence emission spectrum of a Chinese gooseberry (kiwi) leaf is shown (Gross and Ohad, 1983). Fluorescence spectroscopy may thus be used as a tool in the study of chlorophyll organization during the growth and ripening periods of fruit.

VI. Chlorophyll Biosynthesis

The biosynthesis of chlorophyll has been intensively investigated. The results have been recently reviewed by Schneider (1975) and Bogorad (1976).

The biosynthetic pathway for chlorophyll involves multiple steps each

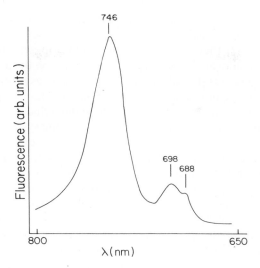

Fig. 1.5 Low temperature fluorescence emission spectrum of an intact leaf of the kiwi plant. Reprinted with permission from (*Photochem. Photobiol.*, **37**, Gross and Ohad), Copyright (1983), Pergamon Journals Ltd.

catalysed by specific enzymes (Fig. 1.6). The first step is the synthesis of the precursor δ-aminolevulinic acid (ALA) (Fig. 1.6(a)). The pyrrole precursor, porphobilinogen (PBG), is formed through condensation of two molecules of ALA (Fig. 1.6 (b)). The cyclic tetrapyrrole is formed through condensation of four PBG molecules with elimination of NH_3. The product is called uroporphyrinogen (urogen), which has one acetic and one propionic acid residue attached on each of the pyrrole rings. A mixture of isomers is obtained of which uroporphyrinogen III is the intermediate in chlorophyll biosynthesis (Fig. 1.6 (c)).

The next steps involve the decarboxylation of the acetic side-chains to methyl groups and the oxidative decarboxylation of the propionic acid residues to vinyl groups at rings A and B. After a subsequent oxidation the macrocycle protoporphyrin is formed (Fig. 1.6 (d)). Protoporphyrin may chelate with metals; in the animal kingdom with Fe leading to the formation of haemoglobin, in plants with magnesium forming Mg protoporphyrin. Through multiple transformations Mg protoporphyrin is transformed into protochlorophyllide with the cyclopentanone ring (E) characteristic for chlorophylls.

The following step, the most investigated, is the conversion of protochlorophyllide into chlorophyllide *a*, which implies the photoreduction of the ring D.

Thus the chromophore to which chlorophyll owes its green colour is formed (Fig. 1.6 (e)). Etiolated seedlings which originate from seeds germinated in the dark are not green as they do not accumulate chlorophyll, but contain small amounts of protochlorophyllide. Through illumination the absorption maximum *in vivo* at 650 nm is shifted within 30 s to 678 nm, as protochlorophyllide is transformed into chlorophyll. In fruits which contain chlorophyll in the pulp the light regime may be sufficient to allow the reduction to take place or they may contain a special adapted reductive enzyme system.

The next step is the esterification of the 7-propionic acid residue of chlorophyllide with phytol, by chlorophyllase. Chlorophyllase is considered to have both biosynthetic and catabolic activity. *In vitro* the equilibrium of the reaction can be manipulated shifting it in one sense or another (Moll and Stegwee, 1978). However the most obvious *in vitro* activity is that chlorophyllase catalyses the removal of phytol from chlorophylls (Holden, 1976). *In vivo* chlorophyllase is known especially as a splitting enzyme for both chlorophylls and pheophytins (Schneider, 1975). In etiolated leaves transferred to light

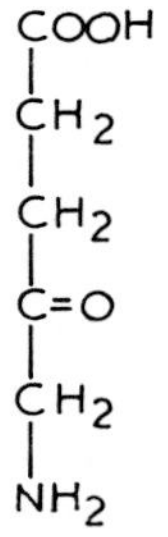

Fig. 1.6(a) δ-Aminolevulinic acid (ALA).

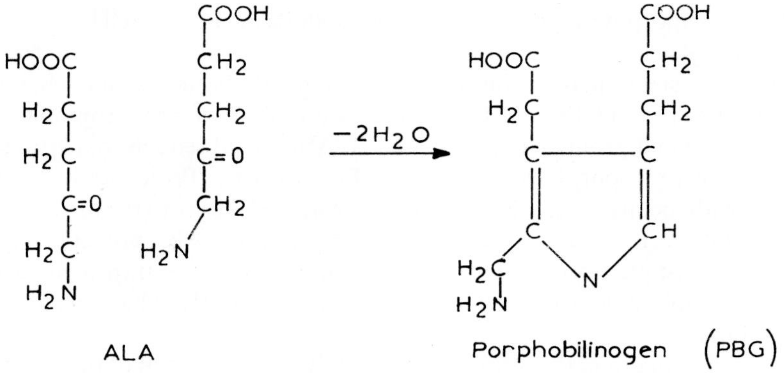

Fig. 1.6(b) Formation of porphobilinogen (PBG).

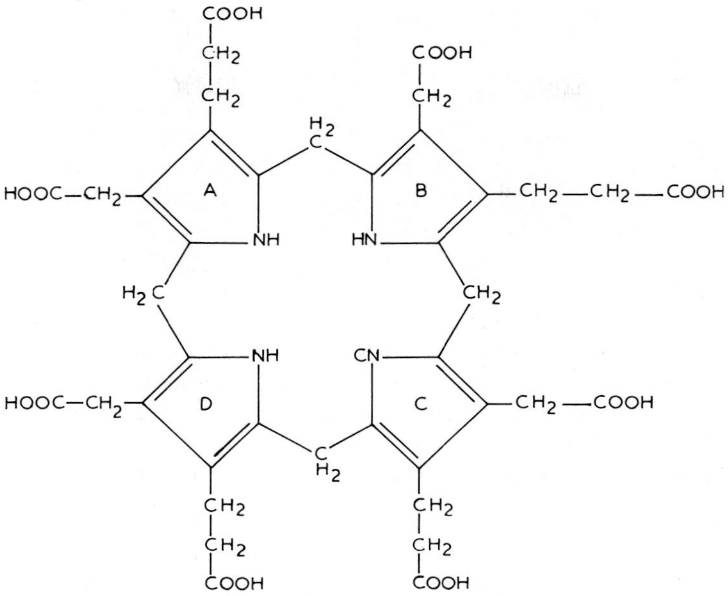

Fig. 1.6(c) Uroporphyrinogen III

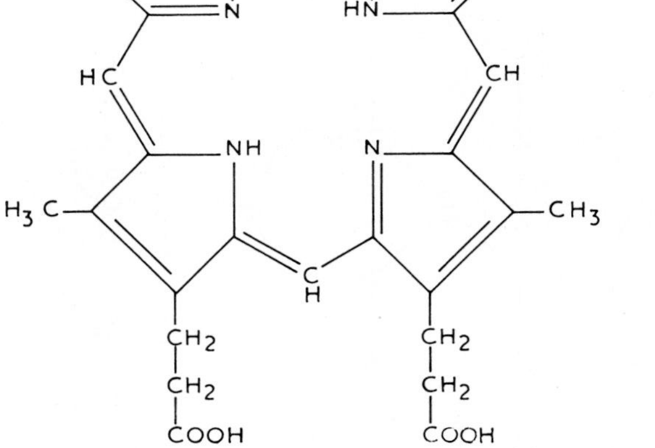

Fig. 1.6(d) Protophorphyrin ix

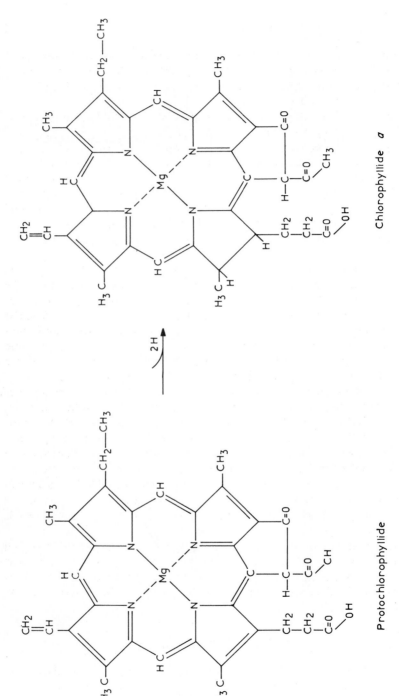

Fig. 1.6(e) Reduction of protochlorophyllide to chlorophyllide *a*.

chlorophyllase activity increased, paralleling the accumulation of chlorophyll (Holden, 1976). In tobacco plants chlorophyllase activity is highest in 51-day old plants and begins to decline before the maximum chlorophyll content is reached (Shimizu and Tanaki, 1963).

Phytylation of chlorophyllide may involve reaction with geranylgeranyl pyrophosphate followed by reduction of the side-chain to a phytyl group. Recently in etiolated leaves an enzyme which was named chlorophyll synthetase was shown to catalyse the esterification of chlorophyllide with geranylgeraniol. Sequential steps leading to chlorophyll-phytol involve the gradual reduction of the esterifying alcohol first to dihydrogeranylgeraniol and then to tetrahydrogeranylgeraniol and finally to phytol. The intermediate chlorophyllide esters were isolated and the activity of the chlorophyll synthetase was partially characterized (Schoch et al., 1977; Rüdiger et al., 1980).

Chloroplast and chromoplasts isolated from pepper fruits (*Capsicum annuum* L.) have been shown to have chlorophyll synthetase activity. The reduction of geranylgeraniol to phytol depends markedly on the concentration of NADPH in the medium (Dogbo et al., 1982). The results confirm the early findings of Costes (1966) that in leaves geranylgeraniol is the main precursor of phytol in chlorophyll *a*. Further investigations will permit the elucidation of the chlorophyllase activity, which in fruits seems to be either totally fortuitous or not yet understood, as will be discussed later.

The last step in the biosynthetic pathway of chlorophyll is the formation of chlorophyll *b*. Chlorophyll *b* is supposed to be formed through the oxidation of the 3-methyl group of the B ring to a formyl group. Indirect evidence for this is the fact that chlorophyll *b* is formed only when chlorophyll *a* is present. Chlorophyllide *b* has never been isolated and there are no plants containing solely chlorophyll *b*. During greening chlorophyll *a* is the first product.

Recently the occurrence of several novel chlorophyll *a* and *b* chromophores in higher plants was reported. The various chromophores differ by the degree of oxidation of their side-chains at the 2 and 4 positions of the macrocycle (Rebeiz et al., 1980).

It is assumed with certainty that all the enzymes necessary for chlorophyll biosynthesis are located in the chloroplast as chlorophyll biosynthesis and accumulation is directly connected to plastid development and differentiation. The changes in amounts of chlorophyll during leaf ontogenesis has been reviewed by Šesták (1977).

VII. Chlorophyll Degradation

While the biosynthetic pathway of chlorophyll is almost completely understood, knowledge of chlorophyll degradation is limited and the catabolism is only outlined hypothetically. Consequently the few comprehensive reviews on the subject are "more an inventory of the lack than of sound knowledge" in the field (Kufner, 1980).

The principal steps would result in the release of chlorophyll from the chlorophyll–protein complex. Chlorophyll and its magnesiumless derivative pheophytin are then split into phytol and chlorophyllides, respectively phaeophorbides. Then follows their oxidation in chlorins and finally the destruction of the porphyrin ring into colourless low molecular compounds (Simpson et al., 1976). The question of the reutilization of these products is also hypothetical.

During fruit ripening and senescence of green leaves in autumn, the green colour disappears as chlorophyl is disintegrated. During summer the chlorophyll content found in leaves is the result of steady chlorophyll synthesis and degradation. During the autumn coloration period preceding leaf desiccation and fall, chlorophyll decays rapidly, producing low levels of pheophytin with only occasional faint traces of pheophorbide and chlorophyllide at the period of most rapid chlorophyll breakdown (Sanger, 1971). In some yellow leaves phytol-containing fragments derivable from ring D have been detected (Park et al., 1973).

Breakdown and synthesis of thylakoid membranes and their lipids in leaves also occur during the natural day–night growth of plants. The turnover however is not visible since the decomposition in the night is recompensed by new synthesis during the day. The biological half-time ($\tau_{1/2}$) has been calculated to have values in the range 2.5–7 days (Lichtenthaler and Grumbach, 1974). In the course of chlorophyll degradation chlorophyll a disappears more rapidly than chlorophyll b, which is already more oxidated than chlorophyll a. Consequently the ratio chlorophyll a/b is continuously shifted to lower values during leaf senescence.

In tobacco leaves, chlorophylls decrease more rapidly than carotenoids during senescence. The chlorophyll a/b ratio decreases only in post-mature leaves (Whitfield and Rowan, 1974). In a recent investigation on chlorophyll degradation during chloroplast disintegration, as the level of chlorophyllide b increased, it was assumed that chlorophyll a is metabolized first to chlorophyll b. It was also proved that the chlorophyll degradation products are not reused in chlorophyll biosynthesis (Grumbach et al., 1982).

The fine structure of chloroplasts during the yellowing of leaves has been studied by Ljubešić (1968). It was found that the chloroplast volume diminishes considerably due to the disappearance of both the plastid stroma and the stroma thylakoids. The grana thylakoids remained unchanged although the level of chlorophyll decreased to 50%. The number and size of plastoglobules increased continuously during ageing, becoming the predominant structure in the yellow leaves Fig. 1.7(a). Through induced regreening the reverse process occurred, the ultrastructure of the chloroplasts in the regreened leaf being similar to that of the normal green leaf Fig. 1.7(b). Chlorophyll degradation in fruits is discussed below.

VIII. Chlorophyll in Fruits

Chlorophyll is present in unripe fruits, which at the beginning of their development contain chloroplasts, in higher amounts in peel than in pulp. During ripening the chloroplasts are gradually disorganized as the thylakoids are destroyed and chlorophylls break down. Although this phenomenon is almost ubiquitous occurring in the great majority of fruits, there are some

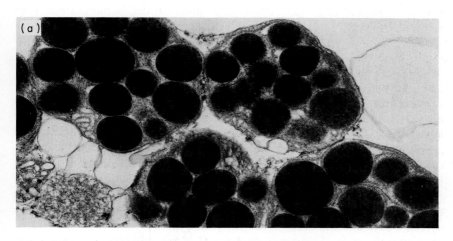

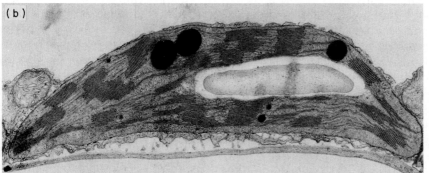

Fig. 1.7 The fine structure of chloroplasts after (a) yellowing and (b) regreening of leaves. From Ljubešić (1968).

exceptions; fruits which retain chlorophyll at the ripe stage. Examples are fruits with diversified coloration in which chlorophylls form the ground colour as in certain apple, pear and gooseberry cultivars. In others they predominate and the fruit appear totally light green. Examples are again apple and pear cultivars, as well as fig, plum and grape cultivars.

Chlorophylls are found in high amounts in the deep green peel of the ripe lime, avocado and watermelon. In other fruits chlorophylls are found in the flesh, as in the green fleshed melon cultivars, or in the light green pericarp of the Chinese gooseberry or the green yellow endocarp of the avocado.

The chlorophyll content, and in many fruits also the ratio chl a/b, varies with the genus, species, cultivars, the environmental factors and the developmental stage. In the fruit itself there are pronounced differences between the chlorophyll content of the peel and the pulp. In fruits with green pulp the chlorophyll content differs at various depths. Some examples are given in Table 1.

IX. Chlorophyll Changes during Fruit Ripening

A systematic study of the pigment changes during fruit ripening began only recently, although seasonal colour changes were described a long time ago in both citrus (Miller *et al.*, 1940) and apples (Haller and Magness, 1944). As was gradually realized that colour and quality are not casually related and fruit colour became to be considered as a quality index, research on the subject was intensified. For commercial purposes the fruit colour rating in categories of ripeness, by physical methods, is intensively used. For a deeper understanding of the colour changes during the ripening process, systematical research was carried out and is still necessary.

In an early study of colour changes in two apple cultivars, chlorophyll was globally assessed during fruit development. In both apple cultivars chlorophyll continually decreased during ripening (Workman, 1963). The disappearance of chlorophylls a and b during the maturation of pears Passe-Crassane was found to be a reaction of the first order. In the process chlorophyll a decreased more rapidly than chlorophyll b (Laval-Martin, 1969). The chlorophyll changes during ripening in three different fruits are shown in Figs 1.8–1.10. In all three fruits in which the colour at ripeness is imparted by chlorophylls, carotenoids or anthocyanins, the amounts of chlorophylls a and b decrease continually.

In the ripe mesocarp of the green-fleshed melon chlorophylls a and b are still present (Fig. 1.8). In unripe Satsuma mandarins the chlorophylls are present in high amounts, masking the carotenoids which are also present in the chloroplasts. Both chlorophylls and carotenoids decrease as the chloroplasts

Table 1. Total chlorophyll content in various fruits

Fruit	Cultivar	Part of Fruit	Ripening stage	Chlorophyll content μg (g fresh wt)$^{-1}$	References
Banana	Dwarf Cavendish	Peel	Unripe	60.0	Gross and Flügel (1982)
Muskmelon	Galia	Peel	Unripe	345.0	Flügel and Gross (1982)
Kumquat	Nagami	Flavedo	Unripe	26.5	Huyskens (1983)
Tangerine	Dancy	Flavedo	Unripe	330.0	Gross (1981)
Pummelo	Goliath	Flavedo	Unripe	90.0	Gross et al. (1983)
Gooseberry	Achilles	Whole	Ripe	21.0	Gross (1982/83)
Grape	Riesling	Whole	Ripe	15.0	Gross (1984)
Pear	Super Trévoux	Peel	Ripe	4.5	Gross (1984)
	Spadona	Peel	Ripe	45.0	
Muskmelon	Crenshaw	Pulp	Unripe	4.0	Reid et al. (1970)
	Persian	Pulp	Unripe	20.0	
Apple	Golden Delicious (1979)	Peel	Harvest ripe	53.0	
Apple	Golden Delicious (1981)	Peel	Harvest ripe	10.0	Henze (1983)
Avocado	Nabal	Peel	Ripe	396.0	Gross and Ohad (1983)
		Outer green pulp	Ripe	176.0	
		Inner yellow pulp	Ripe	8.5	
Muskmelon	Galia	Outer green pulp	Ripe	22.8	Gross and Ohad (1983)
		Inner whitish pulp	Ripe	2.1	
Chinese gooseberry		Outer pulp	Ripe	13.7	Possingham et al. (1980)
		Inner pulp	Ripe	9.1	

are disintegrated. At about 160 days following anthesis, at "colour break", the carotenoids become visible and their enhanced biosynthesis begins. The chlorophylls disappear totally (Fig. 1.9). In strawberries the decrease of both chlorophylls and carotenoids indicates the disintegration of chloroplasts, which is followed by an enhanced anthocyanin synthesis. The fruit colour changes from green, through white, whitish pink to red.

During the early phases of fruit development, the phase of cell multiplication, chlorophyll is intensively synthesized as in developing leaves (Sanger, 1971), but there are no detailed studies on this subject. After the fruit's growth

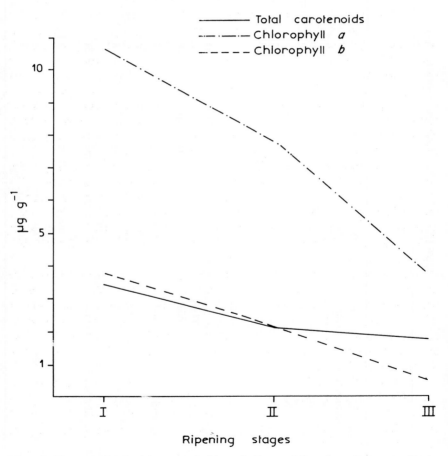

Fig. 1.8 The quantitative changes of chlorophyll a and b and total carotenoids in mesocarp of *Cucumis melo* cv. Galia during ripening. From Flügel and Gross (1982).

ceases, the decrease of the chloroplast pigment begins. The initial increasing biosynthesis may become evident when the pigment content is expressed on a per fruit basis as reported for strawberries (Woodward, 1972), for tomatoes (Laval-Martin, 1974), for oranges (Eilati et al., 1975) or when the pigment analyses are carried out at the very early stages of fruit development, as in the mesocarp of peaches (Lessertois and Monéger, 1978). In the "cherry" tomato the chlorophyll content, expressed as μg per fruit, increased from 23 μg per fruit in the growing fruit to 160 μg per fruit in the mature green. Afterward the content decreased gradually, disappearing almost totally (2 μg per fruit) in the deep red ripe fruit.

The increase of chlorophyll at the early stages of fruit growth may be seen both in oranges and peaches, as shown in Figs 1.11 and 1.12. In these recent studies on pigment changes the ratio chlorophyll a/b was also followed. In most cases it decreased during ripening, but there are fruits in which it increased, proving that in some cases chlorophyll b is more rapidly destroyed than chlorophyll a.

In the study carried out by Eilati et al. (1975) on pigment changes in developing Shamouti oranges, the ratio chl a/b remained almost constant until 200 days after bloom, indicating a proportional decrease of both chlorophylls. Eventually this ratio decreased due to a more rapid destruction of the major component, chlorophyll a.

In the flavedo of other citrus fruits, Hamlin oranges, Robinson tangerines and Marsh grapefruits, the total chlorophyll content decreased and the ratio chl a/b decreased as well (Jahn, 1975). The same trend has been found in pummelo (Gross et al., 1983). In maturing apple chl a/b ratio fluctuates, increasing and then decreasing slightly (Knee, 1972). In Earliglo peaches the ratio a/b increased from 1.0 to 3.4 in the growth period and then decreased abruptly until complete ripeness (Fig. 1.12).

On the other hand in some fruits it has been found that chlorophyll b is more rapidly destroyed. For Dancy tangerine flavedo the ratio chl a/b increased continuously from 2.3 to 3 and finally to 4 before complete disaggregation (Gross, 1981). A preferential degradation of chlorophyll b was also found to occur in Satsuma mandarins ripened on the tree (Shimokawa, 1979). A similar increasing chlorophyll a/b ratio was found in the peels of ripening banana (Gross and Flügel, 1982).

The different destruction rates of chlorophyll a and b may be determined by the different organization of the chlorophyll–protein complexes within the membranes of the thylakoid. More research is needed on chlorophyll biosynthesis and catabolism in fruits. The daily turnover rate, if there is one, has never been investigated.

The destruction of chlorophyll during ripening has been related to an increasing activity of chlorophyllase (Rhodes and Wooltorton, 1967). In

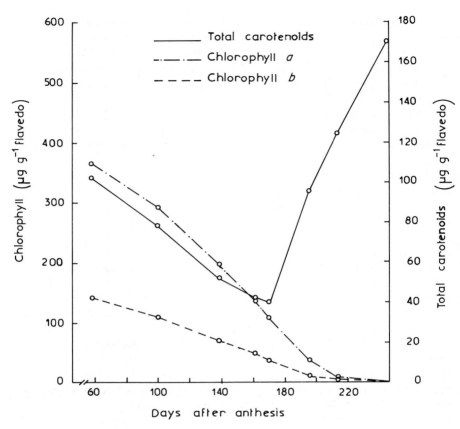

Fig. 1.9 Changes of chlorophylls *a* and *b* and total carotenoids in flavedo of Satsuma mandarins (*Citrus unshui*) during fruit development. From Noga (1981). (Reproduced by courtesy of Dr. G. Noga.)

detached apple fruits stored at 12°C it was found that chlorophyllase activity rose before the commencement of the rise in respiration and continued beyond the peak. As the activity of chlorophyllase increased, the chlorophyll content decreased. Chlorophyllase activity has been more thoroughly investigated in correlation with the effect of the different phytohormones. This will be discussed later.

As in leaves, no intermediate tetrapyrrolic chlorophyll derivatives have been detected in ripening fruit, although an adequate analytical method has been used (Durand and Laval-Martin, 1974). In a pepper cultivar (*Capsicum frutescens*) and in banana peels two green pigments besides chlorophyll *a* and *b* have been detected in low concentrations,. They may be degradation

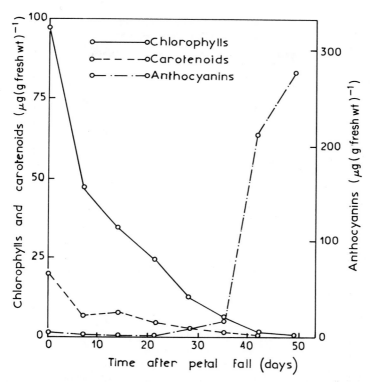

Fig. 1.10 Changes in pigment concentration per unit fresh weight in developing strawberry fruits. From Woodward (1972).

products of the chlorophylls, although their nature is not clear (Schanderl and Lynn, 1966; Lynn et al., 1967).

X. Fruits Containing Chlorophyll in the Pulp when ripe

The special category of fruits which, when ripe, have a pulp of green colour imparted by chlorophyll, will be considered further. These are the avocado (*Persea americana*), a green-fleshed muskmelon (*Cucumis melo*) and the Chinese gooseberry of kiwifruit (*Actinidia chinensis*). In all three the chlorophyll content varies with the depth, the layer adjacent to the peel having the highest chlorophyll content and the darkest green colour. The ultrastructural changes that take place during ripening do not involve the usual

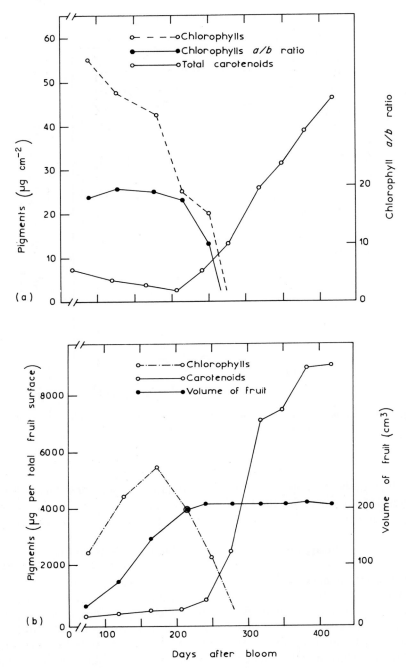

Fig. 1.11 Seasonal changes in peel pigments and fruit volume of Shamouti oranges during ripening. From Eilati *et al.* (1975). *J. Exp. Bot.* **26**, 624 (Oxford University Press).

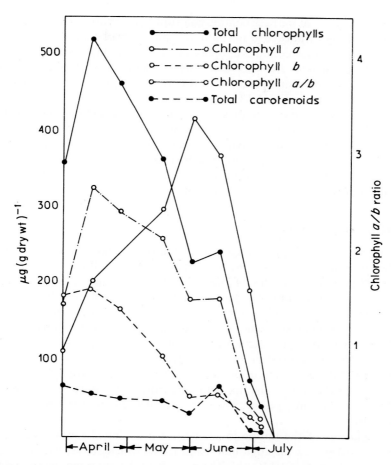

Fig. 1.12 Changes of pigment content and chl a/b ratio in the mesocarp of peach during growth and maturation. Reprinted with permission from (*Phytochemistry*, **17**, Lessertois and Monéger), Copyright (1978), Pergamon Journals Ltd.

transformation of chloroplasts into chromoplasts that occurs in the majority of fruits during the ripening process.

A. Avocado

The variation of chlorophyll levels with depth in the avocado fruits is given in Table 1. The ultrastructure of young and harvest-ripe avocado fruit has been investigated by Cran and Possingham (1973). They found that the skin and outer green layers of the flesh contain chloroplasts with an extensive thylakoid system with highly stacked grana of more than 50 thylakoids. The yellow flesh adjacent to the stone contains etioplasts with a crystalline prolamellar body. The pale green flesh between these layers contains plastids intermediate between chloroplasts and etioplasts, with prominent prolamellar bodies from which grana radiate (Fig. 1.13).

The presence of the various plastid types may be related to the different light conditions within the fruit. In the yellow central layer the conversion of protochlorophyll to chlorophyllide cannot proceed in the absence of light, whereas in the pale green intermediate zone the slow conversion of protochlorophyll into chlorophyll is insufficient to disperse the entire prolamellar body (Figs. 1.13(b) and (c)).

B. Green-fleshed melon

The pigment and plastid changes in the mesocarp and exocarp of a green-fleshed muskmelon *Cucumis melo* cv. Galia have been followed by Flügel and Gross (1982). The changes of the content of chlorophyll *a* and *b* in the mesocarp are shown in Fig. 1.8. Chlorophyll *a* decreased from 10.6 μg to 3.3 μg (g fresh wt)$^{-1}$ and chlorophyll *b* from 3.8 μg to trace amounts. These changes correspond well with the visually recognizable colour change of the mesocarp from pale green to whitish green.

In the deep green exocarp of the unripe fruit the chlorophyll content was very high: 245 μg g^{-1} chlorophyll *a* and 90 μg g^{-1} chlorophyll *b*. As they gradually disappeared in the ripe fruit the carotenoids were no longer masked and the peel appeared yellow. The ultrastructural changes are shown in Figs 1.14–1.16.

The plastids of the unripe inner mesocarp exhibited low stacked grana already swollen, few plastoglobules, and starch which almost did not change at the ripe stage (Fig. 1.14(a)(b). The plastids of the outer mesocarp and that of the exocarp were very similar (Figs 15(a), 16(a). At the unripe stage, both showed plastids with highly stacked grana, up to 120 thylakoids—which disappeared totally in the ripe fruit. The plastids of the ripe fruit contained numerous vesicles which are membrane remnants and increased number of enlarged clustered plastoglobules (Figs 15(b), 16(b)).

According to Sitte (1977), this type of chromoplast found in the ripe fruit is a gerontoplast, a plastid characteristic of a senescent cell. This is based on the fact that the plastid is unable to carry out a *de novo* carotenoid biosynthesis. These pigments are found in decreasing amount as remnant chloroplast carotenoids. Such gerontoplasts are found in senescent leaves (Sitte, 1977).

Highly differentiated chloroplasts with a high stacking degree of thylakoids are generally found in plants grown under low-light conditions or in shade leaves (Wild, 1979; Lichtenthaler, 1981; Meier and Lichtenthaler, 1981). As melons are cultivated under high light intensity it seems that the plastid differentiation is not necessarily dependent on the light conditions. Moreover the plastids of the exocarp and the outer mesocarp are of the same type, although only a fraction of the light incident on the surface passes through the skin. This has been measured in avocado fruits, which also show the same type of plastids in the skin and the outer flesh layer.

C. Chinese Gooseberry (Kiwifruit)

The pigment changes that occur in the pericarp of the Chinese gooseberry (*Actinidia chinensis*) during ripening have been studied by Gross (1982). The fruit has a light green colour imparted by chlorophyll, which does not change visibly during ripening. During fruit ripening the total chlorophyll content decreased from 28.3 μg g^{-1} to 20 μg g^{-1}, about 30% of the chlorophyll being destroyed. Chlorophyll is unevenly distributed, the outer pericarp containing 1.5 times more pigment than the inner pericarp—13.7 μg g^{-1} and 9.1 μg g^{-1} respectively (Possingham et al., 1980). In this latter study the ultrastructure was also investigated (see Fig. 1.17).

The chloroplasts of the outer pericarp layer have a well defined grana and stroma thylakoid system and numerous plastoglobules (Fig. 1.17(a)). The chloroplasts of the inner pericarp (Fig. 1.17(b)) show a massive proliferation of stroma thylakoids, as in the chloroplasts found in the bundle sheaths of C_4-carboxylic acid photosynthetic plants (Kirk and Tilney-Bassett, 1978).

A similarity between the three kinds of fruit belonging to the three different genera *Persea*, *Actinidia* and *Cucumis* is the existence of two kinds of chloroplast at various depths within the fruit flesh. A more pronounced similarity exists between the avocado fruit and the green-fleshed muskmelon. In the outer zone adjacent to the peel they both have a high chlorophyll concentration contained in differentiated chloroplasts with highly stacked grana. Although the Chinese gooseberry also has chlorophyll differently distributed within the pericarp and two kinds of chloroplast, it was found that in this fruit the organization of the chlorophyll–protein complexes in the thylakoids differs totally from that existing in other fruits. This was discovered with the aid of *in vivo* fluorescence spectroscopy described in Section V (Gross and Ohad, 1983).

While the avocado fruit and the green-fleshed cantaloupe have a fluorescence emission spectrum similar to that obtained from green photosynthetic tissue (Fig. 1.5), which indicates a normal organization of the chlorophyll–protein complexes (Figs 1.18 and 1.19), the pattern of the fluorescence emission spectrum of the Chinese gooseberry (kiwifruit) is different (Fig. 1.20). The spectra of both the unripe and ripe kiwifruit are characterized by a

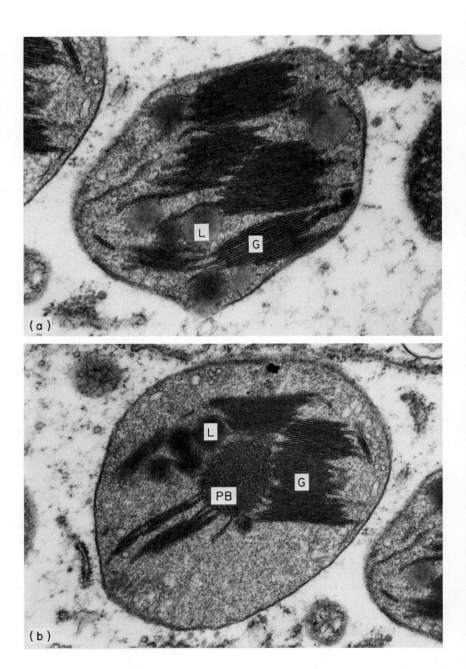

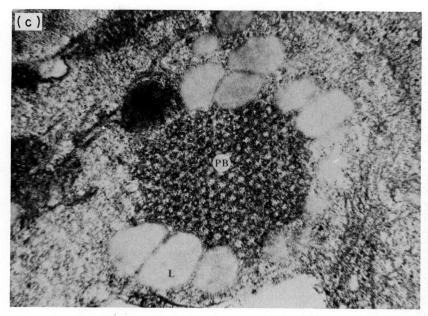

Fig. 1.13 Fine structure of avocado plastids. (a) Chloroplast from the green zone of mature avocado. Tissue taken immediately below the skin. ×41 850. (b) Plastid from the zone intermediate between green and yellow. ×41 850. (c) Plastid (etioplast) from the interior yellow tissue of mature avocado. ×70000. G, grana; L, lipid droplet; PB, prolamellar body. From Cran and Possingham (1973).

higher emission at 686 and 695 nm of the PS II components as compared with that of PS I components (735 nm).

The spectra of different pericarp layers containing decreasing amounts of chlorophyll are also different in unripe and ripe fruits (Figs 1.21 and 1.22). In the ripe fruit the outer layer has a higher fluorescence emission at 740 nm (PS I) as compared to that at 686 and 696 (PS II). A gradual decrease in the fluorescence ratio of PS I to PS II is observed in layers taken towards the interior of the pericarp. The results may be interpreted as a reduction of the PS I antenna and/or a dissociation from the PS II complex, the latter becoming increasingly disorganized.

In the avocado fruit the PS I antennae are still the major emitting complex, even in the innermost endocarp layers. This suggests that the possible disconnection or loss of PS I antenna in *Actinidia* in the corresponding layer is not simply due to the absence of light, since the avocado endocarp is

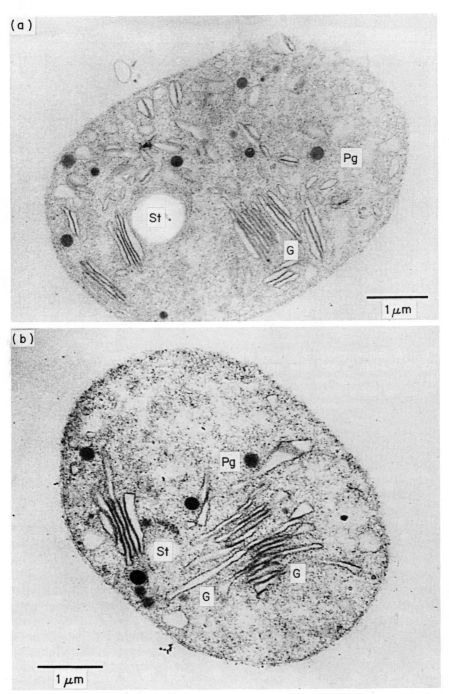

Fig. 1.14 (a) Plastid of the unripe inner mesocarp with low stacked grana (G), few plastoglobules (Pg) and starch (St). (b) Plastid of the ripe inner mesocarp with undifferentiated and swollen grana thylakoids, plastoglobules and starch.

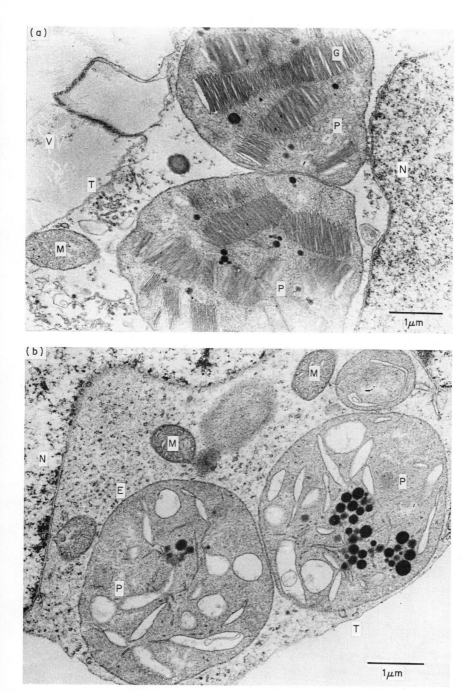

Fig. 1.15 (a) Unripe outer mesocarp layer with plastids (P), nucleus (N), mitochondria (M), vacuole (V) and tonoplast (T). The grana thylakoids are well developed and highly stacked (G). (b) Ripe outer mesocarp layer with plastids (P) plastid envelope (E), mitochondria (M), nucleus and tonoplast (T). Note the predominating plastoglobules in (N) comparison with the unripe stage.

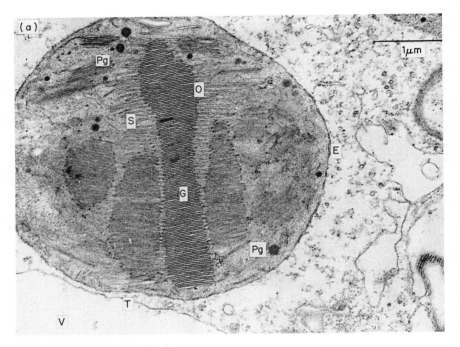

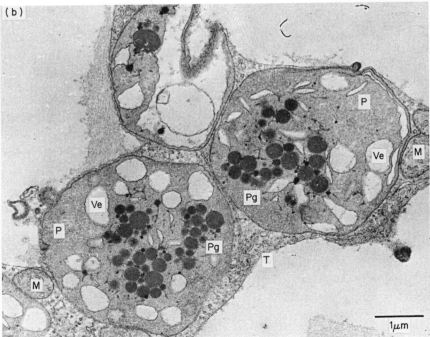

Fig. 1.16 (a) Plastid of the unripe exocarp with highly stacked grana (G), stroma thylakoids (S), few plastoglobules (Pg), and plastid envelope (E)....(b) Ripe exocarp layer with plastids (P), mitochondria (M) swelling vesicles (Ve) and tonoplast (T). Note the multiplied and enlarged plastoglobules (Pg) and the enormously reduced grana thylakoids in comparison with the unripe stage. From Flügel and Gross (1982).

considerably thicker than that of the *Actinidia* pericarp and therefore even less light might penetrate into the interior of this fruit.

XI. Photosynthesis in Developing Fruits

Although developing fruits are dependent on assimilates from the leaves for growth, green fruits containing chlorophyll, and therefore chloroplasts, are capable of photosynthesis. However, the chlorophyll per unit area is much lower in fruits and the photosynthesic activity is correspondingly lower. The photosynthetic rate in Golden Delicious apples measured *in vivo* during a vegetation period decreases continually, paralleling the chlorophyll decrease (Fig. 1.23) (Lenz and Noga, 1982).

Tetley (1931) and Kursanov (1934) demonstrated very early that apples are capable of photosynthesis. Further investigations showed that this is a common feature in many fruits: orange, lemon, avocado. Citrus fruits were later found to be particularly suitable for studies in photosynthesis (Bean and Todd, 1960; Todd *et al.*, 1961; Bean *et al.*, 1963). In developing grapes the photosynthetic rate followed a curve similar to that given in Fig. 1.23 (Geisler and Radler, 1963).

Generally it was concluded that the CO_2 uptake in the fruit is very slight and of little importance compared to the simultaneous uptake of CO_2 by the leaves, and hence can contribute only slightly to the development of the fruit. This does not exclude the possibility that the substances formed directly in the fruit may have qualitative importance (Hansen, 1970). This assumption was confirmed more recently. Jonathan apples grown in light have a different chemical composition from fruits grown in the dark because the photosynthetic activity of the apple contributes directly to its chemical composition (Clijster, 1975). Moreover it was found that as well as the chloroplasts of the peel, the internal tissues of the apple are also capable of photosynthesis, although at a lower degree than in peel. The presence of the internal chloroplasts was related to the degree to which light penetrates the interior cavity and which was unexpectedly high (Phan, 1970). The existence of a special type of chloroplast in the internal apple tissues was confirmed by ultrastructural studies (Phan, 1975; Lenz and Noga, 1982).

In another fruit which also displays plastid dimorphism, the cherry tomato (Laval-Martin, 1974), the photosynthetic properties of the internal and peripheral tissue were investigated (Laval-Martin *et al.*, 1977). The photosynthetic efficiency was found to be highest in the outer region of the fruit and related to the light-dependent activity of two carboxylases: phosphoenolpyruvate carboxylase and ribulosebiphosphate carboxylase, of which the activity of the former was extremely high. The photosynthetic potential and phosphoenolpyruvate carboxylase activity of the fruit tissues were parallel to the chlorophyll content during growth and maturation.

The photosynthetic efficiency of fruit appears commensurate with the CO_2 concentrating mechanism of CAM/CC_4 plants. The presence and activity of phosphoenolpyruvate carboxykinase (PEPCK), phosphoenolpyruvate carboxylase (PEPC) and ribulose-1,5-biphosphate carboxylase, particularly in the internal tissue of developing apple fruit (Blanke *et al.*, 1984) and grape berry indicates the CO_2 reassimilation of respired CO_2 from imported carbon and/or dark CO_2 fixation. In recycling CO_2, it functions as a regulatory mechanism for the photosynthate import between glucogenesis, malate storage and TCA-energy production. Based on the carboxylation potential of PEPCK and PEPC of apple or grape, the internal CO_2 fixation could exceed the fruit's respiration several times. A similar light-dependent high activity of phosphoenolpyruvate carboxylase was found in green grapes which utilize CO_2 very rapidly for the biosynthesis of organic acids (Ribéreau-Gayon, 1967).

In the course of these investigations on fruit photosynthesis, it was observed that additional factors interfere with the efficiency of the CO_2 uptake. A recent study of Valencia oranges showed that fruits exposed to light accumulated CO_2 at a rate which was 25–50% of that of the leaves. The uptake was dependent on fruit size as this is related to changes in stomatal density, on photosynthetically active radiation, on diffusive conductance of the fruit

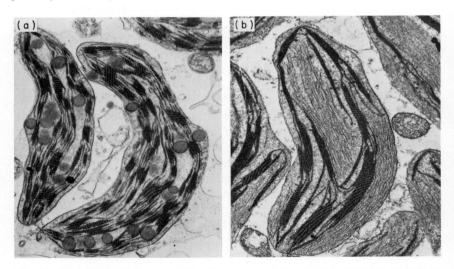

Fig. 1.17 Chloroplasts of Chinese gooseberry pericarp. (a) Section of outer pericarp. × 9600. (b) Section of inner pericarp tissue showing a proliferation of inter-granal membranes. × 13 000. From Possingham *et al.* (1980).

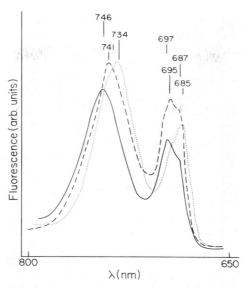

Fig. 1.18 Low temperature fluorescence emission spectra of slices taken from various layers of ripe avocado fruit: (——) green peel; (– – – –) green layer adjacent to the peel; (····) innermost yellow layer.

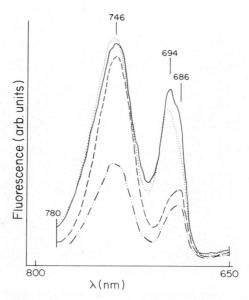

Fig. 1.19 Low temperature fluorescence emission spectra of slices taken from various layers of cantaloupe fruit: (····) green peel; (——) outer layer of the pulp; (– – –), (– .. –) intermediate pulp layers.

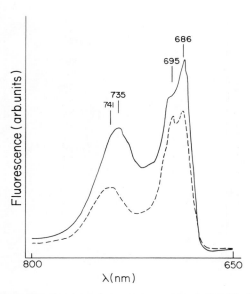

Fig. 1.20 Low temperature fluorescence emission spectra of kiwifruit slices taken from the medium pericarp layer. (———) unripe, and (– – –) ripe fruit.

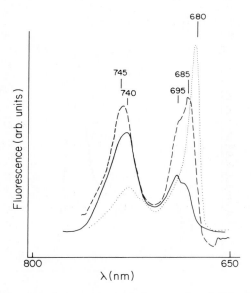

Fig. 1.21 Low temperature fluorescence emission spectra of slices taken from various layers of unripe kiwifruit: (– – –) peel and adjacent pericarp layers; (———) pericarp interine layer; (·····) yellowish green innermost pericarp layer.

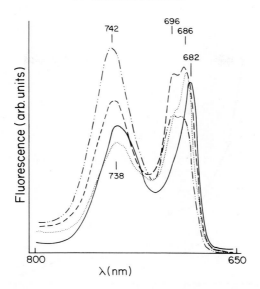

Fig. 1.22 Low temperature fluorescence emission spectra of slices taken from various layers of ripe kiwifruit: (—..—) peel and immediate adjacent pericarp layers; (– – –), (···) two successive layers of the innermost pericarp; (———) yellowish green inner layer of the pericarp. Reprinted with permission from (*Photochem. Photobiol.*, **37**, Gross and Ohad), Copyright (1983), Pergamon Journals Ltd.

epidermis and on soil moisture (Moreshet and Green, 1980). The role of the photosynthesis is to limit losses of CO_2, which is liberated by intense respiration.

XII. The role of Phytohormones in Fruit Ripening

The regulation of the ripening process is due to the interaction of different plant hormones which may have inhibiting or promoting effects. The groups of auxins, cytokinins and gibberellins have senescence delaying effects, whereas ethylene is "the ripening hormone" and abscisic acid accelerates senescence. All the hormones influence the chloroplast–chromoplast transformation and therefore the colour changes of the fruit. This particular effect will be discussed further. It is obvious that other mechanisms are also involved which regulate the pigment biosynthesis.

A. Ethylene

Since ethylene is the hormone which causes the transformation of chloroplasts into chromoplasts, it will be discussed more extensively. In 1924 it was

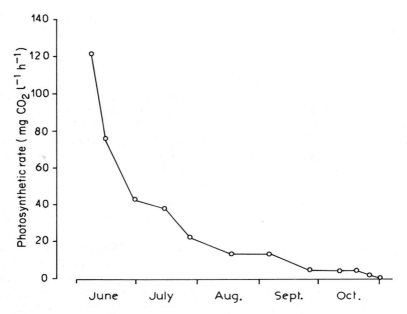

Fig. 1.23 Photosynthetic rates in Golden Delicious apple during a vegetation period. From Lenz and Noga (1982).

discovered that ethylene is responsible for the colour change in citrus which was induced by incomplete combustion products from kerosene stoves (Denny, 1924). Advancements in analytical methods, such as the highly sensitive technique of gas chromatography, permitted accurate measurements of ethylene even in trace amounts. Endogenous ethylene was found to be present in all fruits, both climacteric and non-climacteric (Burg and Burg, 1962; McGlasson, 1970; Reid, 1975). A minimal threshold concentration is necessary to trigger ripening followed by an autocatalytic pronounced increase throughout the process. The effect of ethylene is a drastic decrease in chlorophyll followed by an increase of carotenoid or anthocyanin pigments, depending on the fruit.

The role of plant hormones was elucidated mainly by using exogenous treatments with the respective hormone. Treatments with ethylene were found to be very advantageous, so that it became the most widely used plant growth regulator in agriculture. This was due both to the intensive investigations of its role in plant physiology and to the synthesis of chemicals which evolve ethylene when applied to crops, such as (2-chloroethyl) phosphonic acid (CEPA) known as ethephon or ethrel. Ethylene is successfully used to promote ripening in fruits such as cantaloupe (Kasmire *et al.*, 1970), tomatoes

(Iwahori and Lyons 1970), Japanese plum (Guelfat-Reich and Ben-Arieh, 1975), nectarines (Ben-Arieh and Guelfat-Reich, 1975) and others.

1. Degreening of citrus fruits

Citrus fruits mature internally before an acceptable colour develops in the peel. Ethylene is widely used in degreening citrus fruits, normally after harvest and before packing (Winston, 1955; Fuchs and Cohen, 1969; Grierson *et al.*, 1972; Jahn, 1973, 1974). Two degreening methods have been developed: the short method, in which ethylene is in intermittent contact with the fruit, and the trickle system, in which ethylene is continuously added. The principle underlying both methods is to apply ethylene in an atmosphere high in oxygen and low in CO_2. In Japan a special method has been developed and has been effective, especially for Satsuma mandarins, which consists in sealing the fruits for 15 h with ethylene (Kitagawa *et al.*, 1974).

The effectiveness of the degreening depends on various factors, such as the level of oxygen, in addition to the ethylene (Jahn *et al.*, 1969). The fruit maturity and temperature equally affect the degreening response (Jahn *et al.*, 1973; Jahn, 1974b). The effectiveness of degreening is different in various citrus species. This different effectiveness seems to be related to gas diffusion affecting the rate of release of the absorbed ethylene, which is highest in Valencia oranges and lowest in Satsuma mandarins (Kitagawa *et al.*, 1977).

A factor which affects the rate of degreening is possibly the chl *b* content relative to that of chl *a*. Possibly chl *b* is degraded more slowly than chl *a*. If this is the case, the response to the degreening is predictable when the variations of the ratio chl *a/b* during ripening are known, a fact of which some authors seem so far to be unaware (Winston, 1955; Fuchs and Cohen, 1969). The chlorophyll *a/b* ratio varies with genus, species and cultivar and changes differently during ripening as has already been discussed in Section IX (Jahn, 1975; El-Zeftawi, 1976; Shimokawa, 1976).

2. Regreening of citrus fruits

Another problem related to the colour of citrus fruit is the regreening. This phenomenon occurs after the fruit has developed an acceptable colour before picking and takes place only in certain species. Ultrastructural investigations showed that the increase in chlorophyll level is due to the reversal transformation of chromoplasts into chloroplasts in regreening Valencia oranges (Thomson *et al.*, 1967). These same ultrastructural changes have recently been found to occur in yellowing and regreening lemons and are shown in Fig. 1.24 (a)–(d) (Ljubešić, 1984). That the chloroplast–chromoplast conversion is not irreversible has also been shown in leaves (Fig. 1.7). In leaves the regreening process, introduced by detaching all young parts of the plant, produced the enlargement of the plastids due to formation of large amounts of stroma in which thylakoids were rebuilt. The chlorophyll content reached 80% of that of the normal leaf, but more plastoglobules were present (Ljubešić, 1968).

Regreening occurs particularly with the sweet oranges Valencia and Lane Late Navel, both late-maturing cultivars. As suggested for leaves in which regreening was artificially induced (Dyer and Osborne, 1971), the phenomenon is genetically determined, occurring only in certain species or cultivars. It is also influenced by weather conditions during growth and maturation, in particular by a higher temperature of the root system (Caprio, 1956). Degreening these fruits with ethylene is less effective as compared with other citrus cultivars (Eaks and Dawson, 1979). The influence of nitrogen on regreening is controversial (Jones and Embleton, 1959; Coggins et al., 1981). Other factors such as light intensities, height of the fruit on the tree and hormonal control are also considered to affect regreening (El-Zaftawi, 1977).

The ultrastructure of the flavedo of regreened Valencia oranges has recently been reinvestigated by El-Zeftawi (1978a). As in leaves, regreening involved an increase in plastid lamellae without grana stacks. The absence of grana correlates with the low levels of chlorophylls. Fruits treated with ethephon at the regreening stage did not regreen. It is emphasized that light is the determining factor, but it seems more probable that a change in the hormonal balance must also be taken into account.

3. *The mode of action of ethylene*

Although ethylene is so widely used in triggering and accelerating the ripening process, its mode of action is not fully understood. The ultrastructural changes occurring in normally ripening citrus fruits involve the gradual transformation of chloroplasts into chromoplasts, as was observed in an earlier study on Valencia oranges (Thomson, 1966) and in a more recent one on pumello (Gross et al., 1983) (Fig. 1.26). The ultrastructural changes of plastids in flavedo of ripening grapefruit are analogous (Fig. 1.25). The changes involved the usual transformations: the breakdown of the thylakoid lamellae as chlorophylls disappeared and the formation of chromoplasts in which plastoglobules are the predominating feature.

In pummelo, during conversion of chloroplasts into chromoplasts two types of chromoplasts were formed, one containing only plastoglobules (see Fig. 1.26(c)). The other, which predominated, contained unusually long achlorophyllous membranes that were arranged in concentric circles (Fig. 1.26(d)).

The ultrastructural changes of plastids induced by ethylene are analogous.

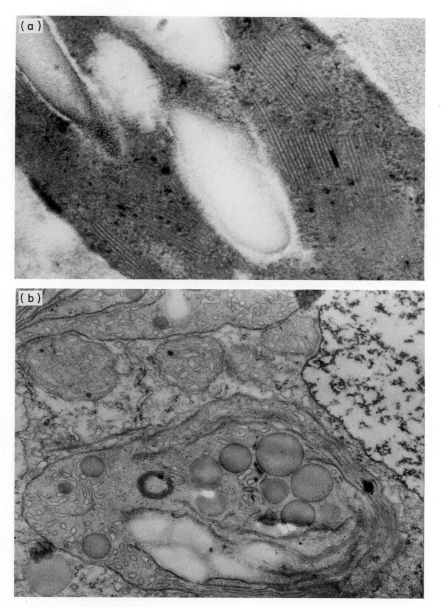

Fig. 1.24 Plastid transformations in yellowing and regreening lemons. (a) Part of a chloroplast from the young green fruit. In very dense stroma grana and starch grains are present. ×45 000. (b) Chromoplasts from the yellow fruit. Large plastoglobules and starch grains can be seen in the middle of the chromoplast. Long single thylakoids and peripheral reticulum are situated on the periphery. ×27 000.

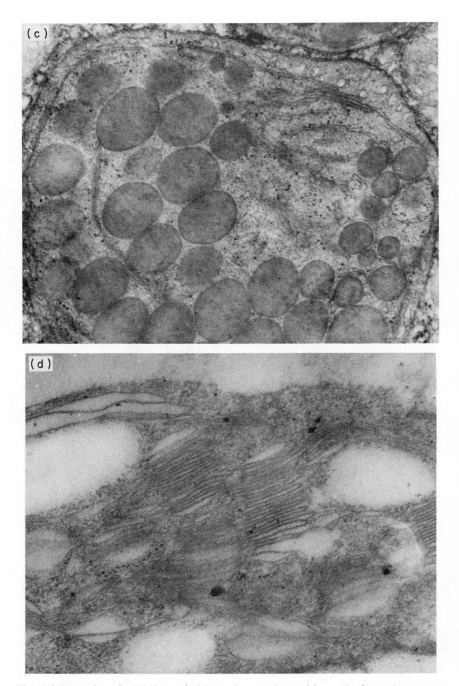

Fig. 1.24 (continued) (c) Part of chloro-chromoplast with newly formed grana at beginning of the regreening process. Plastoglobules are large and numerous, Ribosomes form an array along the thylakoids. ×46 800. (d) Chloroplast from the regreening fruit Stacked grana are located in very dense stroma. ×37 800. From Ljubešić (1984).

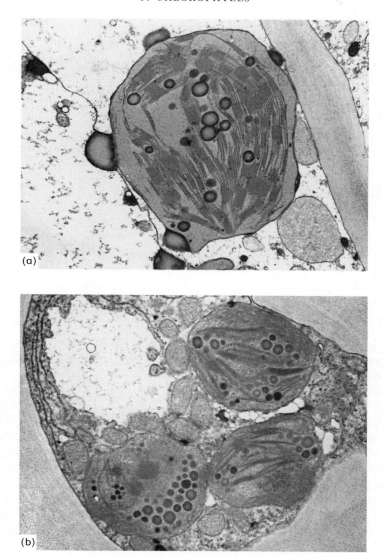

Fig. 1.25

In Satsuma mandarins the rapid reduction in chloroplast size was attributed to ethylene. The disintegration of the inner lamellar membranes was similar to that found in normally ripening fruits. An interesting feature is the appearance of finger-like protuberances and peripheral reticula. The reticulum may be an adaptation for transport of substances necessary for

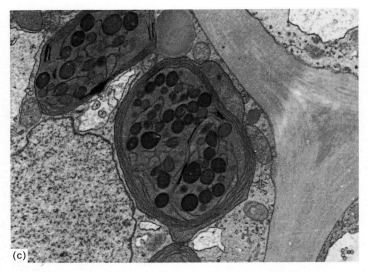

Fig. 1.25 Plastid changes in flavedo of ripening grapefruit. (a) Chloroplast of young fruit. (b) Intermediate plastid of pale yellow fruit. (c) Chromoplast of ripe yellow fruit. Gross and Timberg (1984: unpublished data).

chlorophyll degradation, from cytoplasm to chloroplast (Shimokawa and Sakanoshita, 1977; Shimokawa et al., 1978a).

Purvis (1980) found that in calamondin fruit exogenous ethylene induced a rapid ultrastructural transformation and an increase in chlorophyllase activity. The chloroplasts of the green flavedo showed an extensive thylakoid system with highly stacked grana. After 48 h exposure to ethylene the grana have disintegrated. After 72 h characteristic chromoplasts with osmiophilic globuli were formed. Whether the loss of chlorophyll is causally related to breakdown of the chloroplasts and/or chlorophyllase activity is not readily apparent. Chlorophyllase levels were inversely related to chlorophyll content but electron micrographs showed that internal membranes of chloroplasts were disrupted simultaneously with the decrease in chlorophyll content.

The dramatic increase in chlorophyllase activity which occurs during ripening is difficult to explain. A high chlorophyllase activity may not necessarily cause degreening. In natural degreening and regreening oranges the chlorophyllase levels decreased and increased in parallel with the chlorophyll levels, having alternatively a role in chlorophyll degradation and chlorophyll synthesis (Aljuburi et al., 1979).

Ethylene-enhanced chlorophyllase activity, a 195% increase has been found in another calamondin cultivar (Barmore, 1975). Cycloheximide applied

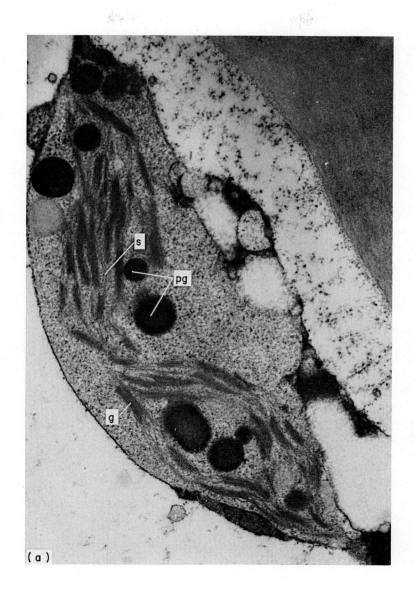

Fig. 1.26 (continued)

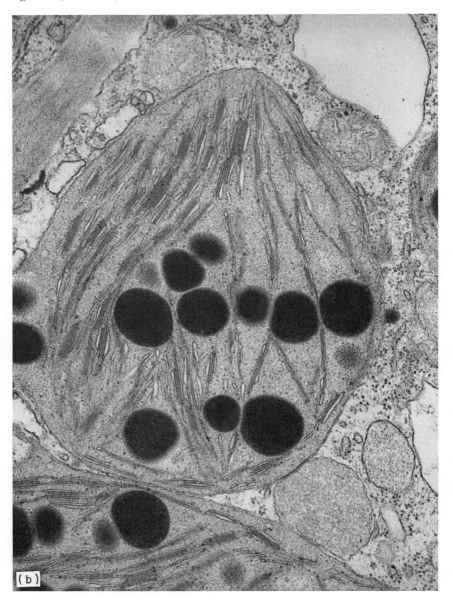

Fig. 1.26 (continued)

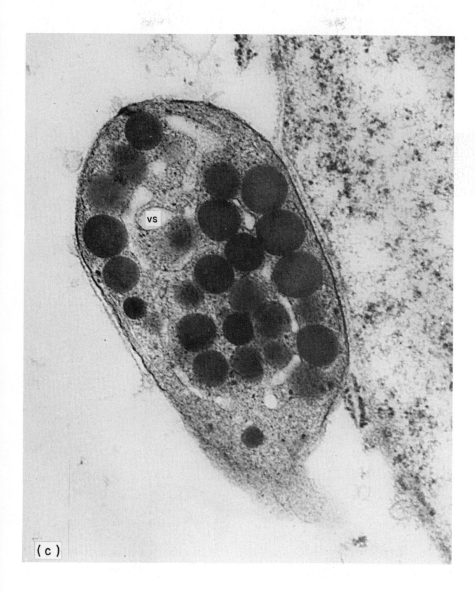

Fig. 1.26 (continued)

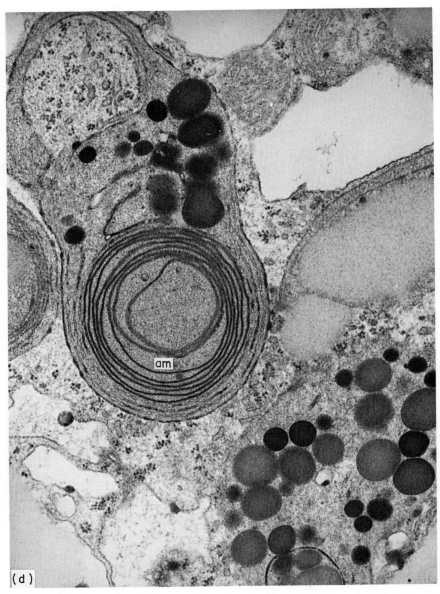

Fig. 1.26 Plastid changes in flavedo of developing pummelo *Citrus grandis* Goliath. (a) Chloroplasts of green fruit with grana thylakoids (g), stroma thylakoids (s) and plastoglobules (pg). ×25 500. (b) Chloroplast showing the beginning of disorganization of the lamellar system as the grana thylakoids, especially at the periphery, are swollen. ×27 000. (c) Chromoplast of ripe fruit of common type showing a massive proliferation of plastoglobules and some vesicles that are swollen membrane remnants (vs). ×50 400. (d) Chromoplast of ripe fruit of uncommon type showing multiple concentric membranes not including plastoglobules. am, achlorophyllous membranes. ×24 300. From Gross *et al.* (1983).

immediately prior to the ethylene treatment reduced the effect of chlorophyll degradation by approximately 80% and the chlorophyllase activity by 60%. Therefore it appears that chlorophyll degradation is partly mediated through increased chlorophyllase synthesis and activation. Ethylene-enhanced chlorophyllase activity also occurred during the degreening of *Citrus unshiu* Marc (Shimokawa *et al.*, 1978b). As previously found, cycloheximide inhibited the ethylene-enhanced chlorophyll degradation and repressed the enhancement of chlorophyllase activity, while chloramphenicol had no effect. A *de novo* cytoplasmic synthesis of chlorophyllase is supposed to take place. However the chloroplast–cytoplasm interrelationship in senescent chloroplasts is still obscure. It is postulated that the peripheral reticulum described in the ethylene-treated fruits may be a shuttle system between chloroplast and cytoplasm (Shimokawa *et al.*, 1978a).

A recent report has investigated the involvement of ethylene in chlorophyll degradation in citrus flavedo of two cultivars Robinson tangerine and calamondin Bianco which are not self-sustaining but require a continuous supply of ethylene (Purvis and Barmore, 1981). When fruit was removed from ethylene the chlorophyllase activity did not change, even though chlorophyll degradation had stopped. It is probable that the level of chlorophyllase is not the regulating factor in chlorophyll degradation. Ethylene may interact with the thylakoid membranes to release chlorophyll molecules or induce another enzyme outside the chloroplast whose activity precedes and influences that of the chlorophyllase. Synthesis of proteins in both cytoplasm and chloroplast might be required in order to maintain a high level of chlorophyllase in the leaves. This could explain the high stability of chlorophylls in isolated chloroplasts (Sabater and Rodriguez, 1978).

B. Gibberellin

Gibberellic acid, known as GA_3, is the most widely distributed from the over sixty-five naturally occurring gibberellins. It attains its highest levels when the fruit is unripe. Its effect is opposite to that of ethylene as it delays maturation, inhibiting the chlorophyll degradation that is the conversion of chloroplasts into chromoplasts.

The effect of gibberellin has been tried on various fruits both attached to and detached from the tree and also on fruit tissue. The citrus fruits have been more extensively investigated (Soost and Burnett, 1961; Coggins and Hield, 1962; Ismail *et al.*, 1967) but the same delaying effect was also observed in other fruits including tomatoes (Abdel-Kader *et al.*, 1966; Dostal and Leopold, 1967) apricots (Abdel-Gawad and Romani, 1974) and bananas (Vendrell and Naito, 1970; Bruinsma *et al.*, 1975). For the banana assays segments of unripe fruits were treated *in vitro*. The pre-harvesting treatment with GA$_3$ was found to be more effective in apricots (Abdel-Gawad and Romani, 1974) and oranges (Goldschmidt and Eilati, 1970). Treated Shamouti oranges remained green for more than six months. GA$_3$ is effective even when applied to a limited portion of the peel. During a month of storage, the treated portion remained green while the rest of fruit developed the typical orange colour (Eilati et al., 1969).

In some of the assays mentioned, treatment with both hormones gibberellin and ethylene allowed their antagonistic effects to be displayed in the chloroplasts (Abdel-Kader et al., 1966; Scott and Leopold, 1968; Goldschmidt et al., 1978). In late maturing citrus cultivars, which are susceptible to regreening, the treatment with exogenous gibberellin intensified the symptom, which is due to an excess of endogenous gibberellin (Rasmussen, 1975). Treatment with potassium gibberellate caused formation of about twice the usual level of chlorophyll during the regreening process. The treated fruit exhibited accelerated rates of protochlorophyllide synthesis, which leads to enhanced regreening through accumulation of higher levels of chlorophylls (Coggins and Lewis, 1962).

The practical use of gibberellin is to extend harvest and to improve storage. For lemon the harvest can be extended into late spring (El-Zeftawi, 1980). The maturation of persimmon fruits can be delayed by gibberellin treatment, which delays colour development (Kitagawa et al., 1966; Testoni, 1979; Gross et al., 1984).

C. Abscisic Acid

At first the only role of abscisic acid (ABA) in fruit maturation was thought to be abscission. The evidence that it has a role as a regulator of fruit ripening is recent (Milborrow, 1974a). The synthesis of ABA from mevalonate occurs in the chloroplast (Milborrow, 1974b).

ABA has been found in different fruits—pears (Rudnicki et al., 1968), apple, avocado, lemon, olive (Milborrow, 1967), tomatoes (Dörffling, 1970), grapes (Coombe and Hale, 1973), peaches (Sandhu and Dhillon, 1982)—in which its level increases during ripening, enhancing the fruit colour. The relationship between ABA and the fruit ripeness was recognized first in pears (Rudnicki et al., 1968).

In the flavedo of citrus fruits large amounts of both free and bound abscisic acid accumulate during natural or ethylene-induced senescence. When mature green harvested Shamouti oranges were treated with ethylene a pronounced increase in ABA occurred, while only 20% of the chlorophyll was destroyed. Increases of ABA upon exposure of different citrus fruits to ethylene occurred also in plastids other than chloroplasts—in leukoplasts and chromoplasts (Goldschmidt et al., 1973; Brisker et al., 1976).

D. Auxins

The principal member of this group is the indole acetic acid (IAA) which was the first plant hormone to be characterized chemically. Different members of the group were detected in fruits; the levels of the auxins fluctuate during development. Their formation precedes that of gibberellins (Abdel-Kader et al., 1966; Nitsch, 1970). The auxins function as senescence retardants in fruits, causing an inhibition of the onset of ripening. When slices cut from green banana fruits were treated with IAA and a synthetic auxin (2,4D), the ripening was delayed and induced ripening with ethylene was prevented (Vendrell, 1969). In Bartlett pears a similar result was obtained. The rate of ripening measured as chlorophyll degradation and softening was progressively inhibited in proportion to the concentrations of the applied auxins, even though it simultaneously stimulated an increase in ethylene (Frenkel and Dick, 1973; Frenkel, 1975).

E. Cytokinins

Cytokinins are derivatives of adenine. Kinetin, 6-furfurylaminopurine, was first isolated as an artefact from autoclaved preparations of nucleic acids and was named because of its activity in stimulating cell division. Cytokinins are found in immature fruit (Nitsch, 1970). The effect of kinetin is analogous to that of gibberellins in delaying ripening, as has been shown in tomatoes (Abdel-Kader et al., 1966) and green banana slices (Wade and Brady, 1971). In banana, although the peel remained green, the pulp was ripe in terms of softness, sugar and flavour volatiles content.

Pre-harvest treatment with N_6-benzyladenine induced the same effect in apricots (Abdel-Gawad and Romani, 1974) and citrus (Goldschmidt et al., 1978). Kinetin seems to lower the level of chlorophyllase activity or to impede its rise, as has been demonstrated in senescent leaves (Sabater and Rodriquez, 1978).

XIII. Other Factors Influencing Fruit Chlorophyll Content

A. Light

Chlorophyll is synthesized only in light. The plant development, implicitly fruit development, is entirely dependent on photosynthesis. Moreover, in some fruits it has been found that the photosynthetic activity of the fruit itself contributes directly to its chemical composition. Fruit grown in darkness are devoid of pigments and have a different composition from fruit grown in the light (Clijsters, 1975).

B. Temperature

At higher temperatures the ripening process is accelerated. In the pear Passe-Crassane the chlorophyll content decreases faster the higher the temperature, but the ratio chl a/b remains unaltered (Laval-Martin, 1969).

Citrus fruits have a special requirement concerning temperature not encountered in other fruits: the conversion of chloroplasts into chromoplasts is induced by rather low temperatures. It has been found that colour break occurs at about 13°C (Stearns and Young, 1942). A study of citrus in controlled environments revealed that day- and night-time temperature are relevant. At temperatures of 20°C by day and 5°C at night more chlorophyll was destroyed than with a regime of 30°C by day and an 10°C at night (Erikson, 1961). The temperature of the roots was found to contribute to the effect. The optimal conditions for a rapid loss of chlorophyll and a good fruit development occurred at 20°C by day, 7°C at night and at a soil temperature of 12°C (Young and Erikson, 1961). It was then understood why enhanced chlorophyll loss occurs in citrus fruits grown in climates with warm days and cool nights. On the other hand fruits grown in tropical and sub-tropical zones have generally a green peel colour when mature, due to the high temperature which prevents chlorophyll breakdown (Kefford and Chandler, 1970; Monselise, 1973; Reuther, 1973; El-Zeftawi, 1978).

The effects of different root temperatures on the colour development in Satsuma mandarins and calamondins were compared. It was found that colour break occurred earlier at the lower temperatures (12–14°C) whereas more than half of the fruits grown at 30°C still contained significant amounts of chlorophyll at maturity (Sonnen *et al.*, 1979). According to Eilati *et al.* (1969), low temperatures inhibit root growth and consequently cause a lack of root hormone supply to the fruit.

In controlled post-harvest studies the effect of temperatures was similar. Fruits stored at 30°C did not lose the green colour, which occurred progressively faster as the temperature was lowered (Wheaton and Stewart, 1973). Different results were obtained with Shamouti oranges: post-harvest degreening of Shamouti was faster at 25°C than at 20°C. Higher temperatures (30–35°C) accelerated the loss of green colour (Cohen, 1978). In Golden Delicious apples stored at different temperatures, chlorophyll disappeared totally within two weeks at 18°C and within two months at 4°C (Gross, unpubl. data).

C. The effect of fertilizers on the external fruit colour

The effect of potassium and nitrogen fertilization has not been thoroughly investigated. In Valencia oranges higher rates of potassium and nitrogen produced more green colour in fruits than lower rates (Reuther and Smith, 1952). Later investigations on the effects of potassium and nitrogen confirmed these results in Valencia oranges and also in other orange cultivars (Reitz and Koo, 1960; Reese and Koo, 1975).

In tomatoes the normal relationship between chlorophylls and carotenoids which occurs during ripening, that is decrease of chlorophylls and increase of carotenoids, was altered by the potassium status of the fruit. At the early and last ripening stages the chlorophyll content was higher in high-K fruit than in the K-deficient fruit. At the intermediate stages, the chlorophyll level of K-deficient fruit was higher than that of normal fruit (Trudel and Ozbun, 1970).

In leaves the effect of nitrogen on chloroplast pigments has been investigated. In chloroplasts of *Hordeum vulgare* a nitrogen deficiency induced a marked reduction of both chlorophylls, which was interpreted as a lack of nitrogen necessary for both the synthesis of the pyrrole ring and the proteins of the chlorophyll–protein complexes (Verbeek and Lichtenthaler, 1973). There are methods in which the estimation of the nitrogen status of the leaves is based on measurements of the leaf chlorophyll content (Thomas and Oerther, 1972). Variations in the chlorophyll concentrations of leaves of different crops were induced by varying the plants' nitrogen status (Thomas and Gausman, 1977).

The effect of high nitrogen supply on fruit colour is considered to be a "detrimental effect", as the chlorophyll content increases and the fruits remain green. This has been observed in oranges, as mentioned above. Also the regreening of Valencia oranges is enhanced by high nitrogen levels (Jones and Embleton, 1959).

In apples the same effect was obtained with various cultivars even when grown in different climatic conditions (Kvale, 1971). Two recent reports confirm the same effect of nitrogen nutrition on the Golden Delicious cultivar (Olsen and Couey, 1975; Hansen, 1980). In the latter report it was found that ample nitrogen availability in the period near fruit harvest may be more detrimental to the colour than high nitrogen during the earlier period of fruit development. A higher nitrogen supply is recommended when production of green fruit is required.

Recently the effects of low and high nitrogen levels on four apple cultivars have been investigated (Lenz and Gross, unpublished results). The results were once more the same; fruits treated with high nitrogen concentration had a higher chlorophyll level. But this may not be interpreted as due to a higher nitrogen source for chlorophyll synthesis, but as a delay of the ripening process, the chloroplast conversion being blocked. The effect may be a change in the pattern of endogenous hormones, probably an increase of gibberellin level.

2
Anthocyanins

I. Introduction

The flavonoids are a very large and widespread group of plant constituents. The water soluble anthocyanins, which are responsible for the various shades of red and blue of many fruits, are one of the major flavonoid classes.

There are many comprehensive textbooks available, two of which may be considered as classics in the field: *Comparative Biochemistry of the Flavonoids* (Harborne, 1967) and *The Flavonoids* (Harborne *et al.*, 1975). The latter has been updated and is available as *The Flavonoids: Advances in Research* (Harborne and Marbry, 1982). A recent book entitled *Anthocyanins as Food Colors* deals specifically with anthocyanins and their practical aspect (Markakis, 1982).

II. Structure

The basic nucleus of the flavonoids is the flavan nucleus, which consists of two aromatic rings linked together by a three-carbon unit (Fig. 2.1.). Flavonoids are divided into eleven different smaller classes of which flavones, flavonols and anthocyanins are the major ones. The classes differ from one another in the oxidation level of the central pyran ring. At the aromatic rings A and B, hydroxyl groups or other groups are substituted at certain positions determined by the biosynthetic origin of the two aromatic rings. In the A ring the positions 5,7 are hydroxylated, and in the B ring the hydroxylation pattern 3′,4′,5′ is that of the cinnamic acids.

The flavonoids exist *in vivo* as glycosides, one or more of their hydroxyl groups being joined by a semiacetal link to a sugar. The term of anthocyanins (from Greek: *anthos*, flower and *kianos*, blue) was first introduced in 1835 by Marquart. In 1932, Willstätter and Everest, proposed the flavylium salt structure and the terms "anthocyanins" for glycosides and anthocyanidins for

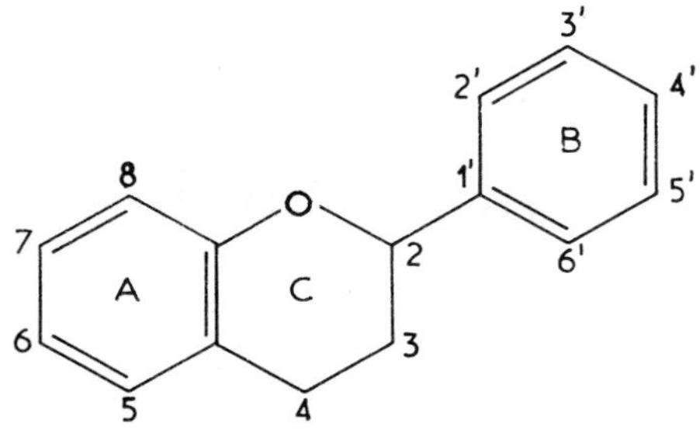

Fig. 2.1 Flavan nucleus.

the aglycones (Willstätter and Everest, 1913). Anthocyanidins and anthocyanins are cations and are presumed to occur in living cells in association with anions of organic acids. In the formulation, the assumed anion is not mentioned.

The elucidation of the chemical structure of the anthocyanins was a struggle due to the difficulties of isolation and separation and the relative lability of the aglycone toward light and alkali. With the advent of the modern methods of chromatography and spectroscopy the obstacles were overcome and the structures of 210 anthocyanins have been determined (Hrazdina, 1982).

The structures of the most common anthocyanidins are given in Fig. 2.2. Their names, given by Willstätter, are derived from the flower sources from which they were first isolated. The most widespread anthocyanidins, the first three compounds in Fig. 2.2, differ in their degree of hydroxylation at the B ring. The methylated derivatives are more widespread in flowers than in fruits. The hydroxyl in position 4' is rarely methylated.

The sugar present in the molecule of the anthocyanins confers a higher solubility and stability. The glycosides are classified according to the nature of the sugar moiety, which is always in the pyranose form. Generally the phenolic group at position 3 is glycosylated, but the hydroxyls at positions 5 and 7 may also form glycosides.

Monosides contain monosaccharides: the pentoses arabinose and xylose or hexoses of which D-glucose is the most common. Others more rarely found are galactose and rhamnose (deoxyhexose). Biosides contain disaccharides: gentiobiose (glucosylglucose ($\beta 1 \rightarrow 6$), sophorose (glucosylglucose ($\beta 1 \rightarrow 2$)), sambubiose (xylosylglucose ($\beta 1 \rightarrow 2$)) and rutinose (rhamnosylglucose ($\alpha 1 \rightarrow 6$)). Triosides contain trisaccharides; they may be linear like gentiotriose or branched like 2^G-xylosylrutinose and 2^G-glycosyl rutinose. The structures of some sugars are shown in Fig. 2.3. In diglycosides two hydroxyls are glycosylated, for example 3,5-diglucosides, 3-rutinoside-5-glucoside and 3-sophoroside-7-glucoside.

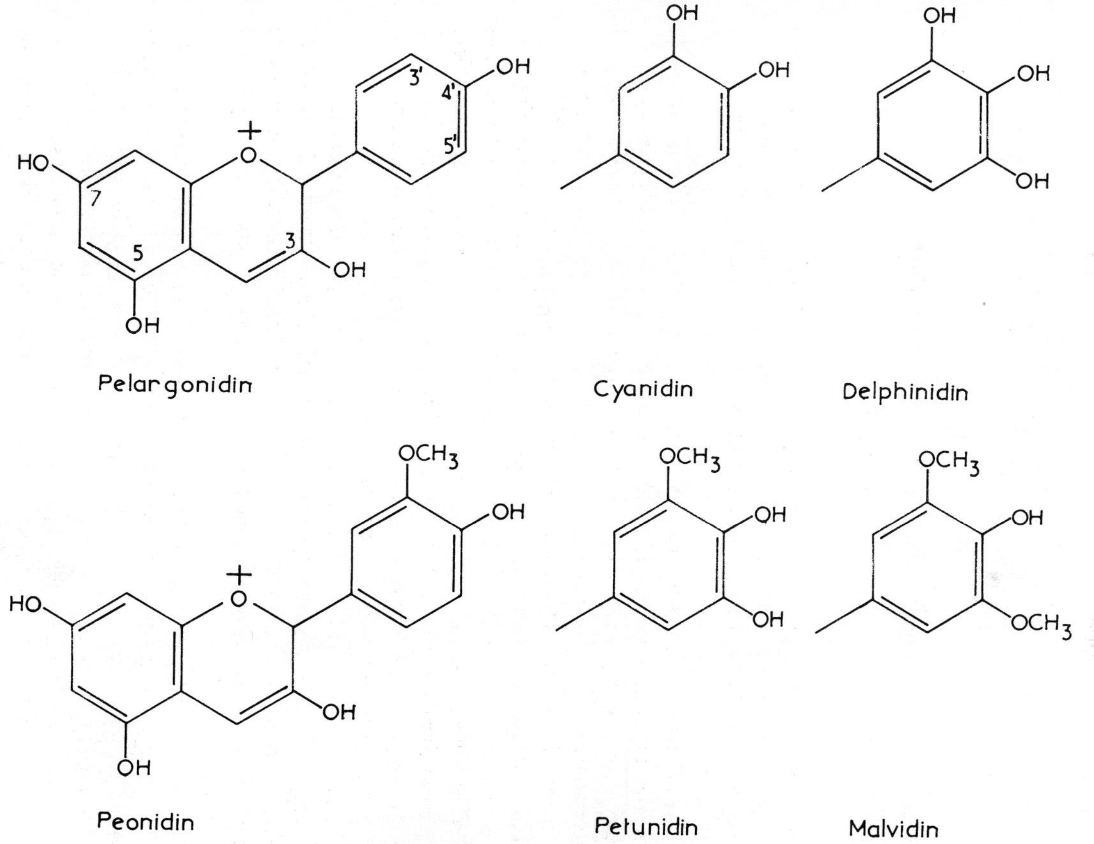

Fig. 2.2 The common anthocyanidins.

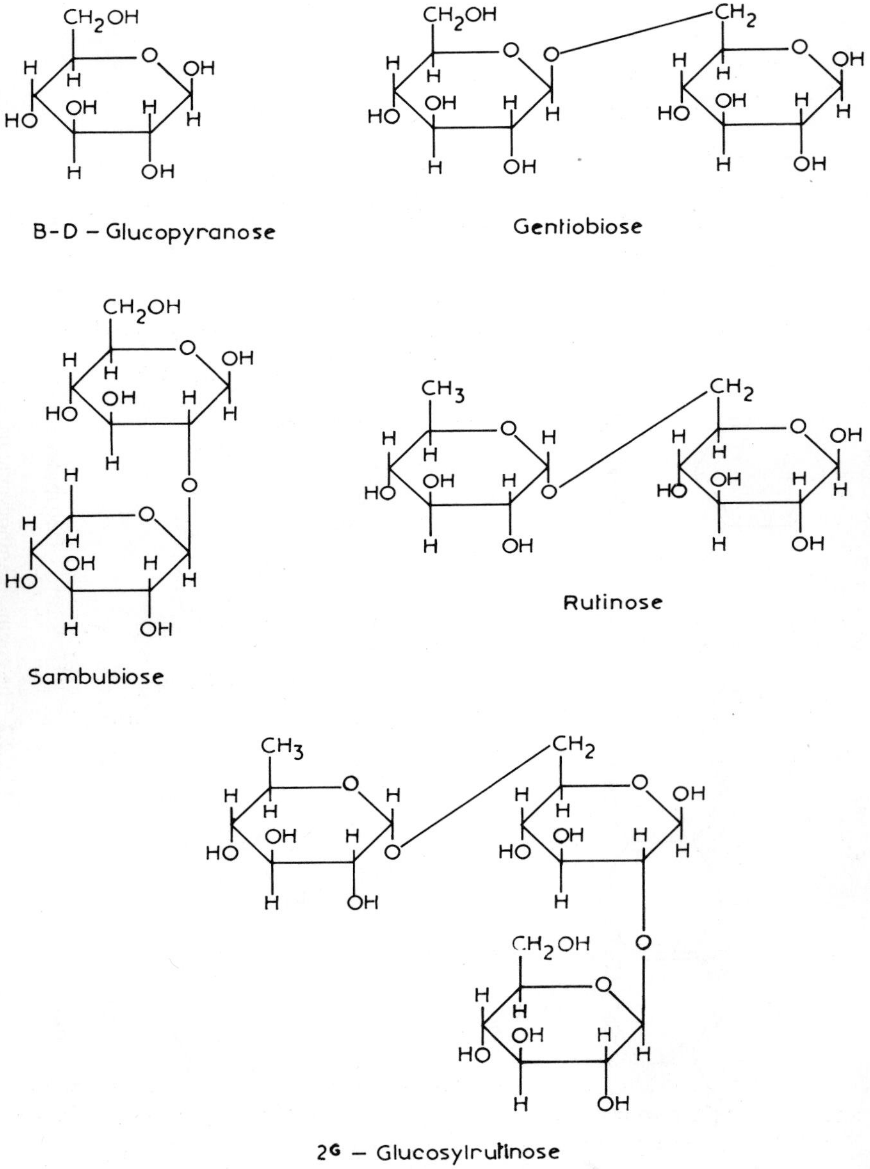

B-D – Glucopyranose

Gentiobiose

Sambubiose

Rutinose

2G – Glucosylrutinose

Fig. 2.3 Some sugars found in anthocyanin pigments.

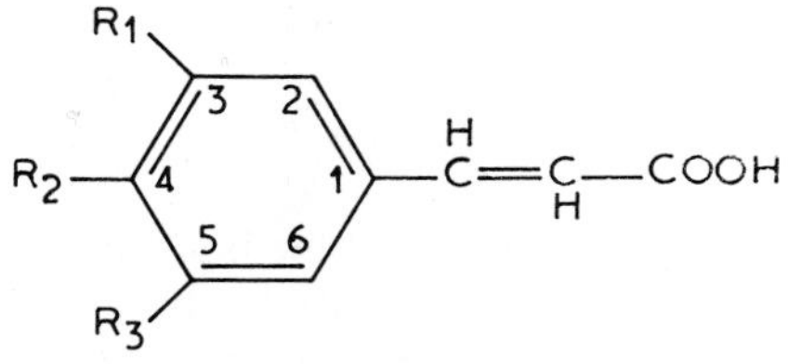

2. ANTHOCYANINS

Cinnamic	($R_1 = R_2 = R_3 = H$)
p-Coumaric	($R_1 = R_3 = H;\ R_2 = OH$)
Caffeic	($R_1 = R_2 = OH;\ R_3 = H$)
Ferulic	($R_1 = OCH_3;\ R_2 = OH;\ R_3 = H$)
Sinapic	($R_1 = R_3 = OCH_3;\ R_2 = OH$)

Fig. 2.4 Structure of cinnamic acids.

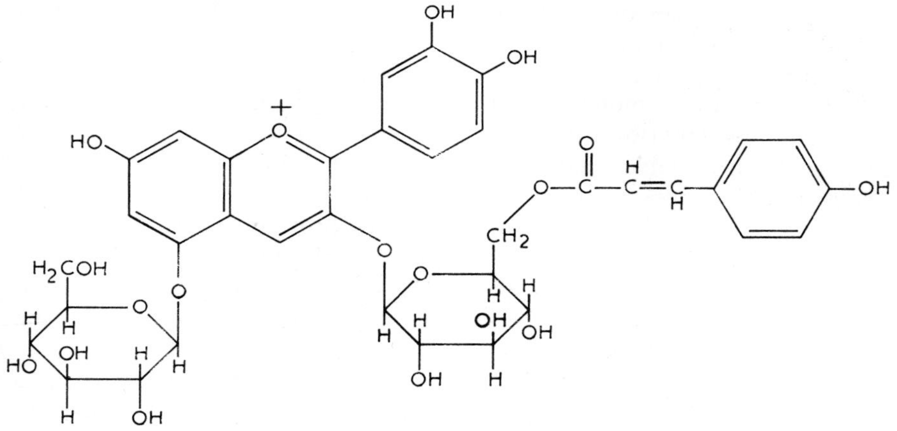

Fig. 2.5 Cyanidin 3-(6-*p*-coumarylglucoside)-5-glucoside.

In some cases the sugars are acylated with acyl groups derived from acetic acid or one of the four cinnamic acids, *p*-coumaric, caffeic, ferulic or sinapic (Fig. 2.4). An example of an acylated anthocyanin is shown in Fig. 2.5. The high number of anthocyanins found in nature is due more to the variation in the nature and position of the sugar and acyl moieties than to the fifteen anthocyanidins so far known.

III. Properties

A. Absorption Spectra

Having a long chromophore of eight conjugated double bonds enhanced by the π electrons of the heterocyclic oxygen of the cation, the anthocyanins appear highly coloured in an acid medium. The electronic absorption spectra of the flavonoids are characterized by two separate bands, one at the longer wavelengths (band I) determined by the B ring conjugation, and the second in the ultraviolet region (band II) determined by the A ring conjugation. The highly coloured anthocyanin salts in acid solution have two main absorption maxima, one in the visible region between 465 and 550 nm, and a less intense one in the ultraviolet range between 270 and 280 nm.

The absorption spectra in the visible region of pelargonidin, cyanidin and delphinidin (in 0.01% HCl in methanol) are shown in Fig. 2.6. The absorption maxima are at 520, 535 and 546 nm; the colour of pelargonidin is scarlet red, of cyanidin crimson and of delphinidin blue-mauve. The absorption maxima vary with the nature of the solvents. Measurements have been standardized using 0.01% HCl in methanol.

The effect of an additional hydroxyl which supplies non-bonding electrons to the existing chromophore of the flavylium system, is a relatively large bathochromic shift (toward the longer wavelengths). Methylation has little or negligible effect. Glycosylation at C-3 produces a hypsochromic shift, the respective glycosides having maxima at about 15 nm shorter. The nature of sugar substitution has no effect, but the position at which the sugar is substituted produces some modification. Substitution of a sugar at C-3 produces a shoulder on the absorption curve at 440 nm; 5- and 3,5-diglycosides have only half the intensity of absorption at 440 nm of the corresponding 3-glycosides.

Acylation of the sugar of anthocyanins by acids of the cinnamic group produce an additional peak in the ultraviolet at 310–335 nm. The spectra of all the anthocyanidins and anthocyanins are tabulated in the literature.

B. Chemical Properties

Anthocyanins behave like indicators in aqueous media. Their structure and thus their colour vary with the pH. At pH 1 they are highly coloured as flavylium cations. With increasing pH the colour gradually fades as colourless pseudobases are formed. In mild alkali blue anhydrobases are formed. The reaction is reversible unless strong alkali is used, which produces irreversible changes. The structural changes that occur in anthocyanins with change of pH are shown in Fig. 2.7.

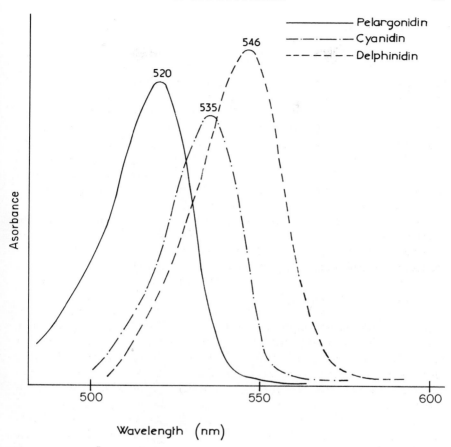

Fig. 2.6 Absorption spectra in the visible region of perlargonidin cyanidin and delphinidin (in 0.01% HCl in methanol).

The structural transformation reaction mechanism of the anthocyanidins has been thoroughly investigated (Timberlake and Bridle, 1975), but is still not completely understood. Recent spectroscopic evidence has proved the existence of eight structural transformation forms (Chen and Hrazdina, 1982).

Anthocyanins containing ortho-dihydroxy groups react with $AlCl_3$ at pH 2–4 to form differently coloured blue chelates which produce a bathochromic shift of 25–35 nm. Anthocyanins form complexes with flavonols which cause a remarkable blueing of the colour, producing a bathochromic shift and a large increase in absorbance. The phenomenon is called co-pigmentation, and can be demonstrated by mixing a solution of anthocyanins with flavonoids,

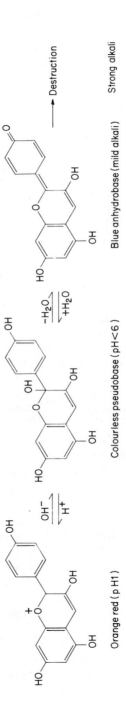

Fig. 2.7 Structural changes of anthocyanins with pH.

varying the pH and their respective concentrations and ratios. The structural aspects of anthocyanin–flavonoid complex formation and its role in plant coloration have recently been investigated (Chen and Hrazdina, 1981). The complex formed between flavonoids and anthocyanins is based mainly on hydrogen bond formation between the carbonyl group of the anthocyanin anhydrobase and aromatic hydroxyl groups of the complex-forming flavonoids, the most effective being the one in the 7-position. These phenomena of variation of colour with the pH, chelation with metals and co-pigmentation occur in plants, contributing to the luxuriant colour variations in flowers and fruits.

Acid hydrolysis of anthocyanins yields anthocyanidin (aglycone) and sugars.

C. Analytical Methods

1. *Non-destructive Methods*

The principle of non-destructive methods was discussed in Chapter 1 (Section III.C.3). These methods can be applied to measuring the anthocyanin colour, since a high correlation has been found to exist between reflectance or transmittance measurements and the anthocyanin content of fruit. Reflectance colour instruments have been used to measure the colour development in cranberries (Sapers *et al.*, 1983), strawberries (Spayd and Morris, 1980), cherries (Drake *et al.*, 1982) and grapes (Watada and Abbott, 1975).

The colour of blueberries can be determined by light transmittance measurements (Dekazos and Birth, 1970; Kushman and Ballinger, 1975). A conveyor for small fruit sorting has been developed based on this technique (Rohrback and McClure, 1978), and a special instrument has been built for separating grapes into specified maturity categories (Ballinger *et al.*, 1978).

D. Pigment Analysis

A complete anthocyanin analysis involves qualitative and quantitative determination. A detailed analytical procedure is described by Francis (in Markakis, 1982).

1. *Extraction*

During analysis some precautions have to be taken, as the anthocyanins are not stable under high light intensities, at high temperatures and in aqueous solution. The pigments are extracted by homogenizing the fruit sample with methanol acidified with 1% HCl. In some cases a quantitative extraction is achieved only after the mixture has been kept overnight at low temperature. If the extract is not clear it must be filtered or centrifuged.

2. *Quantitative Determination*

The determination of anthocyanin content is analogous to that of chlorophyll and is also based on the law of Lambert–Beer, which states that the concentration of a substance is proportional to the optical density (extinction), the length of the light path being l:

$$E = \varepsilon\, C\, L$$

where E is the extinction measured with the spectrophotometer; ε is the molar extinction coefficient, the extinction of a solution containing 1 mol l^{-1} (the values for each pigment are tabulated); C is the molar concentration; and L is 1 as the spectrophotometer cells have a path length of 1 cm.

Concentration in milligrams per litre can be determined by multiplying by the molecular weight of the pigment. The extinction is read at maximum wavelength and V, the dilution factor, is taken into account.

$$C\ (\text{mg } l^{-1}) = \frac{E \times \text{mol. wt} \times 10 \times V}{\varepsilon}$$

When the extract contains other groups of pigments such as chlorophylls and carotenoids as well as anthocyanins, they may be determined simultaneously by reading the extinction at four wavelengths. A method has been developed with suitable equations using the corresponding extinction coefficients (Schmidt-Stohn *et al.*, 1980).

3. *Column Chromatography*

Before applying the extract on paper or on thin-layer adsorbents for separation into individual pigments, the extract must be concentrated *in vacuo* and at low temperature to avoid destruction. A good chromatographic separation is obtained when the extract is first freed from interfering impurities. This may be achieved by column chromatography, as at the same time the pigments are concentrated or even preliminarily separated.

The most common adsorbents used in column chromatography are cation-exchange resins, alumina, polyamide, polyvinylpyrrolidone (PVP), (Polyclar AT) or formophenolic resins. In the past ion-exchange resins (Dowex, Amberlite) yielded variable quantitative results, distorted relative proportions of the individual pigments and incomplete recovery. Such erroneous results might have been due to the use of hydrochloric acid as eluent (Smith and Luh, 1965; Fuleki and Francis, 1968a; von Elbe and Schaller, 1968). Recently the use of ion-exchange resins has been successfully reintroduced for purification of anthocyanins from traces of metals and a preliminary separation of the anthocyanin into groups has been achieved (Pifferi and Vaccari, 1980).

Polyvinylpyrrolidone (PVP) was first used for the isolation and purification of strawberry anthocyanins (Wrolstad and Putnam, 1969; Wrolstad et al., 1970b) and pigments of a grape variety could be purified and separated into individual pigments (Hrazdina, 1970). A polyamide column permits the detection and enrichment of minor components and the separation of acylated anthocyanins with minimal degradation (Strack and Mansell, 1975). A combination of column chromatography on polyamide and polyvinylpyrrolidone gave improved results in separating a complex mixture of acylated pigments (Hrazdina and Franzese, 1974).

Column chromatography on alumina has been used as a rapid method to separate complex-forming pigments (cyanidin-, petunidin and delphinidin-glycosides) on a preparative scale (Birkhofer et al., 1966). Recently purification and concentration of anthocyanins has been achieved on porous polymers, formophenolic resin having the highest adsorption capacity as compared with polyacrylic polymers and PVP (Saquet-Barel et al., 1982).

4. Paper and TLC Chromatography

For separation of the crude anthocyanin extract into individual pigments, paper chromatography was first introduced in 1948 by Bates-Smith and has been increasingly applied since then. Pioneering work was carried out by Geissman and Jurd (1955) and in particular by Harborne (1958), who established and tabulated R_f values in different solvent systems for a great number of pigments (Harborne, 1967).

The concentrated pigment extracted can be applied on a Whatman paper even without preliminary purification (the chromatographic separation being a purification process as well). As the paper is thick, a considerable amount can be applied.

The identification of a pigment is based on its chromatographic and spectrophotometric properties. The relevant chromatographic property is its R_f value (that is the distance covered in a certain solvent system which is determined by the polarity of the molecule). The polarity of the anthocyanidins increases with the number of hydroxyls in the molecule; for example, delphinidin is more polar than cyanidin; petunidin is more polar that peonidin and malvidin. The polarity also depends on the nature and number of the sugar substituents, the acylated monoglucosides being less polar than the non-acylated glucosides, of which the 3,5-diglucosides are the most polar.

The colour of the adsorbed pigment, in UV and visible light, is also indicative of the nature of the chromophore. The R_f value varies with the solvent system used of which a considerable number are listed, the choice of the most appropriate depending on the sample. Generally for rechromatography the solvent system is alternated.

Two-dimensional chromatography gives a sharp separation. With this method, grape anthocyanins were separated into 20 pigments by Ribéreau-Gayon before 1960 (Ribéreau-Gayon, 1982). The anthocyanins of various *Vitis* spp. were separated, identified and their relative proportions determined. Paper electrophoresis was also used for anthocyanin separation (Von Elbe *et al.*, 1969).

Almost at the same time the development of the technique of thin-layer chromatography started. The most adequate adsorbents were found to be silica gel and cellulose (von Hess and Mayer, 1962; Morton, 1967; Conradie and Neethling, 1968; Nybom, 1968; Mullick, 1969). This method has been improved upon and is used today (Barritt and Torre, 1973; Gomkoto, 1977; Torre and Barritt, 1977; Flora, 1978; Prasad and Jha, 1978; Vaccari and Pifferi, 1978; Miniati, 1981).

High pressure liquid chromatography (HPLC) was also found to be suitable for anthocyanins (Adamovics and Steinitz, 1976; Wilkinson *et al.*, 1977; Camire and Clydestale, 1979; Strack *et al.*, 1980). Using this technique, the anthocyanins of grapes, which contain the most complex pigment mixture, were resolved into 22–24 discrete peaks. Problematic pigments, acetic acid acylated pigments and acylated 3,5-diglucosides were successfully separated (Williams *et al.*, 1978; Wulf and Nagel, 1978).

5. *Spectrophotometric Properties*

The spectrophotometric properties of each individual pigment separated by chromatography additionally characterize the substance, each pigment having a characteristic absorption spectrum with two maxima, the major in the visible, the minor in UV.

The presence of *ortho*-dihydroxy-groups in the B-ring of the aglycone is tested by recording the spectrum after addition of $AlCl_3$, which produces a pronounced bathochromic shift. Substitution of a sugar at C-3 produces a shoulder at 440 nm. The ratio $\lambda_{440}/\lambda_{max}$ can be used as a parameter for the glycosylation pattern. Acylation of the sugar moiety produces an additional distinct peak or a shoulder in UV at 310–335 nm.

6. *Identification of the Sugar*

Acid hydrolysis of anthocyanins in boiling water for 30 min yields aglycone and sugar moieties. The anthocyanidins are extracted in amyl alcohol and may be further analysed. The sugars are found in the aqueous phase. After removing the acid by washing with an amine, the sugars are separated chromatographically and identified by standard methods. Exact information on the structure of the anthocyanins can be obtained by "controlled hydrolysis". The kinetics of the hydrolysis can be followed chromatographically at short time intervals, the number of intermediates increasing from simple to complex sugar derivatives.

7. Acyl Groups

The acids of the acylated pigment can be removed by mild alkaline hydrolysis and then identified by standard methods or by HPLC.

8. Quantitative Determination of Individual Pigments

After chromatographic separation into individual pigments, the constituents are determined quantitatively. The usual method is to determine each pigment spectrophotometrically after its elution from the chromatogram. A more rapid method is to determine the colour intensity of each spot of the chromatogram *in situ*. This is done by photodensitometry: as the chromatogram is scanned, the colour density is measured with a densitometer fitted with an automatic integrator. The method was first applied to determine individual anthocyanins in cranberry (Fuleki and Francis, 1968b). The quantitative distribution of anthocyanins has been determined by densitometry in sour cherries (Dekazos, 1970), figs (Puech *et al.*, 1975), *Rubus* spp. (Torre and Barritt, 1977) and muscadine grapes (Flora, 1978).

IV. Biosynthesis of Anthocyanins

The progress in the investigations of the enzymology of the biosynthetic pathway of flavonoids has recently been reviewed (Grisebach, 1980; Ebel and Hahlbroch, 1982).

The two aromatic rings of the flavan skeleton (Fig. 2.1) are derived from two different precursors; acetate units for ring A and phenylpropanoid precursor for ring B, which are joined by a condensation reaction. The active precursor of ring A is malonyl-CoA. This is formed through the carboxylation of acetyl-CoA, by the action of acetyl-CoA carboxylase, in the presence of ATP. The active precursor of ring B is *p*-coumaryl-CoA. This is formed by a series of reactions termed the "general phenylpropanoid metabolism", since the products of this pathway are intermediates not only of flavonoids, but also of other phenolic compounds, benzoic acids, lignin, etc.

The sequence of the reactions in the flavonoid biosynthesis involves the

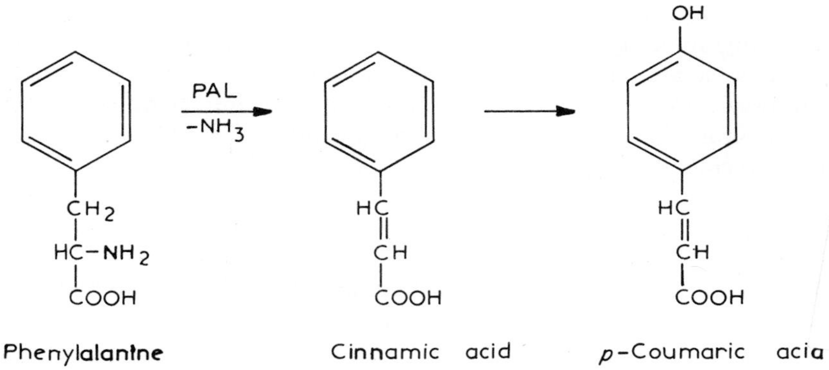

Fig. 2.8 The general phenylpropanoid metabolism.

transformation of phenylalanine into cinnamic acid through elimination of ammonia. The reaction is catalysed by phenylalanine ammonia-lyase (PAL). The enzyme cinnamate-4-hydroxylase catalyses the formation of *p*-coumaric acid. This is transformed into its active form *p*-coumaryl-CoA by the enzyme 4-coumarate-CoA lyase (Fig. 2.8).

PAL has been studied extensively as it is the link between primary metabolism and the phenylpropanoid pathways; however, being involved in all these pathways, it must not be considered a "key enzyme" implicated exclusively in flavonoid biosynthesis. The central reaction of flavonoid biosynthesis is the condensation of the acyl residues from one molecule of 4-coumaryl and three molecules of malonyl-CoA. The scheme outlined in Fig. 2.9 shows that chalcone is the first common precursor for all flavonoids. The enzyme chalcone synthase which catalyses its formation is the key enzyme in flavonoid biosynthesis. The postulated mechanism consists of three successive condensation steps of the 4-coumarate with acetate units derived from malonyl-CoA. The hypothetical intermediate has the side-chain elongated by six carbon atoms, which cyclize to give the ring A (Grisebach, 1980).

The activity of chalcone synthase has been related to anthocyanin biosynthesis (Fritsch and Grisebach, 1975; Hrazdina et al., 1978). The chalcone isomerase isomerizes chalcone into flavonone, which is then converted into flavones, flavonols or anthocyanins (Wellman and Baron, 1974). The specific steps in the biosynthesis of anthocyanins are not yet entirely elucidated. The anthocyanin biosynthesis is supposed to take place via dihydroflavonols (Kho *et al.*, 1977; Kho, 1978; Forkmann, 1980; Grisebach, 1980). The last steps in anthocyanin biosynthesis are glycosylation and acylation. First, the position at C-3 is glycosylated to stabilize the flavylium cation and subsequently the other positions. Several enzymes,

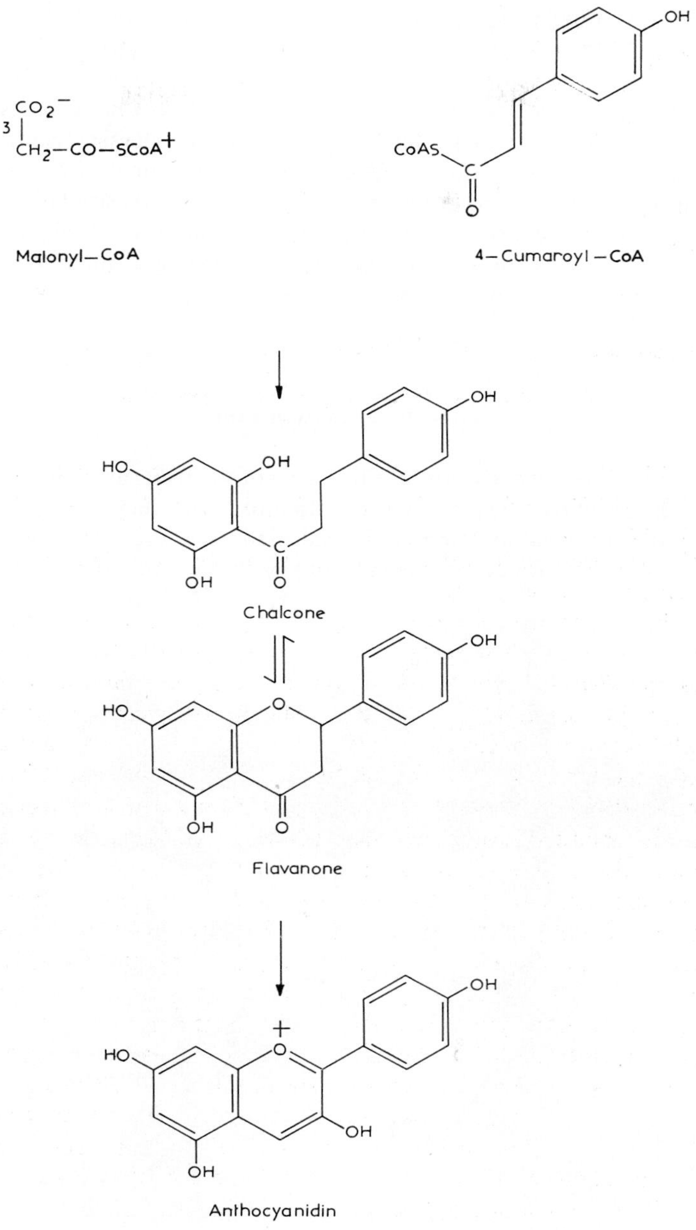

Fig. 2.9 Biosynthetic pathway of anthocyanins.

glucosyltransferases which catalyse glycosylation using uridine diphosphate-D-glucose (UDPG) as glucosyl-donor, have been characterized (Saleh et al., 1976; Kamsteeg et al., 1980).

The enzymes involved in the three stages of flavonoid biosynthesis, flavanone synthase (initial), chalcone–flavonone isomerase (central), and UDP-glucose: anthocaynidin-3-o-glucosyltransferase (final) are cytoplasmic and endoplasmic reticulum-bound enzymes. The biosynthesized pigments are transported into the vacuole where they accumulate (252). In an earlier investigation it was suggested that the enzymes are membrane-bound localized on the inner face of the tonoplast or in the vacuole (Fritsch and Grisebach, 1975).

V. Anthocyanins in Fruits

Anthocyanins are located mainly in the skins of the fruits as in plums, apples, pears, grapes and American cranberries. In most fruits they accumulate in the vacuoles of epidermal and sub-epidermal tissue (see Figs 2.10 and 2.11). In other fruits they are found both in skin and flesh, predominating in the skin as

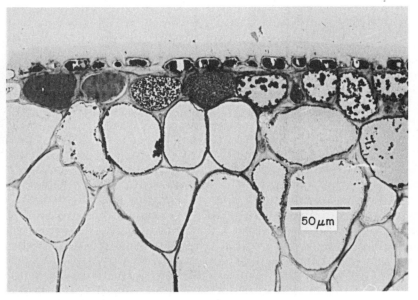

Fig. 2.10 Epoxy section through the skin of fresh Pilgrim cranberry (× 300). Two pigment-bearing cell layers beneath the cuticle are visible. From Sapers et al. (1983).

2. ANTHOCYANINS

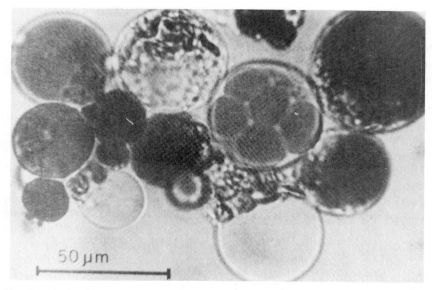

Fig. 2.11 Vacuole preparation from mature DeChaunac grape berry sub-epidermal tissues. From Moskowitz and Hrazdina (1981).

in some sweet cherries, or more evenly distributed as in sour cherries. In berries they are generally distributed throughout the whole fruit as in strawberries, currants and raspberries. An uncommon distribution is that of blood oranges, in which they are found only in the pulp.

The role of anthocyanins in fruits is to attract animals, which contribute to seed dispersal. Nothing is known about the role of the pigments found in the interior of the fruit. The colour differences among various fruits depend on the nature and the concentration of the anthocyanins. Additional factors affecting fruit colour are the pH of cell sap, the co-pigmentation effect determined by the presence of other flavonoids in the vacuole, and the metal chelation determined by the availability of metals which form complexes with the orthohydroxyl groups of the aglycone. The distribution of anthocyanins in plants has recently been surveyed by Hrazdina (1982). An extensive survey on anthocyanin distribution in food plants conducted by Timberlake and Bridle has also recently been published (Timberlake, 1981; Timberlake and Bridle, 1982).

Most of the results reported on anthocyanin distribution are based on early investigations in which the emphasis was on qualitative identification. The few quantitative results deal mainly with the total anthocyanin content.

VI. Characterization of Fruit Anthocyanins

Most fruits contain a mixture of anthocyanins from a simple pattern of only one pigment as in passion fruit (Ishikura and Sugahara, 1979) or two anthocyanins as in peaches (Ishikura, 1975) and pears (Timberlake and Bridle, 1971), to a complex pattern of over twenty pigments as found in some grapes (Williams *et al.*, 1978; Wulf and Nagel, 1978). The large number of over 200 different anthocyanins found in plants is due to the variation of the glycosidic type, as so far only 15 aglycones have been characterized.

In fruits the variation is by far more limited than in flowers. A total of about 50 different anthocyanins are present in the more common fruits surveyed in this book. The aglycones are represented by the six common anthocyanidins (p. 61) of which cyanidin represents 55%, peonidin and delphinidin (each 12%), pelargonidin and malvidin (each about 8%) and petunidin 6%. In some fruits such as apple and red currants, cyanidin is the only aglycone (Timberlake and Bridle, 1971; Øydvin, 1974); others such as grapes and lowbush blueberries contain five of the six common anthocyanidins, pelargonidin being absent (Francis *et al.*, 1966; Williams *et al.*, 1978; Wulf and Nagel, 1978). The most common pigment is cyanidine 3-glucoside, being ubiquitous in all genera and species except one avocado cultivar whose peel becomes intense purple during post-harvest ripening, due to anthocyanins (Prabha *et al.*, 1980). According to Harborne (1972), cyanidin is considered to be the most primitive pigment; in the advanced plants pelargonidin and delphinidin are more widespread. Increasing O-methylation, complex O-glycosylation and acylations with cinnamic acids can be correlated with evolutionary advancement.

Almost all glycosides of the common fruits have sugars linked to position 3 of the aglycone. The glycosidic pattern is dominated by the monosides, the most common sugar being glucose. In grapes *Vitis vinifera* monoglucosides are found exclusively (Wulf and Nagel, 1978); in apples a mixture of monosides, the major pigment containing galactose, the minor pigments containing glucose, arabinose and xylose. The most widespread dioside is 3-rutinoside, which is found in both sour and sweet cherries (Harborne, 1967; Dekazos, 1970), in black and red currants (Øydvin, 1974; LeLous *et al.*, 1975); figs (Puech *et al.*, 1975), olives (Maestro Durán and Vázquez Roncero, 1976), plums (Harborne, 1967) and raspberries (Torre and Barritt, 1977). Rutinoside is accompanied by sophoroside in sour cherries, red currants and raspberries. Sambubioside is found only in red currants and raspberries.

Less widespread are the 3,5-diglucosides, which are found in tropical fruits such as litchi and pomegranate (Du *et al.*, 1975; Prasad and Jha, 1978) and in American grapes (Williams *et al.*, 1978). Triglucosides, more often branched, are present in sour cherries, red currants, raspberries and olives.

Table 2.1

Pattern	Anthocyanidin	Glycoside type
I	One	One, various monosides
II	Various	One monoside
III	Various	Various monosides
IV	One	Various biosides
V	Various	One, various biosides
VI	Various	3,5-Diglucoside
VII	One, various	Various triosides
VIII	Various	Acylated monoside and diosides

Acylated anthocyanins occur rarely in fruits, the main source being the grapes (Williams et al., 1978). An acylated triglucoside has been detected in olives (Maestro Durán and Vázquez Roncero, 1976) and in avocado (Prabha et al., 1980); in apple and pears acylated anthocyanins have only been tentatively identified (Timberlake and Bridle, 1971).

The acylating acids are the most common cinnamic acids, p-coumaric and caffeic acids. Recently, besides these acids, acetic acid was identified with the aid of HPLC method, as the acylating acid of five anthocyanin 3-glucosides in the American Cabernet-Sauvignon (Wulf and Nagel, 1978). Acetic acid-acylated anthocyanins have also been separated with conventional methods in various *Vitis* spp. (Anderson et al., 1970; Fong et al., 1971). Acetic acid-acylated pigments are difficult to identify with conventional methods. Reinvestigations with HPLC would show if acetic acid is only confined to *Vitis* or if it is also present in other fruits (in which it has not yet been detected due to the lack of an adequate technique).

Based on these above mentioned qualitative data it is difficult to establish general patterns of distribution. However, a tentative systematization would comprise the major patterns shown in Table 2.1. Many fruits may be characterized by more than one pattern, which merge into each other. The respective patterns are given in the tables on anthocyanin distribution in the Appendix.

A. Quantitative Distribution of Anthocyanins in Fruits

Data referring to the quantitative determination of total anthocyanins and the relative proportions of the individual anthocyanins in fruits are very scarce (perhaps due to the lack of an adequate method). The rapid modern method of HPLC should be applied to fill all the gaps found in the distribution tables and to update the old data.

Table 2.2 *Total Anthocyanin Content in Some Fruits*

Fruit	mg $(100\,g\,fresh\,wt)^{-1}$	References
Apple		
Cranberry	45–100	Sapers et al., 1983
Currant, red	16	Øydvin, 1974
Cherry, sour	45	Dekazos, 1970
Grape, muscadine	40–403	Flora, 1978
Raspberry	20–60	Torre and Barritt, 1977
Strawberry	45–70	Wrolstad et al., 1970

Examples showing the anthocyanin levels in some fruits are given in Table 2.2. Data on relative proportions are given in the tables in the Appendix.

B. Importance of Anthocyanins in Taxonomy

Specific anthocyanin patterns characterize different species and cultivars. In some cases a common pattern is found in all species of a sole genus. The majority of examples available are based on qualitative surveys.

The Rosaceae family has been extensively investigated. Fifteen *Malus* spp. were found to have the same pattern, containing as main pigment cyanidin 3-galactoside, which is not found in other species (Harborne, 1967). The two *Vaccinium* spp. *macrocarpon* and *angustifolium* both have a very similar pattern characterized by the presence of three monoglycosides of all the five anthocyanidins excepting pelargonidin. A recent extended survey has proved that the pattern is not so uniform and consistent as was assumed in earlier surveys (Ballinger *et al.*, 1979).

The anthocyanin pattern of the sour cherry *Prunus cerasus* differs from the sweet cherry *P. avium* in containing, in addition to the more common pigments, both cyanidin 3-sophoroside and as a major pigment cyanidin 3-(2^G-glucosylrutinoside).

A pattern similar to that found in the sour cherry is found in the *Rubus* spp. in the raspberry, which contains additionally pelargonidin glycosides as minor pigments. Such a pattern is also found in *Ribes rubrum*, which differs in containing additional sambubioside and another branched triglycoside, which in turn are absent in *Ribes nigrum*.

The family of Vitaceae, namely the genus *Vitis*, is largely cultivated, and the anthocyanins of the grapes responsible for the colour of the red wines have been thoroughly investigated (Ribéreau-Gayon, 1982). It was found that each species has a characteristic pattern. The European *Vitis vinifera* can be distinguished from the American species by the absence of anthocyanidin 3,5-

diglucosides. A practical method for differentiating red wines of *V. vinifera* and those of hybrids has been developed based on this fact (Ribéreau-Gayon, 1982).

The anthocyanins are useful taxonomic markers at the familial and generic levels. Of additional taxonomic use may be the exact differentiation of the different fruit cultivars based on a detailed characterization of their anthocyanin pattern. Intercultivar variations exist both in the relative proportions of the major pigments and in the presence or absence of minor pigments, which is even more significant. This has been illustrated in varieties of sour cherries (von Elbe *et al.*, 1969) and in more recent and complete reports on red currants (Øydvin, 1974), raspberries (Barritt and Torre, 1975) and grapes (Williams *et al.*, 1978).

Two cultivars of red currants are lacking two pigments of the general pattern of six pigments which characterizes the others. One of the four pigments present, a branched triglycoside, rose to a level as high as 70% as compared to 30% in the cultivars with all six glycosides (The tables are reproduced in the Appendix.)

A survey of 37 red raspberry cultivars and selections showed that they can be separated into two groups. Nine cultivars contained two major anthocyanins, cyanidin 3-glucoside and cyanidin 3-sophoroside, and 27 cultivars contained in addition cyanidin 3-rutinoside and cyanidin 3-glucosylrutinoside. Related pelargonidin glycosides were detected in most cultivars and cyanidin 3,5-diglucoside was present in 13 cultivars. Each cultivar can be distinguished from the other by the relative proportions of the individual pigments of the anthocyanin pattern (Barritt and Torre, 1975).

VII. Pigment Changes During Ripening

During maturation the anthocyanins are synthesized at an increasing rate, especially near maturity, reaching a maximum in the fully ripe fruit. This is well illustrated in Fig. 1.10 in ripening strawberries. It can be seen that anthocyanin production was very low until 35 days after petal fall, when chlorophyll and carotenoid are almost at their lowest levels. It then increased sharply, reaching nearly 75% of the final concentration in seven days.

Following the qualitative and quantitative changes during maturation has a double impact. As the anthocyanin content increases gradually during maturation the total anthocyanin content is considered to be an index of maturity, besides being one of the most important quality parameters (Dekazos and Birth, 1970; Kushman and Ballinger, 1975; Watada and Abbott, 1975; Drake *et al.*, 1982). The second important implication of great interest is that by following systematically the anthocyanin pattern changes during ripening, we may achieve a deeper insight into the biosynthetic mechanism. It is known from earlier studies, especially on flowers (Harborne, 1967), that during development the sequences are characterized by a progression from lesser to greater complexity, both in anthocyanidins and glycosylation pattern.

A few recent reports deal with the systematic investigation of anthocyanins during ripening. In developing Montmorency sour cherries the pigment changes were followed at two week intervals by determining the absorption spectra of the acidulated methanolic pigment extracts at four different stages of maturity. This is shown in Fig. 2.12. The total anthocyanin content increased continuously during ripening, from 2 to 43.6 mg (100 g fresh wt)$^{-1}$, a development analogous to that occurring in strawberry (Fig. 1.10). The anthocyanin pattern contained, even at the most unripe stage, all four pigments present in the fully ripe fruit (the two anthocyanidins may be artefacts). Each individual pigment followed the general increasing trend, while their relative percentages remained almost unchanged (see Table 8 in Appendix).

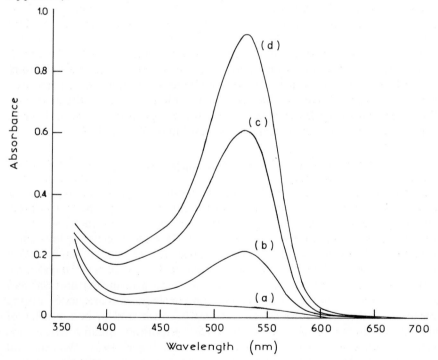

Fig. 2.12 The absorption spectra of anthocyanin pigments from Montmorency sour cherries in MeOH (1% HCl) at four stages of maturity: (a) very immature; (b) immature; (c) partially mature; (d) mature. From Dekazos (1970).

2. ANTHOCYANINS

The development of anthocyanins in grapes is characterized by a rapid increase followed by a slow accumulation and a final decrease at the end of maturation (Ribéreau-Gayon, 1982).

In red currant Red Dutch cv. the anthocyanin changes investigated at two maturity stages, the pink unripe and the dark fully ripe, showed a different development from that found in sour cherries (Øydvin, 1974). The total anthocyanin content almost doubled during ripening. As cyanidin is the only aglycone, the changes affect the glycoside pattern. Whereas in the unripe fruit sambubioside was the main pigment (39%), in ripe fruit it decreased to 22% of total pigments. The most striking feature at the last ripening stage was an enhanced synthesis of branched triosides, 2^G-xylosylrutinoside and 2^G-glucosylrutinoside, rising to 53% of the total glycosides (see Tables A.12(b) in the Appendix).

In the litchi, the red pigmentation is due to a mixture of cyanidin and pelargonidin glycosides. The changes in pigmentation pattern during ripening followed the outlined biosynthetic sequences; in the early stages cyanidin glycosides appeared and subsequently pelargonidin 3-rhamnoside and pelargonidin 3,5-diglucoside became dominant in the ripe fruit (Prasad and Jha, 1978; see Table A.23 in the Appendix).

In raspberry cv. Meeker the total anthocyanin content increased four-fold in the ripe fruit. The relative proportion of the four major pigments, which were all found in the unripe fruit, did not vary appreciably with maturity. Whereas cyanidin 3-sophoroside decreased, cyanidin 3-glucoside and cyanidin 3-glucosylrutinoside increased. Pelargonidin glycosides in trace amounts were detectable in the ripe stage (Barritt and Torre, 1975; see Table A.43(c) in the Appendix).

The anthocyanins of Manzanillo olives have recently been characterized as being a mixture of cyanidin glycosides: one monoside, two diosides and three triosides (Maestro Durán and Vázquez Roncero, 1976). The changes in the levels of these three groups were followed during ripening from October to the beginning of January. Before attaining their final relative proportions, which were constant during the last five weeks at monosides 15%, diosides 60% and triosides 25%, the fluctuations in their respective trends deviated from the outlined sequence. The level of the diosides had a maximum before mid-season and then decreased, whereas the monoglucosides remained at about the same level during the entire season. The triosides however followed an unexpected development, being highest (50%) at the beginning, decreasing to a minimum (10%) before mid-season and subsequently rising again to the final level (see also Table A.29 in the Appendix).

Evidently additional comprehensive systematic research, now facilitated by the availability of the HPLC method, will contribute to a better understanding of the subject in the future.

VIII. Factors Influencing Anthocyanin Levels in Fruits

A. Light

The amount of anthocyanin formation in plants depends on numerous environmental factors such as light, temperature, water condition, exogenous sugar, wounding and infections. The most important of all is the effect of light. Farmers were not only aware of the correlation between colour and the maturity stages of fruits, but observed for example in apples, that apples were redder on the sunny side (Magness, 1928). The fruit colour of grapes is also markedly affected by the amount of solar radiation that a grape cluster receives. This applies particularly to low-pigmented cultivars, like Tokay and Emperor (Weaver and McCune, 1960).

The studies on apple colour continued in order to find the optimal light regimes for its development (Sigelman and Hendricks, 1958; Looney, 1968; Proctor and Creasy, 1971). In the last study, the effect of supplementary artificial light in addition to the available natural light was considered and the conditions of maximal efficiency were established. Subsequent investigations on colour stimulation in apples with supplementary light were extended on several cultivars. The minimum energy requirement for anthocyanin synthesis varied considerably with cultivar and changed during the seasons (Proctor, 1974).

The effect of light on anthocyanin synthesis is expressed in the activation of the different enzymes involved in their biosynthesis. In spite of phenylalanine ammonia lyase (PAL) not being a key enzyme in this biosynthetic pathway, it was found that anthocyanin synthesis is directly influenced by PAL. PAL is stimulated by all the regulatory factors mentioned above (Camm and Towers, 1973).

In a series of fruits such as strawberry and apples, a direct correlation could be established between anthocyanin accumulation during ripening and PAL activity. Moreover, in the apple skin, PAL activity was observed only in the red parts of the peel (Aoki *et al.*, 1970; Hyodo, 1971). In ripening sweet cherries, anthocyanin synthesis and PAL activity did not appear to be directly related (Melin *et al.*, 1977). In Jonathan apples anthocyanin synthesis appeared to be regulated by the level of PAL activity, which in turn was initiated by light. Although not the critical enzyme in the anthocyanin synthesis in apple skin, it was found, however, to be rate-limiting (Faragher and Chalmers, 1977). A close relationship between PAL activity and anthocyanin synthesis was found in Red Spy apples (Tan, 1979). In the absence of light anthocyanin synthesis did not take place. The regulation of PAL levels seems to be due to a phenylalanine ammonia lyase inactivity system (PAL-IS) that has recently been isolated from different plants and is also present in apples (Creasy, 1976).

Relationships between the other enzymes which are involved in flavonoid biosynthesis and anthocyanin accumulation in fruits have not yet been reported.

B. Temperature

It is known that most anthocyanin-containing fruits develop a higher coloration in cooler regions. This was confirmed in the early study on apple colour (Magness, 1928). In the investigation by Tan (1979) on the relationship and interaction between PAL and PAL-IS and anthocyanin in apples, different temperature regimes were considered. Low temperatures of 6°C stimulated both anthocyanin synthesis, respectively PAL activity greatly. Fruits at alternating temperatures of 6°C and 18°C in light produced anthocyanin levels twice as high as fruits at constant 18°C, the levels of PAL and PAL-IS varying accordingly.

The influence of day and night temperature on anthocyanin synthesis in the apple skin has recently been investigated thoroughly (Diener and Naumann, 1981). The optimal temperature regimes changed with maturity. Unripe fruits necessitated lower temperature regimes (12°C day and 2°C night), whereas for more mature fruits high temperature values were more effective (24°C day and night). No difference was observed for different cultivars. It is suggested that it is possible to find the most efficient temperature regime for each ripening stage.

It was observed that in controlled temperature investigations, red grapes tend to develop the highest fruit coloration at 15–25°C day and 10–20°C night temperature (Kliewer and Torres, 1972). At higher temperatures (35°C day and 30°C night) the anthocyanin biosynthesis is completely inhibited in Tokay grapes and greatly reduced in the more deeply coloured Cardinal and Pinot noir cultivars (Kliewer, 1970).

The inability of a grape cultivar to form anthocyanin in fruits at high temperatures or in the absence of light may be related to the anthocyanin pattern present in the grapes. It has been assumed that the presence of anthocyanidins containing a B ring unsubstituted at C-3', or the absence of acylated anthocyanins could be responsible for the higher sensitivity to light and temperature effects (Kliewer, 1977).

C. Nutritional Effects

1. *Sugar*

It is obvious that anthocyanin synthesis requires the presence of free sugar. It has been found that the presence of sugar has a triggering effect on anthocyanin accumulation (Thiman et al., 1951; Pirrie and Mullins, 1976). Measuring the changes of sugar and anthocyanins in the skin of developing grape berries, a close relationship between sugar and anthocyanin content was found, the accumulation of sugar preceding the rise in anthocyanin content (Pirrie and Mullins, 1977, 1980). In litchi, the sugar metabolism contributed to the red pigmentation. Rhamnose was detectable in the peel in high amounts during active ripening, while in over-ripe fruits it was not detected, probably having been utilized during the biosynthesis of the respective glycoside (Prasad and Jha, 1978).

2. Nitrogen

Based on early investigation, the effect of nitrogen is associated with poor anthocyanin accumulation in fruits. It is not known whether the effect is indirect, producing increasing vegetative growth which hinders the penetration of light, delaying fruit maturity; or direct, as more nitrogen-containing substances such as amino acids and proteins are synthesized, interfering with the biosynthesis of sugars and anthocyanins.

In grapes it was found by Kliewer (1977) that high nitrogen doses reduced coloration, delaying maturity. According to Kliewer this is due to the diversion of photosynthate from carbohydrate accumulation to amino acid and protein synthesis.

D. The Effect of Phytohormones

The characteristic effects of each phytohormone in inhibiting or promoting ripening are described in Chapter 1, Section XII.

1. Ethylene

Ethylene, the ripening hormone, increases the rate of anthocyanin accumulation, these pigments being naturally associated with fruit maturation. The response to ethephon treatment of grapes (Vitis vinifera) has been found to vary with the time of application and the cultivar (Hale et al., 1970; Weaver and Pool, 1971). In apples, too, ethylene promotes anthocyanin accumulation (Chalmers and Faragher, 1977). A combination of ethephon and auxin enhanced the red colour, particularly of fruit treated previously with succinic acid 2,2-dimethylhydrazide (Edgerton and Blanpied, 1970).

In addition to apples, other fruits such as black currants, cherries, raspberries and olives gave good results after treatment with ethephon. Generally the responses vary with the date of treatment, the environmental conditions and the cultivars (Pecheur and Ribaillier, 1975). In figs, ethephon treatment promoted the rate of anthocyanin accumulation, but the total amount was 40% lower than in normal fruits which ripened 18 days later. Etiolated figs did not synthesize anthocyanins, indicating that the anthocyanin-synthesizing system has an obligatory light requirement (Puech *et al.*, 1976). Ethephon was also found to be effective for cranberries (Craker, 1971; Bramlage *et al.*, 1972; Rigbey *et al.*, 1977). In a recent experiment on McFarlin cranberry, ethephon treatment produced enhanced colour, but the fruit was devoid of keeping quality, being suitable only for processing and not for the fresh market (Shawa, 1979).

The mode of action of ethylene in promoting anthocyanin accumulation is probably to stimulate the activity of the enzymes involved in anthocyanin biosynthesis. The activity of phenylalanine ammonia-lyase is stimulated by light, ethylene and fruit ripening (Camm and Towers, 1973). However, ethylene can only stimulate PAL activity which is light-induced. When this condition is not fulfilled as in the etiolated figs, the anthocyanin synthesis is hindered.

2. *Other Phytohormones*

Gibberellin, which delays maturation, is extensively used in sweet cherry production to increase the processing season (Drake *et al.*, 1978). Cytokinins, which generally delay senescence, have an opposite effect on olives. Exogenous treatment induced anthocyanin accumulation, without affecting other parameters of maturation (Shulman and Lavee, 1973). Accordingly, it was found that the levels of endogenous cytokinins were considerably higher in black mature olives than in green fruits (Shulman and Lavee, 1973).

3
Carotenoids

I. Introduction

The yellow, orange and red colours of a great many fruits are due to the presence of carotenoids. Carotenoids are among the most widespread and important natural pigments. Together with chlorophylls they are found in all organisms capable of photosynthesis. Unlike chlorophylls and anthocyanins they are found also in various animals. It is estimated that about 100 million tonnes are produced annually in nature. Since carotenoids are synthesized only in plants, these constitute the source of all animal carotenoids. Animal products such as milk, butter and egg yolk contain carotenoids.

Carotenoids derive their name from the main representative of their group β-carotene, which was isolated from carrots (*Daucus carota*) by Wackenroder in 1831. Six years later Berzelius named the pigments of yellow leaves xanthophylls. In 1906 Tswett achieved the first separation of leaf pigments by inventing the chromatographic method. He separated four carotenoid pigments from two chlorophylls and recognized that the yellow pigments form a family (Tswett, 1906). A year later the carotenoids were characterized as isoprenoid derivatives (Willstätter and Mieg, 1907).

In 1930 Karrer elucidated the structure of β-carotene and vitamin A. Owing to the work of his school (Karrer, Kuhn and Lederer) the structure of more carotenoids was established as Karrer recognized their symmetrical structure. Ruzicka's isoprene rule has been of great help. Zechmeister introduced the concept of polyene, while Kuhn demonstrated the relationship between colour and conjugation in the chain of the double bonds.

In 1934 Zechmeister published the first monograph on the carotenoids (Zechmeister, 1934). In the same year Lederer published three monographs on carotenoids, one dealing specifically with plant carotenoids (Lederer, 1934). Further work followed by Karrer and Jucker (1948) and the comparative biochemistry of the carotenoids was investigated by Goodwin (1952). Meanwhile the number of known carotenoids had increased from 15 to 80.

This was possible due to the improvement in methods of isolation, the development of adsorption chromatography by Kuhn's school and by Strain (see Chapter 1, Section III.C.4(a)), the use of oxidative degradation methods and the development of synthetic methods.

Obviously in such an unsaturated substance a very large number of geometrical isomers of *cis-* and *trans-* configuration are possible, the subject being examined by Zechmeister (1962). After 1960 the new separation method of thin-layer chromatography allowed rapid progress in the field.

The availability of modern physical methods such as absorption, nuclear magnetic resonance (NMR) and mass spectroscopy (MS) brought a revolutionary advance. X-ray crystallography, optical rotatory dispersion (ORD) and circular dichroism (CD) are used in stereochemistry. At present the structures of approximately 450 carotenoids are known (Liaaen-Jensen, 1980). The structures of the carotenoids that were known before 1975 have been listed by Straub (1976).

Recent publications concerned with the carotenoids include the very comprehensive monograph by Isler (1971a). Carotenoid biochemistry is covered by Goodwin (1980), and Bauernfeind (1983) has emphasized the practical aspects of the use of carotenoids. Carotenoids are also dealt with by Hulme (1970), Goodwin (1976) and Czygan (1980). Additionally, the published main lectures of the carotenoid symposia up-date the most recent aspects. The first symposium was held as early as 1963 (Lang, 1963). The International Union of Pure and Applied Chemistry (IUPAC) publish the proceedings of their carotenoid symposia in *Pure and Applied Chemistry* every few years (IUPAC, 1967, 1969, 1973, 1976, 1979; Britton and Goodwin, 1982).

II. Definitions and Nomenclature

Carotenoids are isoprenoid polyenes formed by the joining together of eight C_5 isoprene units:

$$CH=C-CH=CH_2$$
$$|$$
$$CH_3$$

The isoprene units are linked in a regular head to tail manner (Fig. 3.1), except in the centre of the molecule where the order is inverted tail to tail, so that molecule is symmetrical. As a result the two methyl groups near the centre of the polyene chain are separated by six C atoms and the other methyl groups by five C atoms as illustrated in lycopene, the prototype. From its basic structure almost all other carotenoids can be formally derived by hydrogenation, cyclization, oxidation or any combination of these processes.

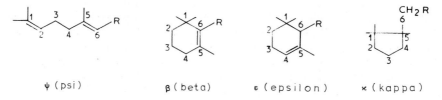

Fig. 3.1 Joining together of isoprenoid units.

According to Karrer, the numbering of the molecule from the end to the centre is 1 to 15, for the additional methyl groups it is 16 to 20 and for the symmetrical part it is 1′ to 20′. The former rules outlined by Karrer have been completed by IUPAC (Isler, 1971b; IUPAC, 1974).

The following discussion uses the trivial names, complemented by the new

Fig. 3.2 Some end-group designations of carotenes.

semi-systematic names. We quote one of the IUPAC rules necessary for further understanding: "The name of a specific hydrocarbon is constructed by adding two Greek letters as prefixes to the stem name carotene, these prefixes being characteristic of the two C_9 end groups: acyclic ψ (psi), cyclohexene β, ε (beta, epsilon), cyclopentane κ (kappa), etc." (Fig. 3.2).

III. Structure and Classification

The carotenoids can be divided into carotenes, which are hydrocarbons $C_{40}H_{56}$, and their oxygenated derivatives, xanthophylls. The most frequently found functions are hydroxy- (monols, diols and polyols), epoxy- (5,6 and 5,8-epoxides), methoxy, aldehyde, oxo, carboxy and ester. An hydroxyl substituent occurs mostly at C-3 in the ionone ring and a carbonyl sustituent occurs at C-4 in the β-ionone ring. In the majority of the cyclic carotenoids the 5,6- and 5',6'-double bonds are the most susceptible to epoxidation. The unconjugated double bond in the α-ionone ε ring does not undergo epoxidation.

Allenic carotenoids have a $C{=}C{=}C$ grouping at one end of the central chain, and acetylenic carotenoids have a $-C{\equiv}C-$ bond in position 7,8 and/or 7',8'. Two more carotenoid types are the degraded carotenoids with less than C_{40} in the skeleton of the apocarotenoids, and the higher carotenoids in which the carbon skeleton contains more than 40 carbon atoms, C_{45} and C_{50}.

Another classification divides the carotenoids into acyclic, and mono- and bicyclic carotenoids (alicyclic). In the following figures the structural formulae of the most widespread carotenoids found in fruits are shown. The acyclic carotenes are phytoene, the first C_{40} compound in carotenoid biosynthesis with three conjugated double bonds, followed by more unsaturated compounds phytofluene, ζ-carotene, neurosporene and lycopene (see Fig. 3.3).

Figure 3.4 shows β-carotene, the most widespread bicyclic carotene, and its hydroxy derivatives, the monol cryptoxanthin and the diol zeaxanthin, together with their various 5,6- and 5,8-monoepoxides and diepoxides—antheraxanthin and violaxanthin being the most important. In Fig. 3.5, α-carotene and its monohydroxy derivative and dihydroxy derivative—the widespread lutein and its epoxide, are shown. Other carotenoids are the allenic triol, the widespread neoxanthin which also contains a 5',6'-epoxide group, and neochrome, which has the same formula and a 5',8'-epoxide group (Fig. 3.6).

Unique carotenoids found in the fruits of certain genera will be discussed later.

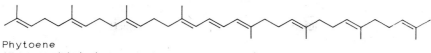

Phytoene
(7,8,11,12,7',8',11',12'-Octahydro-ψ,ψ-carotene)

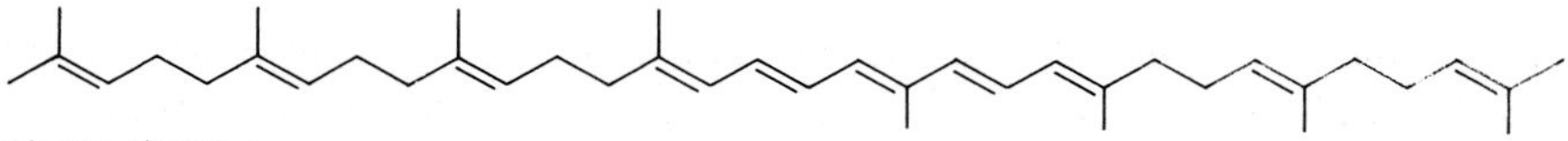

Phytofluene
(7,8,11,12,7',8'-Hexahydro-ψ,ψ-carotene)

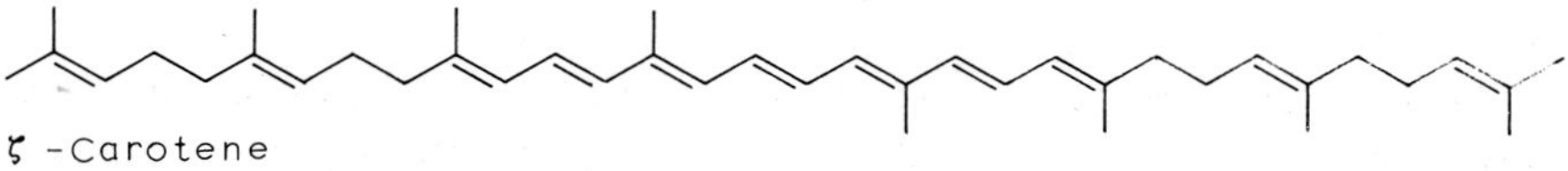

ζ-Carotene
(7,8,7',8'-Tetrahydro-ψ,ψ-carotene)

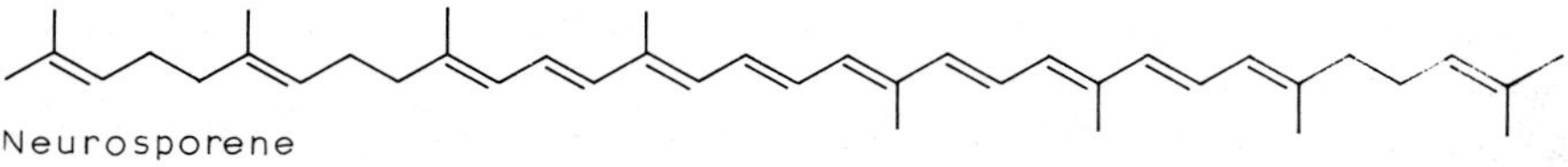

Neurosporene
(7,8-Dihydro-ψ,ψ-carotene)

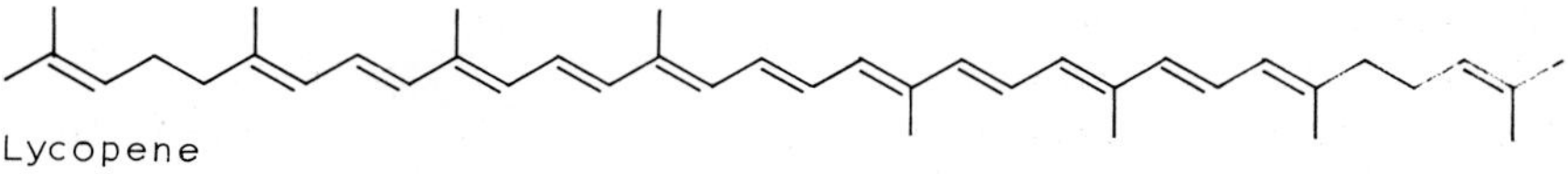

Lycopene
(ψ,ψ-Carotene)

Fig. 3.3 Acyclic carotenes.

IV. Physical Properties

A. Crystallization

Carotenoids crystallize in various forms, coloured from orange red to dark violet. Crystallization of carotenoids requires some practice, 1 mg being the minimum required quantity. The chromatographically pure pigment is crystallized from suitable solvent pairs. Carotenes are dissolved in a minimal volume of benzene or light petroleum to which methanol or ethanol are added in drops and the solution is left in the dark, allowing the pigment to crystallize. In a similar way, xanthophylls are dissolved in the minimum amount of methanol and a few drops of water or light petroleum are added (Britton and Goodwin, 1971).

(a)

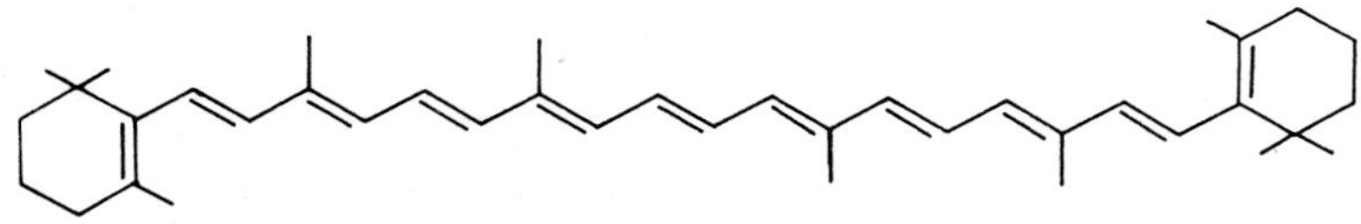

β-Carotene
(β,β-Carotene)

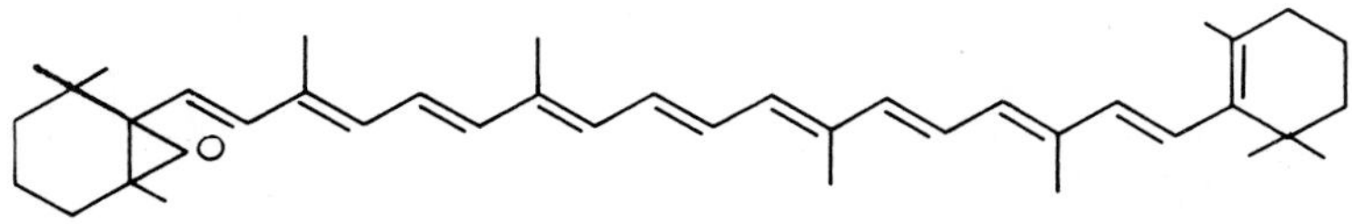

β-Carotene-5,6-epoxide
(5,6-Epoxy-5,6-dihydro-β,β-carotene)

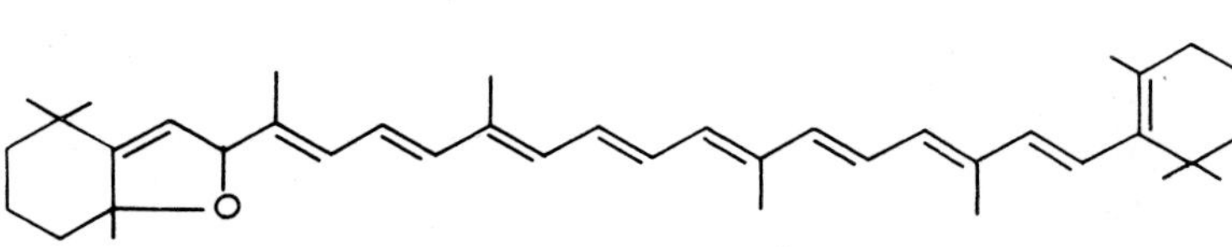

Mutatochrome
(5,8-Epoxy-5,8-dihydro-β,β-carotene)

(b)

Cryptoxanthin
(β,β-Caroten-3-ol)

Cryptoxanthin-5,6-epoxide
(5,6-Epoxy-5,6-dihydro-β,β-caroten-3-ol)

Cryptoflavin
(5,8-Epoxy-5,8-dihydro-β,β-caroten-3-ol)

Cryptoxanthin diepoxide
(5,6,5',6'-Diepoxy-5,6,5',6'-tetrahydro-β,β-caroten-3-ol)

Cryptoxanthin-5,6,5',8'-diepoxide
(5,6,5',8'-Diepoxy-5,6,5',8'-tetrahydro-β,β-caroten-3-ol)

(c)

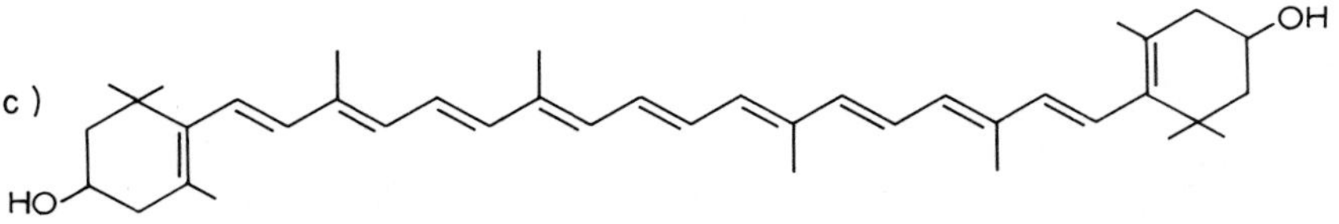

Zeaxanthin
(β,β-Carotene-3,3'-diol)

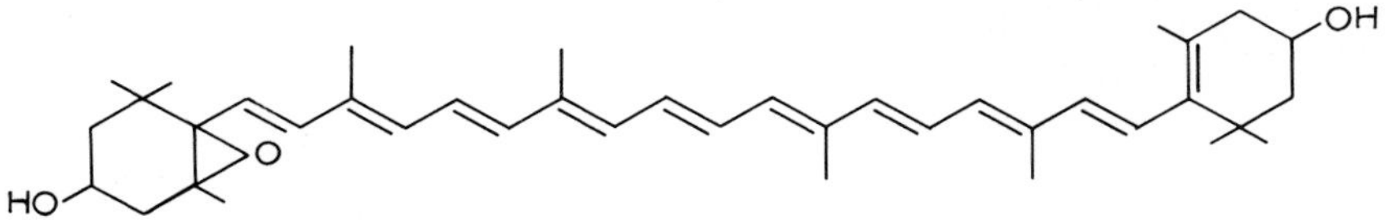

Antheraxanthin
(5,6-Epoxy-5,6-dihydro-β,β-carotene-3,3'-diol)

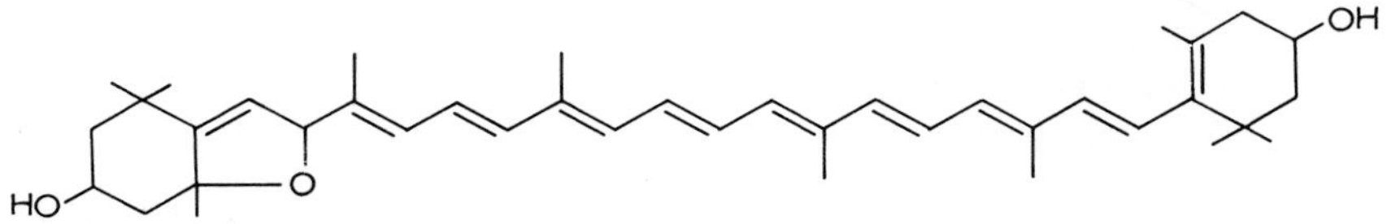

Mutatoxanthin
(5,8-Epoxy-5,8-dihydro-β,β-carotene-3,3'-diol)

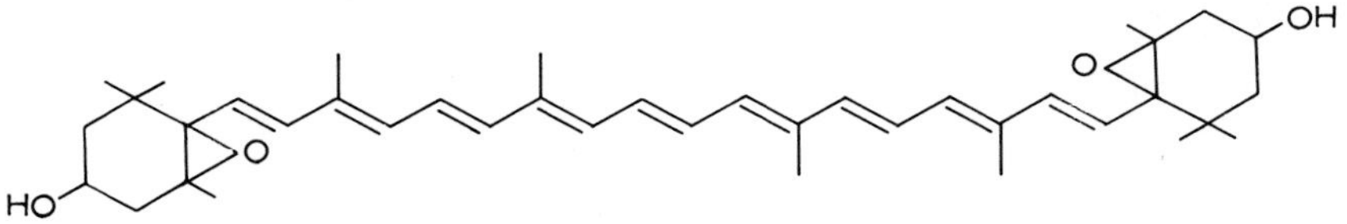

Violaxanthin
(5,6,5',6'-Diepoxy-5,6,5',6'-tetrahydro-β,β-carotene-3,3'-diol)

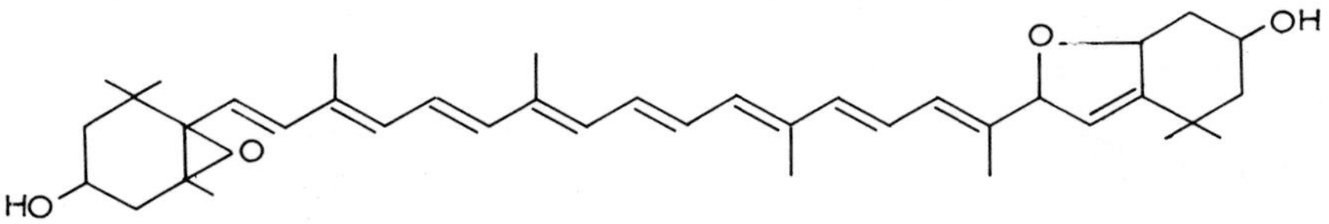

Luteoxanthin
(5,6,5',8'-Diepoxy-5,6,5',8'-tetrahydro-β,β-carotene-3,3'-diol)

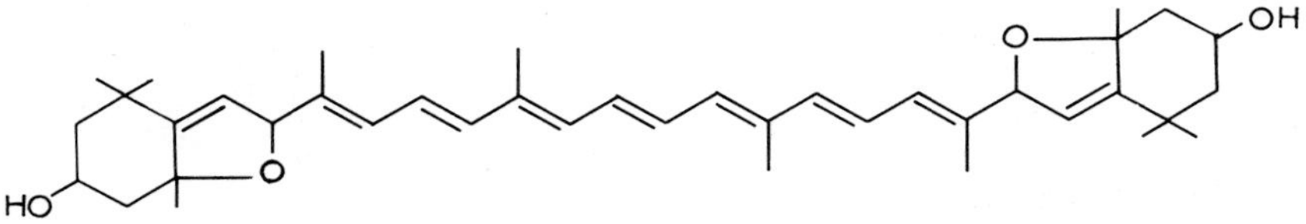

Auroxanthin
(5,8,5',8'-Diepoxy-5,8,5',8'-tetrahydro-β,β-carotene-3,3'-diol)

Fig. 3.4 Carotenoids of the β-ionone series.

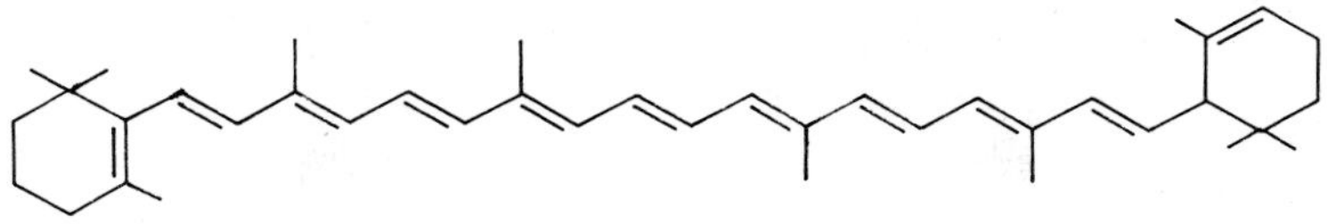

α-Carotene
(β,ε-Carotene)

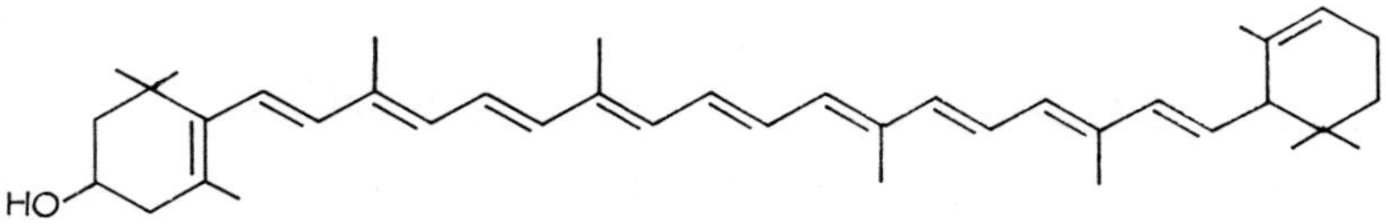

3-Hydroxy-α-carotene
(β,ε-Caroten-3-ol)

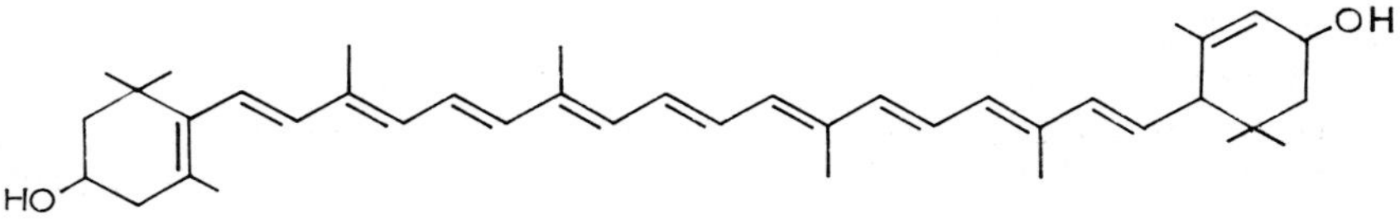

Lutein
(β,ε-Carotene-3,3'-diol)

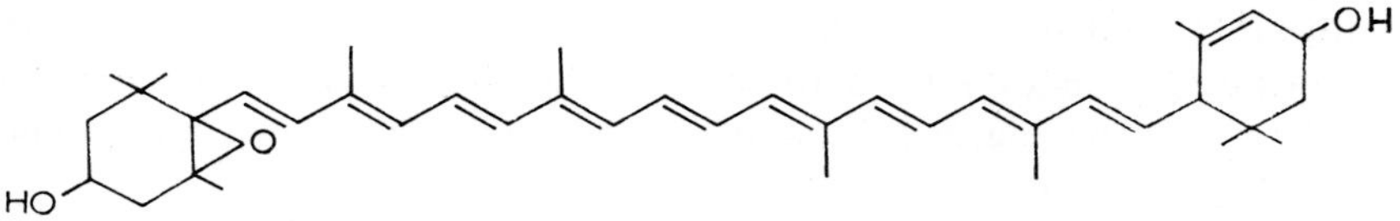

Isolutein
(5,6-Epoxy-5,6-dihydro-β,ε-carotene-3,3'-diol)

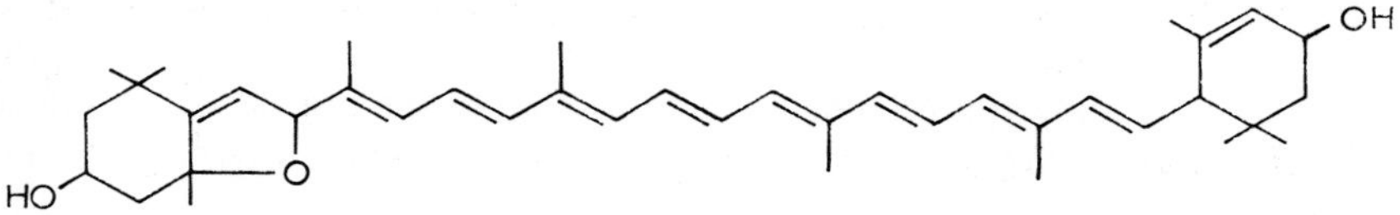

Chrysanthemaxanthin
(5,8-Epoxy-5,8-dihydro-β,ε-carotene-3,3'-diol)

Fig. 3.5 Carotenoids of the ε-ionone series.

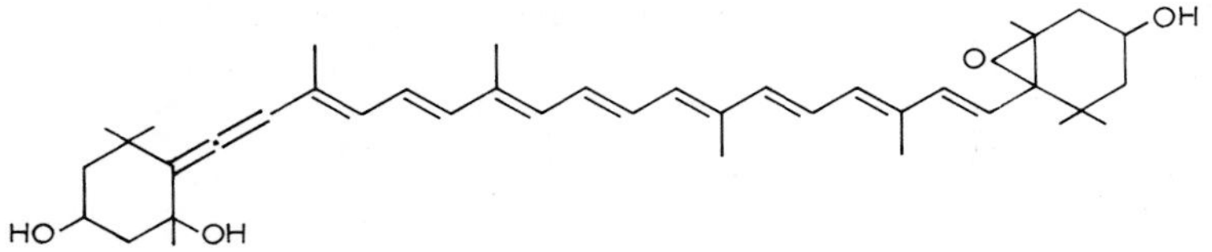

Neoxanthin
(5,6-Epoxy-6,7-didehydro-5,6,5',6'-tetrahydro-β,β-carotene-3,5,3'-triol)

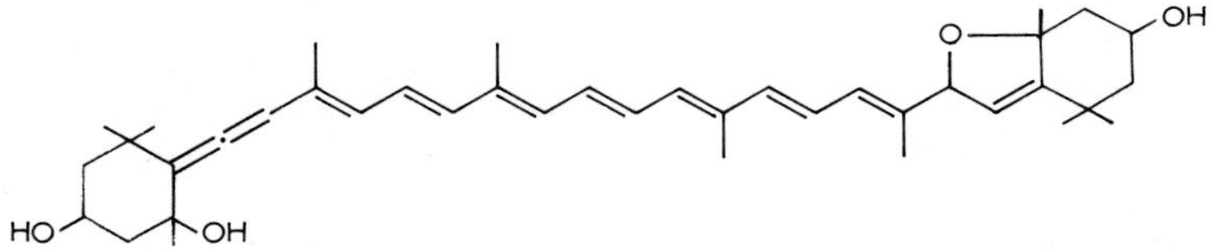

Neochrome
(5,8'-Epoxy-6,7-didehydro-5,6,5',8'-tetrahydro-β,β-carotene-3,5,3'-triol)

Fig. 3.6 Neoxanthin and neochrome.

B. Spectroscopic Properties

1. *Ultraviolet and Visible Light Absorption*

Each carotenoid is characterized by its electronic absorption spectrum. The position of the absorption bands, usually three, is a function of the number of conjugated double bonds. The positions of the absorption maxima are affected by the length of the chromophore, the nature of the double bond in the chain or in the ring, and the taking out of conjunction of one double bond in the ring or its elimination through epoxidation.

A ring closure as in β-carotene produces a less marked fine structure. The introduction of a carbonyl group produces the loss of fine structure, the spectrum having only one broad absorption maximum. The bathochromic shift produced by an in-chain carbonyl is about 25 nm, whereas that produced by an in-ring carbonyl is about 7 nm. The influence of substituents in the spectrum is negligible. The solvents used influence the position of the absorption maxima: petroleum ether and ethanol have little or no effect; chloroform and benzene produce a shift of +15 nm and carbon disulphide produces a shift of +35 nm.

Absorption maxima of some common carotenoids (excepting the uncommon β-citraurinene) that illustrate the differences in absorption spectra caused by differences in structure are given in Table 3.1. The absorption spectra of some carotenoids are shown in Fig. 3.7.

3. CAROTENOIDS

Table 3.1 *Absorption maxima of some common carotenoids of fruits*

Carotenoid	Conjugated Double Bonds		Abs. max. (nm)		
	In Chain	In Ring			
Phytofluene	5		331,	348,	367
ζ-Carotene	7		380,	400,	424
Lycopene	11		447,	472,	504
α-Carotene	9	1	423,	444,	473
β-Carotene	9	2	425,	450,	478
Zeaxanthin	9	2	427,	451,	480
Antheraxanthin	9	1	420,	444,	472
Mutatoxanthin	8	1	404,	427,	453
Violaxanthin	9		418,	440,	470
β-Citraurin	8 (+1 C=O)	1		456	
β-Citraurinene	8	1	402,	424,	450

Carotenes recorded in petroleum ether, xanthophylls in ethanol. From Davies (1976); Farin *et al.* (1983).

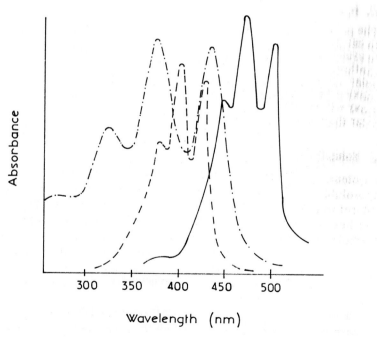

Fig. 3.7 Absorption spectra of some carotenoids: (——) lycopene; (– – –) ζ-carotene (in petroleum ether); (– · – · –) persicaxanthin (in ethanol).

2. *Relationship between Absorption Spectra of Carotenoids and their Colour*

The chromophores of phytoene (three conjugated double bonds) and phytofluene (five conjugated double bonds) are not long enough to provide colour. Both substances have their absorption spectra in the ultraviolet (UV) region with phytofluene exhibiting a green–blue UV fluorescence. Other carotenoids have their absorption bands within the limits of visible light, 400–700 nm. They absorb mainly in the blue (430–470 nm), but also in the blue-green (470–500 nm) and green (500–530 nm) regions of the spectrum and their colour is determined by the light they reflect (or transmit, when in solution). ζ-Carotene, having a chromophore of seven conjugated double bonds with absorption bands mainly in the violet region of the spectrum, exhibits a light yellow colour. With the displacement of the absorption bands of a pigment toward regions of longer wavelengths, hue increases from yellow, through orange to red.

C. Polarity

The polarity of the carotenoids is directly related to their chemical structure. In carotenes polarity increases with the number of conjugated double bonds. In cyclic carotenes, the β-ionone ring is more polar than the ε-ionone ring. In xanthophylls the polarity of the functional groups increases from the less polar monoepoxy to the most polar hydroxy group in the order 5,6-epoxy < 5,6,5′,6′-diepoxy < 5,8-epoxy < 5,8,5′,8′-diepoxy < oxo < monohydroxy < dioxo < dihydroxy < trihydroxy. Xanthophyll esters are slightly more polar than carotenes.

D. Solubility

Carotenoids are considered lipids and therefore are soluble in other lipids – liposoluble – and in fat solvents such as acetone, alcohol, diethyl ether and chloroform. The carotenes are soluble in apolar solvents like petroleum ether and hexane, whereas the xanthophylls disolve best in polar solvents like alcohols, ethanol and methanol.

V. Stereochemistry

The stereochemistry of the carotenoids involves geometrical isomerism about the carbon–carbon double bond and absolute configuration at an assymetric carbon.

A. Geometrical Isomerism

Owing to the double bonds in the molecules, all carotenoids present the phenomenon of *cis–trans* isomerization (Zechmeister, 1962). Naturally occurring *cis*- and poly-*cis* isomers have been isolated. Phytoene, the first C_{40} compound, the precursor of carotenes, has the 15-*cis* configuration. Other *cis*-isomers isolated may be artefacts formed by stereomutation (*cis–trans* isomerization) of the naturally preponderant *trans*-isomers.

The number of N possible stereoisomers of a product with n double bonds is calculated to be $N = 2^n$. In carotenoids the number is much more restricted because the double bonds of the acyclic polyene chain are of two types. In the first type the *cis* configuration is sterically unhindered as in the 15,15′ double bond. In the second type steric hindrance is found between —H and —CH_3; this type includes all double bonds adjacent to a methyl side-chain.

Stereomutation occurs under differing conditions. The isomerization process depends upon the structure and configuration of the carotenoid, the most stable isomer being the all-*trans* form; *trans–cis* isomerization increases with temperature and light intensity and is induced by the presence of acids. Since the identification of a carotenoid is linked to its stereochemistry, a standard method of stereomutation has been worked out (see Section VII.G). The absorption spectrum of a *cis*-isomer presents a subsidiary peak in the near ultraviolet, the *cis*-peak. Generally it is located at a distance of 143 nm from the longest wavelength maximum (e.g. 473 − 143 = 330 nm). The main absorption maxima are also slightly shifted (2–5 nm) toward shorter wavelengths, the *cis*-isomer having a lighter hue (Fig. 3.8).

B. Absolute Configuration

About half of the naturally occurring carotenoids possess at least one asymmetric end carbon and are therefore optically active. A carbon atom to which four different substituents are attached is called an asymmetric atom or in modern terminology a chiral centre. Absolute configuration is the description of the position or order of the different substituents in space. To designate the absolute configuration the general R and S convention (the sequence rule) is used (Weedon, 1971). The most conspicuous physical property of dissymetric substances is their optical activity or optical rotatory power; that is their ability to rotate the plane of polarized light.

Useful tools for the determination of the absolute configuration are measurements of optical rotatory dispersion (ORD) and circular dichroism (CD), the latter being increasingly used in routine analyses. The spectra are compared with the respective spectra of synthesized chiral model carotenoids, suitable derivatives and degradation products.

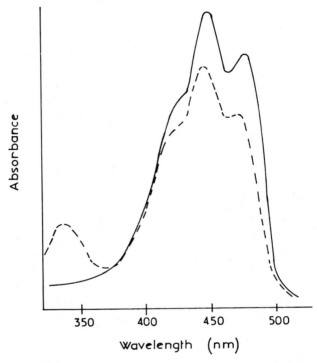

Fig. 3.8 Absorption spectra of (———) *trans-* and (– – –) *cis-β*-carotene.

According to Bartlett *et al.* (1969), the optically active natural carotenoids may be classified into three types: homodichiral, containing two identical chiral end-groups like zeaxanthin (3R, 3′R)-β,β-carotene-3, 3′-diol (Aasen *et al.*, 1971); heterodichiral containing two different chiral end-groups like lutein (6R, 3′R, 6′R)-β,ε-carotene-3,3′-diol (Andrewes *et al.*, 1974); and monochiral, with one chiral and one achiral end-group like α-carotene (6R′)-β,ε-carotene. The absolute configuration of carotenoid epoxides has been established and it has been found that after the acid-catalysed isomerization of 5,6-epoxides to the 5,8-furanoxides, the absolute configuration at C-5 is maintained (Goodfellow *et al.*, 1973).

So far, 32 chiral end-groups have been encountered in natural carotenoids. They are listed by Liaaen-Jensen (1980). Generally it has been found that individual chiral carotenoids have the same absolute configuration regardless of the biological source. This was recently refuted by new findings which were reviewed by Eugster (1983). Of the ten possible stereoisomers of ε,ε-carotene-diol, which has four chiral centres, six isomers have been isolated from natural sources. One of them is lactucaxanthin, isolated from Compositae (*Lactuca sativa*) (Siefermann-Harms *et al.*, 1981); another tunaxanthin, of which three different isomers are found in marine fish (Rønneberg *et al.*, 1978). An isomer of lutein, 3′-epilutein (3′R,6′S-lutein) has been found in the goldfish (Buchecker *et al.*, 1978) and also in egg yolk and flowers of *Caltha palustris* (Buchecker and Eugster, 1979).

The absolute configuration has enormous implications in the study of biogenesis. In order to elucidate the mechanism of biosynthesis it is essential to know both the chirality of the studied compound and the stereochemistry of its biosynthesis.

VI. Chemical Reactions

Since natural carotenoids contain a maximum of three elements (C, H and O), the functional groups are oxygen functions. Their characterization can be carried out on a spectroscopic scale requiring only small amounts of sample, and is usually a matter of routine analysis. Tests that are useful for the determination of carotenoid structure will be discussed further. The chemistry of carotenoids is discussed in detail in Liaaen-Jensen (1971) and Moss and Weedon (1976).

A. Epoxide Test

Isomerization of the 5,6-epoxides into 5,8-epoxides occurs rapidly, even in the presence of trace amounts of dilute acids (Fig. 3.9). The reaction is very fast and the rearrangement produces a hypsochromic shift in the visible absorption spectrum as the chromophore is shortened. A shift of 20 nm is indicative of one epoxide group, whereas a shift of 40 nm indicates the presence of two such groups (Fig. 3.10).

The isomerization of violaxanthin into auroxanthin involves a shift of 40 nm. The chromophore of auroxanthin is similar to that of ζ-carotene with seven conjugated double bonds. The reaction also produces a characteristic colour that varies from green for 5,6-epoxides to ink-blue for 5,8-epoxides. The latter group, of course, gives only the colour reaction. A rapid epoxide test is to expose a thin-layer chromatogram to HCl gas for a few minutes.

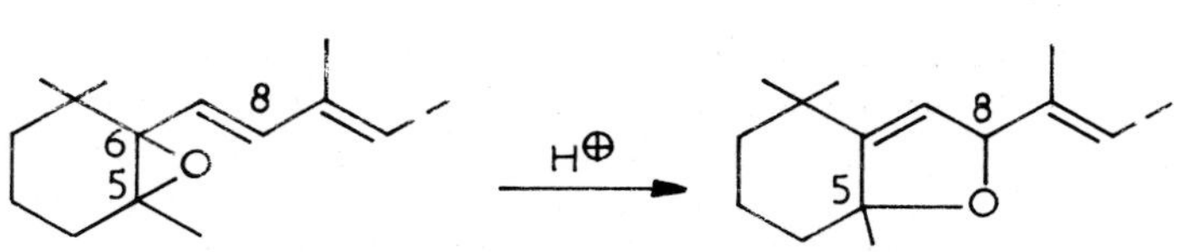

Fig. 3.9 Isomerization of 5,6 epoxide into 5,8 epoxide.

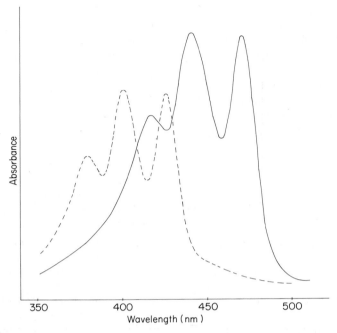

Fig. 3.10 Conversion of violaxanthin (———) into auroxanthin (- - -) by HCl.

B. Carbonyl Reduction Test

The reduction of the carbonyl function can be carried out with metal hydrides—lithium aluminium hydride in ethyl ether or sodium borohydride in ethanol. Both the rate of reduction under standardized conditions and the magnitude of the hypsochromic shift may be of diagnostic value. A hypsochromic shift of about 6–7 nm is indicative of an in-ring carbonyl. A shift of 25–35 nm indicates an in-chain carbonyl as in citraurin. After reduction the typical fine structure reappears (Fig. 3.11). Upon chromatography, reduction products may easily be recognized by their increased polarity and their lighter colour.

C. Hydroxyl Test

1. *Esterification*

Primary and secondary hydroxy groups are acetylated at room temperature by acetic anhydride in dry pyridine, while tertiary hydroxy groups do not react

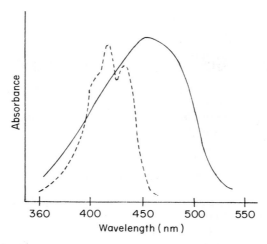

Fig. 3.11 Reduction of citraurin (——) to citraurol (– – –) with NaBH$_4$.

under these conditions. Following the course of the reaction chromatographically, the number of acetylable hydroxyls is indicated by the number of intermediary acetates observed. A monol gives one monoacetate, a diol one intermediary acetate if the molecule is symmetrical or a mixture of several if not, and a triol, three monoacetates and three diacetates.

2. *Silylation*

Tertiary hydroxyl groups as well as primary and secondary hydroxyls can be converted into trimethylethers. Hydroxyls which are sterically hindered (in position 6 and 6') cannot be silylated. Tertiary hydroxyl groups may be determined through silylation of the fully acetylated product.

3. *Allylic Hydroxyl Test (Dehydration and Etherification)*

Allylic hydroxyl groups can be dehydrated by treatment with acidified chloroform. The introduction of an additional double bond produces a bathochromic shift due to the extension of the chromophore. Allylic hydroxyl groups undergo methylation more easily than other alcohols. By treatment with methanol and HCl at room temperature, the methyl ether is easily formed. Being less polar than the alcohol, it can be separated chromatographically.

D. Carotenol Esters

For the investigation of the carotenol esters the saponification is of course omitted. The hydrolysis of the esters proceeds at room temperature in a weak (1–5%) alkaline solution. The kinetics of the reaction can be followed chromatographically as significant changes in polarity occur, the free xanthophylls being much more polar than the apolar esters. The mobility of the esters depends upon the nature of its fatty acid constituents. The nature of the carboxylic acid liberated through saponification can be determined.

VII. Analytical Methods

Analytical methods have been described by Britton and Goodwin (1971) and De Ritter and Purcell (1981), while perhaps the most complete coverage is given by Davies (1976). The analysis of carotenoids is complicated because of their instability—their tendency to undergo stereomutation, their photo- and thermolability and the readiness with which they undergo oxidation. All analytical operations must be carried out in dim light in an inert atmosphere (under nitrogen or vacuum). The solvents used must be purified, the temperature not above 40°C and samples stored at about $-20°C$ under nitrogen.

Isolation of carotenoids involves extraction, saponification and separation. The identification of the separated pigment is based on both spectroscopic and chromatographic properties and on chemical tests. Modern techniques, in particular mass spectrometry, have become indispensable tools for structural elucidation.

A. Extraction

Owing to the diversity of the biological material in which carotenoids are found, no standard technique can be outlined. Prior to extraction the material must be ground or cut into small pieces to facilitate complete extraction. Since the carotenoids are liposoluble they are extracted with organic solvents. All solvents used must be pure, free of oxidizing compounds, acids or halogens. Diethyl ether, which may contain peroxides, must be freshly prepared peroxide-free before use.

Prior to extraction, antioxidants and neutralizing agents are added. For carotenoid analysis of fruit in our laboratory butylated hydroxytoluene (BHT), *c.* 1%, is used as antioxidant. Tris-buffer [(trishydroxymethyl)-aminomethane] was found to be a better neutralizing agent than calcium carbonate or magnesium carbonate used in the earlier investigations, especially for unripe or very acidic fruit.

For fresh tissue, which contains a high percentage of water, water-miscible solvents (i.e. acetone, methanol, ethanol, usually a mixture of acetone and methanol) are used. Dry material may be extracted with water-immiscible solvents such as diethyl ether and benzene. Generally the extraction is carried out in a blender where grinding and extraction occur simultaneously. The best is the Ultra Turrax type in which the motor is mounted above the blades.

The homogenate is filtered under vacuum through a filter paper coated with filter aid in a Büchner funnel. The filter aid is diatomaceous earth (celite). From the water-containing crude extract the pigments are transferred to an appropriate water-immiscible solvent (diethyl ether, petroleum ether) by adding enough saturated NaCl solution. Other water-soluble non-carotenoid pigments are thus removed. Great care must be taken to remove all traces of acetone prior to saponification to prevent formation of artefacts produced by aldol condensations between acetone and carotenals in the presence of alkali (Stewart and Wheaton, 1973).

When chlorophylls are present, they can be determined in the epiphase, diethyl ether, using the equations of Smith and Benitez (Section III.C.1). If chlorophylls are absent, total carotenoids can be estimated spectrometrically.

B. Saponification

Saponification is a purification procedure which removes unwanted lipids and chlorophylls. It does not affect carotenoids, which are generally alkali-stable. When alkali-labile pigments (astaxanthin, fucoxanthin) are present or carotenoid esters are to be analysed, this step is omitted. Saponification is carried out by adding sufficient KOH in methanol or ethanol to have an overall KOH concentration between 5 and 10%. There are different alternative saponification methods: boiling the alcoholic lipid solution for 30 min or keeping it overnight at room temperature under N_2. Fruit extracts saponified for three hours with occasional heating up to 40°C gave the same results as the overnight procedure (our laboratory). Subsequently the saponified extract, mixed with ether in a separation funnel, is washed free of alkali, dried and evaporated *in vacuo*. Water is removed by azeotropic distillation with absolute ethanol, which is preferable to drying with sodium sulphate. The unsaponifiable matter is taken up in a suitable solvent and total carotenoids are determined.

C. Quantitative Determination

The standard method of quantitative determination is spectrophotometric and similar to the determination of chlorophylls and anthocyanins (Davies, 1976). The optical density of a known volume of carotenoid solution is read at the wavelength of maximal absorption. The amount of carotenoid (x) dissolved in V (known volume in ml) is given by the expression:

$$x = \frac{EV}{E_{1\,cm}^{1\%}} \times 100$$

E is the extinction of the solution and $E_{1\,cm}^{1\%}$ is the extinction coefficient of a 1% solution in a 1 cm light pass cell. The extinction coefficients of most carotenoids have been tabulated (Davies, 1976). For the determination of a mixture of carotenoids the $E_{1\,cm}^{1\%}$ of β-carotene = 2500 is used and the result is expressed as β-carotene.

D. Removal of Sterols

Sterols which are very abundant in some fruits can be removed by precipitation in different solvents. After determination of total carotenoids the extract is evaporated *in vacuo* and the unsaponifiable matter is dissolved in a minimum volume of methanol, petroleum ether or acetone and kept overnight at $-20°C$. The sterols are removed by centrifugation in a refrigerated centrifuge.

E. Phase Separation

Separation into epiphasic and hypophasic carotenoids, according to polarity, may be achieved by solvent partition between petroleum ether and aqueous methanol (85–95%). When the petroleum ether pigment extract is shaken in a separation funnel with an equal volume of aqueous methanol, the epiphase (PE) contains the chlorophylls, carotenes and the total esterified xanthophylls. Monoketones and monols are evenly distributed in both phases. Diols and polyols are hypophasic.

A method based on liquid–liquid partition is Craig's countercurrent distribution which was introduced and then applied by Curl in all his investigations on fruit carotenoids. The pigments were first separated into fractions of different polarity. Afterwards each fraction was submitted to column chromatography (Curl, 1953). The method was not used further in carotenoid separation.

F. Chromatography

Chromatography is the most important technique for the separation and purification of carotenoids. Introduced first by Tswett in 1906 it has since been developed and perfected (Tswett, 1906).

1. *Column Chromatography*

This technique is used for separation on a large scale, for preparative work or for a preliminary separation of carotenoids into different polarity groups. Most procedures are based on adsorption chromatography. The adsorbents used are chosen according to the types of carotenoids to be separated: for carotenes more polar adsorbents such as calcium hydroxide or activated alumina; for carotenoids of intermediate polarity, calcium carbonate or magnesium oxide; for strongly polar carotenoids cellulose or sucrose. There are three basic methods of column chromatography separation: (1) zone chromatography where the solvent is run through the column until the zones are separated, the adsorbent is extruded and the coloured zones are cut out and eluted; (2) the stepwise elution method where the zones are eluted from the column sequentially, in order of increasing adsorption, with solvents of increasing polarity; and (3) the gradient elution method which gives sharper separation than the stepwise (discontinuous) elution, used especially in automated column chromatography.

2. *Thin-layer Chromatography (TLC)*

The methods of thin-layer and adsorbent-loaded paper chromatography have both been applied successfully in carotenoid separation (Jensen and Liaaen-Jensen, 1959; Lang, 1963; Randerath, 1963; Stahl, 1965). For thin-layer chromatography a large number of adsorbents is available and suitable solvents are used singly or in mixtures. For paper chromatography kieselguhr, aluminium oxide, and silica gel are the most used adsorbents. For thin-layer chromatography on glass plate, the chromatogram is developed in a closed tank in an ascending manner. For paper chromatography the radial development of circular papers held horizontally between two Petri dishes is used (Jensen and Liaaen-Jensen, 1959). Both methods permit rapid and sharp separation of carotenoids and are generally applied on a micro and semimicro scale. Substances can be detected at trace levels. Thicker thin layers up to 1.5 mm can be used also on a preparative scale.

Most methods are based on adsorption, but partition chromatography has also been successfully applied. Partition chromatography is done on layers impregnated with paraffin or triglyceride, and the mobile phase may contain water (Randerath, 1963). Since carotenoids are coloured, staining methods are not necessary. Colourless fluorescent carotenoids like phytofluene are detected in UV light (360 nm). The adsorbent with the separated carotenoid is scraped from the plates directly into the eluent, filtered and subsequently analysed.

In our laboratory, during a period of about ten years, a combination of column with thin-layer chromatography was used for the separation of carotenoids in fruits (Gross *et al.*, 1971, 1978). Recently a new method for the rapid separation of citrus carotenoids by thin-layer chromatography was developed by Gross (1980). The separation of the carotenoids is achieved by successive thin-layer chromatographic separation on two different adsorbents. The first chromatography on silica gel G developed with the solvent system acetone–petroleum ether (30:70) gives a preliminary fractionation into groups of different polarity (hydrocarbons, monols, diols and polyols). A further separation of each group into individual carotenoids is obtained by rechromatography on MgO–kieselguhr (1:1, w/w). The same solvent system is used, the acetone percentage being increased according to the polarity of each group (from 4 to 30%).

Figure 3.12 represents a chromatogram on silica gel of a saponified carotenoid extract of Dancy tangerine juice. (Solvent system: acetone-petroleum ether (30:70), plate dimensions 10 cm × 20 cm). Some bands are a mixture of several carotenoids or of *cis–trans* isomers and others are single pigments. The citrus apocarotenals are separated as single pigments so that a further separation on MgO developing with an acetone-containing solvent system which would give the artefacts described by Stewart (1973), is not necessary (J. Gross, pers. obs.).

The method is not restricted to citrus fruit only, but may be applied to any carotenoid extract of plant or animal origin. It has been applied successfully in the analysis of various fruits, as will be seen later.

3. *High-performance Liquid Chromatography (HPLC)*

Besides the HPLC methods used for separation of green plant pigments (chlorophylls and carotenoids), HPLC can also be used exclusively for carotenoids. It was first explored by Stewart and Wheaton (1971), the efficiency of the method being tried out on citrus peel carotenoids, typical examples of highly complex mixtures. A two-step procedure was developed, separating carotenes on magnesium oxide and xanthophylls on zinc carbonate. Subsequently Stewart developed an HPLC method for the determination of provitamin A in orange juice, α,β-carotene and β-cryptoxanthin and a two-step method to follow the carotenoid changes in citrus juice during fruit maturation (Stewart, 1977a, b). Recently citrus carotenoids were separated by reversed-phase (rp-HPLC) in a one-step procedure (Noga and Lenz, 1983).

HPLC coupled with a UV-VIS scanning spectrophotometer was found to be suitable for the analysis of *cis–trans* isomerization mixtures (Fiksdahl *et al.*, 1978). HPLC was also tested for identification of aldehyde, ketone, 5,6-epoxy, 5,8-epoxy and *cis* double bonds. To identify functional groups the chromatograms were recorded before and after carrying out the chemical test (Matus *et*

3. CAROTENOIDS

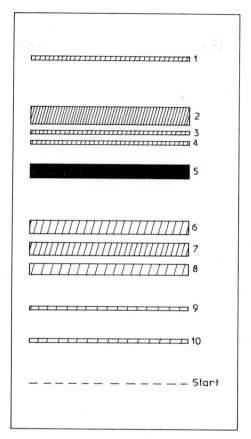

Fig. 3.12 TLC of saponified carotenoid extract of Dancy tangerine juice. Adsorbent, silica gel G; solvent system, acetone–petroleum ether (30:70). Band 1, carotenes; band 2, monol (cryptoxanthin); bands 3 and 4, cryptoxanthin epoxides; band 5, citraurin (found only in peels); band 6, diols (lutein, zeaxanthin and mutatoxanthin); band 7, diol epoxide (two isomers of antheraxanthin); band 8, diol diepoxides (two isomers of violaxanthin); bands 9 and 10, polyols (neoxanthin isomers).

al., 1981). A number of eleven cis-isomers of β,β-carotene could be successfully separated on alumina (Vecchi et al., 1981).

Two independent methods have been developed for the separation of the tomato carotenoids by HPLC and rp-HPLC (Zakaria et al., 1979; Cabibel et al., 1981). In the latter method identification is achieved by comparing both the retention time and the visible spectra (by a stopped flow scanning method) of the peaks. Reversed phase HPLC was applied to the separation of α- from β-carotene in fruit and vegetables (Bushway and Wilson, 1982).

Recently the characterization of the HPLC separated carotenoids from green tobacco leaf extracts has been facilitated by the aid of a microcomputer which collects spectral data from a wavelength-scanning microprocessor-controlled HPLC detector (Stewart *et al.*, 1983). The HPLC methods of plant pigments have recently been reviewed (Taylor and Ikawa, 1980; Schwartz and von Elbe, 1983).

G. Identification of Carotenoids

The carotenoids can be identified on the basis of their chromatographic and spectrophotometric behaviour. The adsorption affinity is determined by the length of the chromophore and by the polarity of the functional groups, as discussed previously. According to Davies (1976), due to the numerous factors which influence the mobility of the pigments on TLC, the R_f value *per se* is of little value.

The absorption spectrum of each chromatographically pure pigment is recorded and compared with values from the literature (Davies, 1976). The slightly different values that may appear in the literature are inherent to the spectrophotometer's accuracy, which varies generally with ± 2 nm. For quantitative determination of each pigment the extinction is read at the maximal absorption and the reported extinction coefficient is used in calculations. For functional groups the described chemical tests are carried out on spectroscopic scale.

For determining the nature of the geometric isomer, stereomutation is induced by iodine-catalysed photoisomerization. To a petroleum ether solution of the pigment, a few drops of an iodine solution in PE are added and the mixture is exposed to light. If the pigment is an all-*trans* isomer a hypsochromic shift occurs, while a *cis*-isomer gives a bathochromic shift. The isomers may be separated chromatographically as a *cis*-isomer differs from the parent all-*trans* isomer in its adsorption affinity and electronic spectrum.

When the pigment is finally identified the ultimate test to confirm the identification is the co-chromatography test. The unknown pigment is mixed with an authentic carotenoid and chromatographed. It is a clear indication that they are identical if the two pigments do not separate on different adsorbents when developed with different solvent mixtures.

Additional physical techniques for structure determination are mass spectroscopy, infrared spectroscopy, proton magnetic resonance spectroscopy, ^{13}C nuclear magnetic resonance (NMR) spectroscopy, optical rotatory dispersion and circular dichroism measurements, of which many may be carried out on a microgram scale.

VIII. Biosynthesis of Carotenoids

Carotenoid biosynthesis is discussed both in the books mentioned earlier and in the IUPAC series of proceedings of carotenoid symposia (IUPAC, 1969, 1976, 1979), and also in Goodwin (1952, 1982), Hulme (1970/71), Isler (1971a) and Czygan (1980). More recently carotenoid biosynthesis was reviewed comprehensively by Spurgeon and Porter (1983) and carotenoid biosynthesis in higher plants was discussed by Britton (1982).

Carotenoid biosynthesis involves a series of stages: (1) formation of mevalonic acid; (2) formation of geranylgeranyl pyrophosphate; (3) formation of phytoene; (4) desaturation of phytoene; (5) cyclization; (6) formation of xanthophylls. The biosynthetic pathway of carotenoids from mevalonate is common to other isoprenoids so that the elucidation of the mechanism involved at the earlier steps was facilitated by drawing an analogy with sterol biosynthesis, phytoene formation being similar to squalene formation.

Genetic studies, inhibition studies, techniques with labelled precursors and incubations *in vitro* with enzyme systems isolated from plant tissue, especially from tomato and pepper have all been used to elucidate the overall pathways and their stereochemistry. Multiple reports in the field have proved the existence of all the precursors involved and the supposed mechanisms in different organisms.

A. Formation of Mevalonic Acid

The biological isoprene precursor of carotenoids and all terpenoids is isopentenyl pyrophosphate (IPP). It is formed from mevalonic acid (MVA) which is first transformed into MVA-5-pyrophosphate in the presence of adenosine triphosphate (ATP) (Fig. 3.13). Mevalonic acid itself is formed through the condensation of three molecules of acetyl-CoA, via acetoacetyl-CoA, and β-hydroxy-β-methyl-glutaryl-CoA, the latter being reduced to MVA. Incorporation of labelled acetate and CO_2 into lycopene by tomato slices was demonstrated by Braithwaite and Goodwin (1950). Isopentenyl

Fig. 3.13 Conversion of mevalonic acid into isopentenyl pyrophosphate.

pyrophosphate is readily incorporated by cell-free systems from tomato fruits (IUPAC, 1969; Spurgeon and Porter, 1983), spinach leaves (Subbarayan et al., 1970) and chromoplasts isolated from pepper fruits (Camara et al., 1982).

B. Formation of Geranylgeranyl Pyrophosphate and Formation of Phytoene

The chain elongation from isopentenyl pyrophosphate (C_5) to geranylgeranyl pyrophosphate (C_{20}), which is catalysed by the enzyme prenyl-transferase, is shown in Fig. 3.14. Through the condensation of two molecules of GGPP phytoene, the first C_{40}-carotene is formed via the intermediate prephytoene pyrophosphate (Altman et al., 1972). In higher plants 15-cis-phytoene is generally found, the trans-isomer occurring in other organisms. Their formation depends on the stereochemistry of hydrogen removal at C_{15}. The enzyme system, the phytoene synthetase complex, first isolated from tomato fruits, was partially purified and characterized by Porter et al. (in IUPAC, 1979).

C. Desaturation of Phytoene

From phytoene, according to the Porter–Lincoln pathway, lycopene is formed through stepwise dehydrogenation. Through each dehydrogenation sequence the chromophore is extended by two double bonds alternately from either side. Phytoene is converted into lycopene, via phytofluene, ζ-carotene and neurosporene (see Fig. 3.3). β-Carotene is formed subsequently by cyclization of lycopene. The scheme was based on the pattern of pigment distribution in various tomato mutants at a time when the structure of all the pigment involved had not yet been elucidated (Porter and Lincoln, 1950). Since all the coloured carotenes are trans-isomers, a cis–trans isomerization must occur at the phytoene or phytofluene level.

In tomato all steps involving isomerization of cis-phytofluene into trans-phytofluene and four dehydrogenations were demonstrated (Qureshi et al., 1974). It is supposed that the removal of hydrogen is carried out by different dehydrogenases. Cis–trans isomerization at the phytoene level occurs also in pepper (Camara et al., 1980).

D. Formation of Cyclic Carotenes

The scheme of the biosynthesis of cyclic carotenes through cyclization of one or both end-groups of lycopene as suggested by Porter and Lincoln is given in Fig. 3.15. The cyclization mechanism was elucidated, as can be seen in Fig. 3.16. The same intermediate (carbonium ion) gives rise to both ε- and β-rings by elimination of different protons, showing that these rings are formed

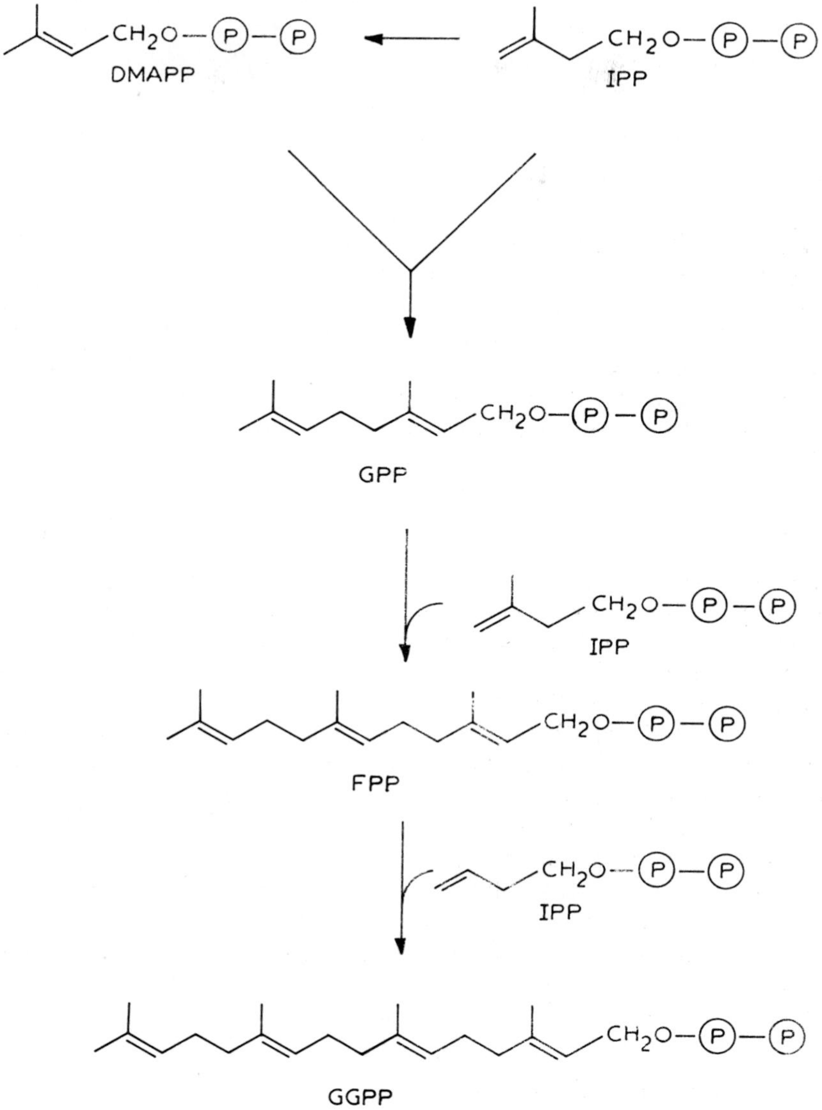

Formation of GGPP from IPP.

Fig. 3.14 Conversion of isopentenyl pyrophosphate into geranylgeranyl pyrophosphate.

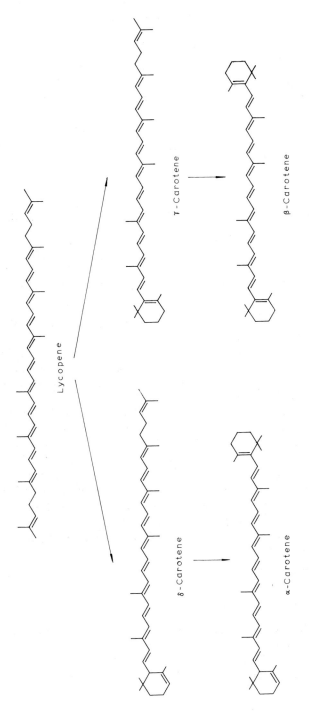

Fig. 3.15 Biosynthesis of monocyclic and bicyclic carotenes from lycopene.

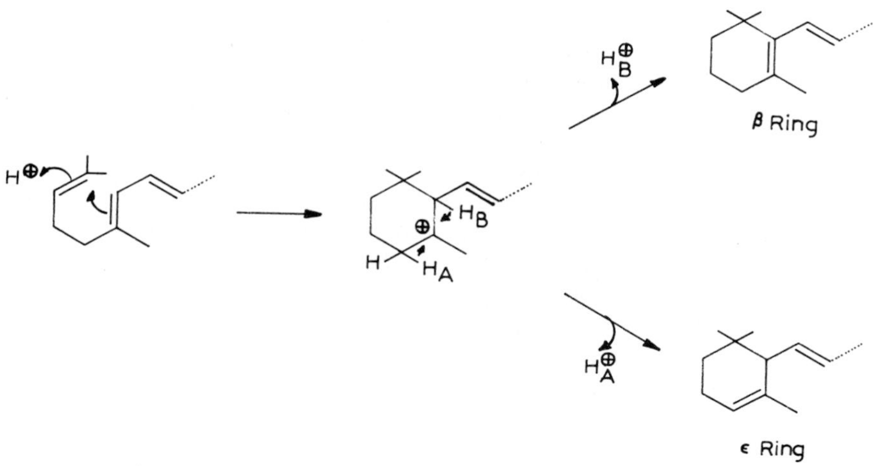

Fig. 3.16 Cyclization mechanism of β- and ε-ring formation.

independently of each other and not through isomerization of one into another (Williams et al., 1967).

The monocyclic intermediates of α- and β-carotene are δ-carotene and γ-carotene respectively. Cyclization may also occur at the neurospurene level. Again studies with tomato enzymes confirmed the scheme outlined (Kushwaha et al., 1970; Papastephanou et al., 1973).

Additional evidence that lycopene is the major cyclization substrate came from *in vitro* studies with cyclization inhibitors such as CPTA (2-(4-chlorophenylthio)-triethylamine hydrochloride) and nicotine, both of which cause accumulation of lycopene as cyclization is inhibited. The effect of CPTA was studied on citrus fruits, which after treatment developed an intense red colour like that of a ripe red tomato (Yokoyama et al., 1971, 1972).

The effect of nicotine was also studied in higher plants, *Cucurbita* species (Howes, 1974) and banana leaves (Gross and Costes, 1976). As in all *in vivo* studies, the latter study indicated that different enzymes may be involved in the synthesis of δ- and γ-carotene as well as for the second cyclization step, as they react differently to inhibition.

E. Formation of Xanthophylls

Insertion of the oxygen function, hydroxy and epoxy groups occurs after cyclization as a late step in biosynthesis when corresponding carotenes are oxidized. This was demonstrated through inhibition studies, especially with a strain of *Flavobacterium* that normally synthesized large amounts of zeaxanthin. In the presence of nicotine lycopene accumulates. On removal of nicotine lycopene is transformed into β-carotene under anaerobic conditions and into zeaxanthin in the presence of oxygen (McDermott et al., 1974). The oxygen is derived from molecular oxygen and not from water (Yamamoto et al., 1962), oxidation probably being catalysed by a mixed function oxidase (Goodwin, 1982) with retention of the absolute configuration (Milborrow, 1982).

Epoxy carotenoids are formed by the oxidation of the corresponding carotene or xanthophyll, e.g. violaxanthin is formed from epoxidation of zeaxanthin via antheraxanthin, the monoepoxide. The mechanism is not known (Spurgeon and Porter, 1983) and the source of oxygen in violaxanthin seems to be different, molecular oxygen or water (Takeguchi and Yamamoto, 1968), depending on the site where the biosynthesis occurs in the chloroplast (Costes et al., 1979). Natural 5,8-epoxides are formed through rearrangement of 5,6-epoxides in a slightly acid medium. Allenic carotenoids are formed through a rearrangement of the epoxide group according to a mechanism proposed for neoxanthin formation from violaxanthin (Davies et al., 1970; Isler, 1971a). The stereochemistry of allene biosynthesis has recently been established (Swift and Milborrow, 1981).

IX. Function of Carotenoids

A. Photofunctions

The photofunctions of carotenoids has been reviewed comprehensively by Krinsky (in Isler, 1971a and IUPAC, 1979) and updated by Mathis and Schenk (in Britton and Goodwin, 1982). The universal presence of carotenoids in photosynthetic organisms as chlorophyll–carotenoid complexes located in the photosynthetic membranes has a justification. From the variety of functions which have been attributed to these pigments, two important photofunctions have been clearly established: (1) their roles as accessory pigments in photosynthesis and (2) as protective agents of the photosynthetic apparatus against the potential damage of visible light.

1. *Role in Photosynthesis*

Chlorophylls play the major role in photosynthesis. For maximum light utilization they are aided by accessory pigments that absorb at different wavelengths from the chlorophylls—and transfer the absorbed light to them. Evidence that carotenoids can transfer energy to chlorophylls was found by measuring the enhancement of chlorophyll fluorescence on illumination with light of the wavelength absorbed by carotenoids.

The mechanism of energy transfer from carotenoids to chlorophylls is still not well understood, but the transfer is efficient only when both molecules are close together in the pigment–protein complex (Goedheer, 1979). Recently it was proved that both β-carotene and xanthophylls have the ability to transfer absorbed energy to chlorophylls (Siefermann-Harms, 1981). Earlier it was thought that only β-carotene was involved (Goedheer, 1979).

2. Photoprotection

The damage caused by visible radiation to cells can be altered by the presence of endogenous carotenoids. This was first observed in the purple photosynthetic bacterium *Rhodopseudomonas sphaeroides*, when the wild-type strain was compared with a blue–green mutant which lacked carotenoids. In the presence of air and light the mutant stopped growing, its chlorophyll destroyed and the cells killed, whereas the wild-type continued to grow. It could be proved that the chlorophyll was the endogenous photosensitizer responsible for the damage (Sistrom *et al.*, 1956).

Photosensitization involves harmful photochemical reactions initiated by the excited sensitizer in the triplet state. In photosynthetic plants the sensitizer is the chlorophyll, which in the excited state can either initiate redox reactions, with formation of free radicals, or transfer the energy to oxygen, forming the highly reactive singlet oxygen $^1O_2$. Carotenoids protect the organisms by either quenching directly the excess of energy of excited chlorophyll or quenching the singlet oxygen.

Photoprotection is related to the length of the chromophore and a minimum of nine conjugated double bonds is necessary (Mathews-Roth *et al.*, 1974). In addition to bacteria and algae, in higher plants a number of photosensitive mutants from maize and sunflower (*Helianthus annuus*) have been described (Walles, 1972; Mustárdy *et al.*, 1976).

The action of some herbicides is based on the photoprotective function of carotenoids. When carotenoid biosynthesis is blocked by herbicides, the protection is no longer provided and the plant is killed in high light intensities (Britton, 1979).

B. The Violaxanthin Cycle

Reversible changes in the violaxanthin content of leaves held under light and dark conditions was first observed by Sapozhnikov (Sapozhnikov *et al.*, 1957). The light-dependent conversion of xanthophylls is due to the "violaxanthin cycle", which involves de-epoxidation of violaxanthin in the light and re-epoxidation of the formed zeaxanthin in the dark, via antheraxanthin:

$$\text{violaxanthin} \underset{\text{dark}}{\overset{\text{light}}{\rightleftarrows}} \text{antheraxanthin} \underset{\text{dark}}{\overset{\text{light}}{\rightleftarrows}} \text{zeaxanthin}$$

The biochemistry of the reactions was investigated in detail mainly by Hager (1975) and Yamamoto (in IUPAC, 1979). Two enzymes are involved: the violaxanthin de-epoxidase situated on the inner side of the thylakoid membrane, and the epoxidase situated on the outer side of the membrane, so that the cycle arrangement is transmembraneous—which implies also the transport of the substrates across the membrane.

The enzyme de-epoxidase located on the inner surface is activated by the drop in pH within the thylakoid, due to photosynthetic electron transport. It is presumed to proceed by a reductive-dehydration mechanism. The epoxidation is catalysed by a mixed function oxygenase in the presence of oxygen and NADPH, its activity being independent of light.

Although numerous links between the violaxanthin cycle and photosynthetic processes have been established, a firm conclusion regarding its function has not yet been drawn.

C. Biological Function—Role of Carotenoids in Nutrition

1. *Provitamins A*

One of the most important physiological functions of carotenoids is to act as vitamin A precursors in animal organisms. Almost all animal species are able to convert enzymatically plant carotenoids of a special structure into vitamin A. Vitamin A is the primary alcohol retinol (shown below).

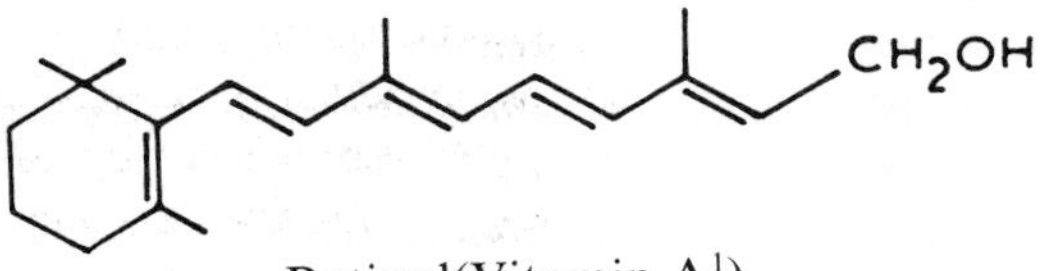

Retinol(Vitamin A^1)

A condition *sine qua non* for provitamin A activity of a carotenoid precursor is the unsubstituted β-ionone ring and unchanged side-chain. Chain lengthening decreases activity. β-Carotene, with two β-ionone rings, is the provitamin A with highest activity. All other carotenoids with provitamin activity have the intact β-ionone ring at only one-half of the molecule, e.g. α-carotene, γ-carotene, 5,6- and 5,8-monoepoxides of β-carotene, cryptoxanthin and its monoepoxides and the β-apocarotenals. The *cis*-isomers have a lower potency than the *trans*-isomers. Vitamin A$_2$ is 3-dehydroretinol and has about one-half of the vitamin activity of retinol. Carotenoids possessing a 3,4-dehydro-β-ionine ring act as provitamin A$_2$, as found by Gross (1967) (see also Budowski et al., 1963; Budowski and Gross, 1965; Gross and Budowski, 1966). An exception is the finding of provitamin A activity of astaxanthin and canthaxanthin in fish (Gross and Budowski, 1966).

Forty carotenoids are listed as vitamin A precursors (Bauernfeind, 1981). Of these the following sixteen are found in fruits: β-carotene α-carotene, γ-carotene, β-zeacarotene, β-carotene, 5,6-monoepoxide, mutatochrome, β-semicarotenone, cryptoxanthin, cryptoxanthin 5,6-monoepoxide, cryptoflavin, β-apo-8'-carotenal, β-apo-10'-carotenal, β-apo-12'-carotenal, β-apo-14'-carotenal, sintaxanthin and cryptocapsin.

Theoretically β-carotene, the most widespread provitamin, must give rise through central cleavage to two molecules of vitamin A. The conversion of β-carotene occurs mainly in the intestinal mucosa from which a soluble enzyme has been isolated. The enzyme which catalyses the central cleavage of β-carotene is 15,15'-dioxygenase, occurring in different species of animals (Lakshmanan *et al.*, 1972). This enzyme yields two molecules of retinal, which is subsequently reduced to retinol. The newly formed retinol is then esterified with long fatty acids, transported in the lymph and stored in the liver.

Although one molecule of β-carotene may be converted into two molecules of vitamin A, bioassays show that only one-half of the molecule is active, thus 0.6 μg carotene = 0.3 μg vitamin A = 1 IU vitamin A. This fact was explained *inter alia* as a result of incomplete resorption. Another acceptable explanation has been offered by Glover and Redfearn (1954). They postulated that β-carotene is degraded through oxidation at one or the other end of the polyene chain. The intermediates are the homologous carotenals. The metabolism scheme is given in Fig. 3.17.

The sources of man's vitamin A are dietary, both animal food which provides vitamin A and plant food—fruit and vegetables—which provide the provitamins A. The vitamin A activity depends on the nature and amount of biologically active carotenoids, their state of isomerization, their stability in the gastrointestinal tract and their digestibility. The FAO/WHO recommendation of 1967 was that one-sixth of dietary carotenes is metabolically available as vitamin A (FAO/WHO, 1967).

In 1974 the NAS-Recommended Dietary Allowances Panel introduced the term "retinol equivalent" to express vitamin A content in addition to international unit activity (1 IU = 0.3 μg retinol). By definition

1 retinol equivalent = 1 μg retinol
                        = 6 μg β-carotene
                        = 12 μg other provitamin A carotenoid
                        = 3.33 IU vitamin A activity from retinol
                        = 10 IU vitamin A activity from β-carotene
                        = 20 IU vitamin A activity from other provitamin A carotenoids

For the calculation of total retinol equivalents corresponding to β-carotene and other provitamins in μg, the following equation is used:

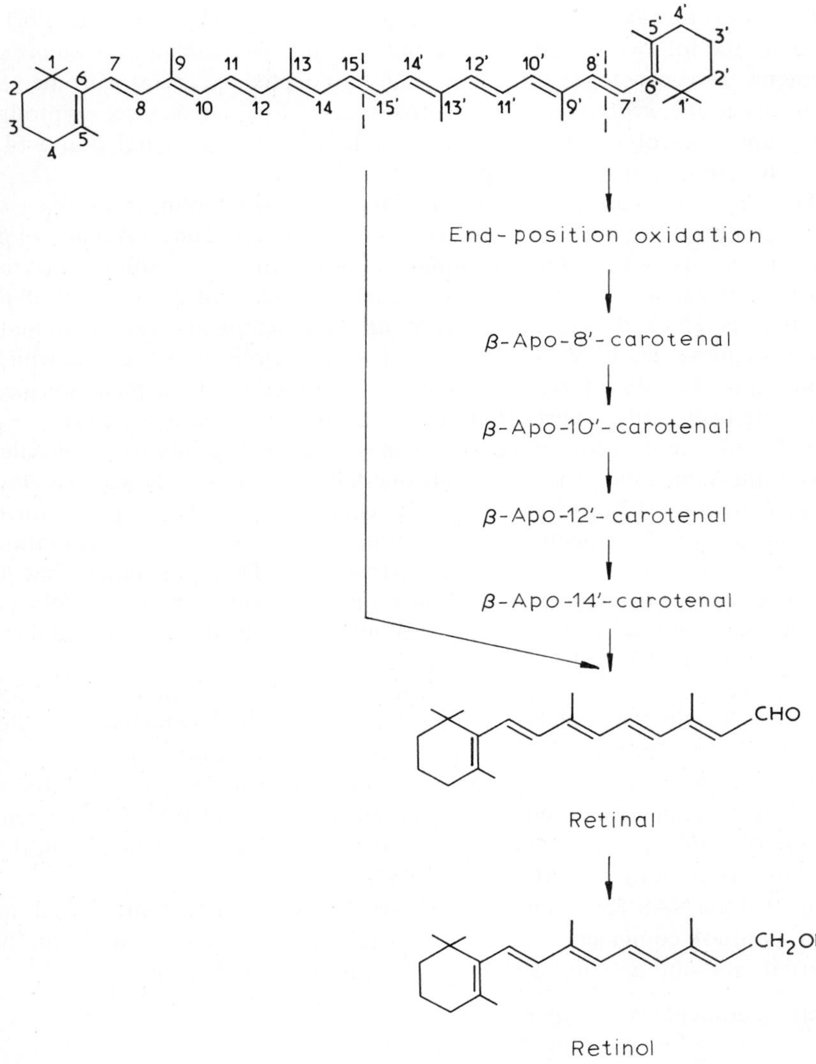

120   PIGMENTS IN FRUITS

[Structure of β-carotene with numbered carbons]

↓

End-position oxidation

↓

β-Apo-8'-carotenal

↓

β-Apo-10'-carotenal

↓

β-Apo-12'-carotenal

↓

β-Apo-14'-carotenal

↓

[Structure with CHO group]

Retinal

↓

[Structure with CH₂OH group]

Retinol

Fig. 3.17 Metabolism schemes of β-carotene to vitamin A.

$$\frac{\text{Total retinol}}{\text{equivalents}} = \frac{\mu g\ \beta\text{-carotene}}{6} + \frac{\mu g\ \text{other provitamin A carotenoids}}{12}$$

Total retinol equivalents content of the tabulated fruits in the Appendix are calculated accordingly.

Fruits and vegetables account approximately for 40% of the total vitamin A value of the food.

The recommended daily allowance is 1000 retinol equivalents. Most data on β-carotene content in food reported in food consumption tables are erroneously high as they are based on methods which determine the total carotenes without separation of β-carotene from other provitamins A which have lower activity. This is also pointed out by Simpson and Chichester (1981), who recently discussed the nutritional significance of carotenoids. The future standardization of new adequate methods of provitamin A determination and its mode of expression is essential.

The carotenoids ingested from food by man are unselectively absorbed (Bauernfeind, 1981). They are found in fat, blood, milk and additionally in retina, heart, liver, pancreas, nerve, bone, marrow, adrenals, seminal vesicles, corpora lutea and placenta. If high-carotene food is consumed excessively a high carotenoid content is found in blood, "carotenemia", and in the skin, which becomes yellow in "carotenodermia", but are not considered as harmful as not producing hypervitaminosis A (Bauernfeind, 1981).

X. Use of Carotenoids

A. Colorants

The use of carotenoids as colorants has been comprehensively discussed by Bauernfeind (1981). Some natural plant pigments used as food colours are annatto (bixin, a C_{25} diapocarotenoid being the main pigment), paprika extracts (capsanthin and capsorubin, see p. 133), alfalfa and tagetes extracts (xanthophylls, mainly lutein), tomato extracts and carrot extracts. These natural extracts are gradually being replaced by synthetic carotenoids, namely β-carotene, β-apo-8'-carotenal and canthaxanthin (4,4'-diketo-β-carotene). β-Carotene, which has vitamin A value, is the most recommended, producing yellow to orange colours; the other two colorants produce a red hue. These carotenoids are added directly to food for human consumption. Other carotenoids called "pigmenters", when added to animal feed, will colour either the body tissue (skin and fat) or animal products such as milk, eggs, butter and cheese.

B. Medical use

The medical applications of carotenoids have been reviewed recently in Bauernfeind (1981) and Britton and Goodwin (1982). The primary medical use of carotenoids is the prevention or correction of vitamin A deficiency in man. The adverse cutaneous response to sunlight in patients with erythropoietic protoporphyria (EPP) has been found to be ameliorated by oral ingestion of β-carotene. Porphyria is related to overproduction *in vivo* of endogenously synthesized porphyrin molecules. These molecules, with a structure similar to chlorophyll, act as endogenous photosensitizers, producing the sensitivity to visible light. The therapeutic effect of β-carotene is based on the same mechanism as photoprotection in plants. Quenching of singlet oxygen $^1O_2$ inhibits the peroxide formation which damages the lipid membranes.

Recently it was found that carotenoid pigments can prevent or slow down the growth of induced skin tumours (Britton and Goodwin, 1982). In another recent review it was concluded that the risk of cancer in human beings may be correlated inversely both with the level of retinol in blood and with the dietary intake of β-carotene (Peto *et al.*, 1981).

XI. Localization of Carotenoids

The carotenoids are located in plastids, chloroplasts and chromoplasts. Chloroplasts are ubiquitous in photosynthetic tissue, especially in leaves, and also in all unripe fruits and some ripe fruits, whereas chromoplasts are found in flowers and ripe fruits, to which they impart colours ranging from yellow to orange to red.

A. Chloroplasts

The pattern of chloroplast carotenoids is universally uniform containing four basic "chloroplast carotenoids", one carotene and three xanthophylls, a possible indication that all higher plants have evolved from a common ancestor (Strain, 1966). Additional carotenoids are found as minor pigments: α-carotene, α- and β-cryptoxanthin, isolutein (lutein 5,6-epoxide) zeaxanthin and antheraxanthin (zeaxanthin 5,6-epoxide). The last two pigments may be found in the 'violaxanthin cycle' (Hager, 1975; IUPAC, 1979). Recently an unusual carotenoid lactucaxanthin (ϵ,ϵ-carotene) was found in the leaves of the lettuce (*Lactuca sativa* (Siefermann-Harms *et al.*, 1981). The carotenoids of leaves have been extensively investigated and reviewed by Strain (1966).

Although the four chloroplast carotenoids are qualitatively the same, various carotenoid patterns are found in different species as a result of the innumerable variations in both the relative percentages of the four main pigments and the nature and relative percentages of the minor pigments. The approximate levels of chloroplast carotenoids are: lutein, which predominates, *c.* 40–50%; β-carotene, *c.* 20–30%; violaxanthin, up to 20%; and neoxanthin, *c.* 10–20%.

In Table 3.1 the quantitative distribution of carotenoids in leaves and green

Table 3.1 *Quantitative carotenoid distribution in the leaves and green peels of banana and avocado fruits, given as the percentage of total carotenoids*

	Banana		Avocado	
Carotenoid pattern	Leaves[a]	Peels[b]	Leaves[c]	Peels[c]
α-Carotene	9.0	13.7	16.0	6.1
β-Carotene	22.6	20.4	11.3	10.8
γ-Carotene	—	—	0.5	—
α-Cryptoxanthin	—	—	0.5	0.4
β-Cryptoxanthin	—	—	0.5	1.4
Lutein	40.9	40.2	51.0	55.8
Isolutein	—	3.7	1.7	6.4
Chrysanthemaxanthin	—	—	2.0	4.2
Zeaxanthin	—	—	1.6	—
Antheraxanthin	5.9	1.4	0.3	—
Luteoxanthin	—	0.7	1.6	1.5
Violaxanthin	7.9	9.5	5.0	3.1
Neoxanthin	13.7	10.4	7.5	9.8
Neochrome	—	—	0.5	0.9

[a] From Gross and Costes (1976).
[b] From Gross and Flügel (1982).
[c] From Gross et al. (1973).

peels of banana and avocado fruits are shown, illustrating both the complexity of the pattern and the interspecies differences.

The occurrence of different types of chloroplasts is a phenomenon generally observed in plants, determined by environmental conditions or associated with a particular type of cell (Thomson and Whatley, 1980). In fruits in which the plastid ultrastructure has been investigated, it is possible that the occurrence of chloroplasts of different types with thylakoids of various stacking degrees is determined ultimately by the prevailing light regime (Lichtenthaler, 1981; Meier and Lichtenthaler, 1981). It is to be expected that fruits with chloroplasts both in peel and pulp, or even within the pericarp, may be of different types. So far there are few fruits in which the ultrastructure has been thoroughly investigated and even fewer in which the corresponding carotenoids have been identified.

Different chloroplast types found within a fruit have been described for avocado, Chinese gooseberry, apple and green-fleshed melon (Chapter 1, Section X.A.,B.,C.). The chloroplast carotenoids found in the peel and pulp of the green-fleshed melon were investigated separately by Flügel and Gross

(1982) and marked quantitative differences were found. Laval-Martin found that the two kinds of chloroplasts in the outer regions and in the jelly, the inner part of the pericarp of "cherry" tomatoes, produced different carotenoid patterns (Laval-Martin, 1974, 1975).

The carotenoid distribution also varies within the chloroplasts. The relative percentages of the carotenoids within the chloroplast membranes is different in the green thylakoids from in the yellow chloroplast envelope, which is devoid of chlorophylls. Lutein is the predominant pigment in the thylakoids, whereas in the envelopes the violaxanthin level is almost three times higher than that of lutein (Jeffrey *et al.*, 1974). This was recently refuted, as evidence was presented that the chloroplast envelope does not contain carotenoids and quinones *in vivo* (Grumbach, 1984). Moreover, even throughout the thylakoid membrane, the carotenoids are unevenly distributed: the main carotenoid pigment in the reaction centres of photosystem I and photosystem II is β-carotene (Thornber *et al.*, 1979) and the light-harvesting chl a/b–protein complex is enriched in lutein and neoxanthin (Lichtenthaler *et al.*, 1982).

1. Biosynthetic Capabilities of Chloroplasts

Chloroplasts are capable of synthesizing chloroplast carotenoids using photosynthetically fixed CO_2. *In vivo* studies with an enzyme system from spinach chloroplasts demonstrated the conversion of labelled isopentenyl pyrophosphate and phytoene to acyclic carotenes (Subbarayan *et al.*, 1970). Recently it was demonstrated that chloroplasts can incorporate $^{14}CO_2$ into β-carotene (Bickel and Schultz, 1976).

As well as labelled CO_2, other precursors such as phosphoglycerate, acetate and mevalonate are capable of crossing the chloroplast envelope and have been incorporated into β-carotene (Grumbach and Forn, 1980).

Although the chloroplast autonomy in acetyl-coenzyme-A formation has been proved, the exact pathways of the formation of acetyl-coenzyme-A and mevalonic acid from photosynthetically fixed CO_2 have not yet been elucidated. The multiple proposed possibilities have been discussed by Goodwin (1982).

The epoxidation of zeaxanthin into violaxanthin is catalysed by the chloroplast envelope membrane (Costes *et al.*, 1979).

B. Chromoplasts

Chromoplasts have been reviewed by Sitte *et al.* (1980). Chromoplasts are different shaped plastids with an envelope consisting of two membranes and a stroma matrix containing ribosomes and filaments of DNA. As revealed by electron microscopy the internal structure is very different and so far four types of chromoplasts occurring in plants have been characterized: globulous, tubulous, membranous and crystallous. In many cases chromoplasts containing elements of different types may also be found. A list of the occurrence of the various chromoplast types found in plant organs is given in Sitte *et al.* (1980).

The most common type of chromoplast is the globulous chromoplast, which is considered to be the oldest in terms of evolution. The globulous chromoplasts are lens-shaped or spheroidal, containing carotenoid-carrying lipid droplets termed plastoglobules, which are of different sizes. The tubulous chromoplasts contain tubules which may or may not be associated with plastoglobules. The membranous chromoplasts contain numerous concentric membranes and are typical of flower chromoplasts. Crystallous chromoplasts contain carotenoid crystals which may determine the chromoplasts' shape. The crystals consist of pure pigment, and because they are large and coloured they were first observed a hundred years ago using a light microscope. In some cases the pigments may crystallize as small crystallites up to 100 being formed per chromoplast. Globulous chromoplasts occur in pepper (Kirk and Juniper, 1967), oranges (Thomson, 1966; Thomson *et al.*, 1967), muskmelon exocarp (p. 36), pummelo (p. 51), kumquat flavedo (Huyskens *et al.*, 1985), pears and plums (Sitte *et al.*, 1980). Tubulous chromoplasts occur in pepper (Spurr and Harris, 1968; Simpson and Lee, 1976). Membraneous chromoplasts have been observed in pepper (Spurr and Harris, 1968), pummelo (p. 52) and tomato (Harris and Spurr, 1969a).

Chromoplasts containing crystals were observed in the pulp of "Ruby Red" grapefruit using the light microscope some twenty years ago. They were described as needle-like and lycopene was the major pigment. As the amount of β-carotene increased platelet-like chromoplasts were formed (Purcell *et al.*, 1963). In the tomato, chromoplasts containing lycopene crystals were described in numerous reports (Ben-Shaul and Naftali, 1969; Laval-Martin, 1974; Mohr, 1979).

Ultrastructural studies on fruit chromoplasts are very incomplete, the vast majority of species having not yet been investigated. Even scarcer are reports about ultrastructure of both peel and pulp, or the exocarp and mesocarp of the same fruit. In fruits in which this research has been carried out, net chromoplast dimorphism has generally been observed. In *Asparagus* fruit, globular chromoplasts were confined to the epidermis and tubulous (fibrillar) chromoplasts were found in the inner regions of the pericarp (Simpson *et al.*, 1977). In "cherry" tomato, two kinds of chromoplasts were observed: in the inner part of the pericarp globular chromoplasts containing mainly β-carotene, in the outer part, chromoplasts containing voluminous sheets of lycopene (Laval-Martin, 1974). It is to be expected that dimorphism is a general phenomenon in all fruits as the carotenoid pattern differs quantitatively in all fruits between peel and pulp, and in many of them it also differs qualitatively.

This raises the question of a possible correlation between the fine structure and carotenoid pattern. As will be discussed later the carotenoid patterns found in fruits by far exceed the number of chromoplast types, but when chromoplast dimorphism exists in the fruit, the carotenoid pattern will certainly be different in each chromoplast type. According to Sitte *et al.* (1980), variations in the type and amounts of carotenoids, polar lipids and proteins may be the key to the development of each of the different types of chromoplasts.

1. *Biosynthetic Capabilities of Chromoplasts*

In higher plants, the chromoplasts are the site of a wide range of chromoplast carotenoids possessing remarkable metabolic activity. They contain all the carotenogenic enzymes and are therefore autonomous from this point of view.

The pathways in carotenoid biosynthesis were elucidated especially through investigations using enzyme systems isolated from tomato fruit plastids (Braithwaite and Goodwin, 1960; Kushwaha *et al.*, 1970; Papastephanou *et al.*, 1973; Qureshi *et al.*, 1974; Spurgeon and Porter, 1983). In the earlier investigations, relatively crude preparations were used. More recently *in vitro* experiments have been carried out with purified chromoplasts from pepper and daffodil flowers (*Narcissus pseudonarcissus*) which were both highly active in carotenoid biosynthesis (Beyer *et al.*, 1980; Camara and Monéger, 1982). The autonomy of the chromoplasts in carotenoid biosynthesis was proved as they were capable of synthesizing mevalonic acid. But this does not exclude the possibility that during chloroplast–chromoplast transformation precursors for carotenoid biosynthesis may be translocated also from extraplastidic sites into the plastid as the membrane permeability to mevalonate increased sharply at this stage (Schneider *et al.*, 1977).

The sites of the enzymes within the plastids have been investigated in both pepper and daffodils, the results suggesting that carotenoid biosynthesis is a compartmentalized process. Both investigations reported similar results, the stroma alone was the site of phytoene synthesis while the subsequent steps of desaturation and cyclization occurred only in the membrane fraction.

XII. Carotenoids in Fruits

A. Main Characteristics

Fruit carotenoids have been reviewed by Goodwin (1952, 1970, 1982) and more recently in Bauernfeind (1981). In this section older data will be updated, including both carotenoid distributions in fruits which were not previously investigated and reinvestigations with modern methods. The distribution tables for fruit carotenoids, both qualitative and quantitative, are presented in the Appendix. The vitamin A value expressed as retinol equivalents (RE) is calculated.

The carotenoid distribution in fruit is extremely complex and subject to considerable variation. Carotenoid patterns characteristic of each species and variety and even of each ripening stage occur. This pattern varies from the relatively simple, three main pigments in the pulp of banana fruit, to 50 carotenoids in citrus fruit juice. The total carotenoid content is also subject to considerable fluctuations in the different species: banana 0.8 $\mu g\ g^{-1}$, pulp of mandarin 14 $\mu g\ g^{-1}$, apricots 35 $\mu g\ g^{-1}$, tomato 54 $\mu g\ g^{-1}$, red pepper 150 $\mu g\ g^{-1}$, mandarin peel 300 $\mu g\ g^{-1}$, persimmon peel 490 $\mu g\ g^{-1}$ (complete data are given in the tables). One characteristic of fruit carotenoids is that a higher rate of biosynthesis occurs in the peel than in the flesh, so that the former contains high carotenoid levels, and in many cases a pattern differing from the pulp. In ripe fruit the xanthophylls are found mainly in esterified form. The carotenoid biosynthesis is autonomous in most of the fruit and thus continues even after the fruit has been separated from its parent plant.

B. Qualitative Distribution

Of the 450 carotenoids occurring in nature only about 70 are found in the fruits reviewed in this volume. These are the provitamins A listed on p. 119 and additionally the widespread diols and their epoxides and some triols and tetrols (structure p. 96, 132). There are also fruit containing unknown carotenoids which have still to be identified (Goodwin, 1982).

1. Unique Carotenoids

Some fruits have a special biosynthetic capacity, producing carotenoids which are found exclusively in the respective genus, although rarely they may be found in other genera as well. This is exemplified by citrus and pepper fruits.

(a) *Citrus C_{30}-apocarotenoids.* The citrus fruits of various shades of yellow, orange and red are the most important carotenogenic fruit. Besides the common carotenoids they contain a group of C_{30}-apocarotenoids which is genus-specific. Their structure is given in Fig. 3.18. The most abundant is β-citraurin (3-hydroxy-8′-apo-β-caroten-8′-al) and was first isolated and identified by Zechmeister and Tuzson (1937). Having a red–orange colour, it contributes to the reddish tint of some citrus fruit. Along with citraurin, β-

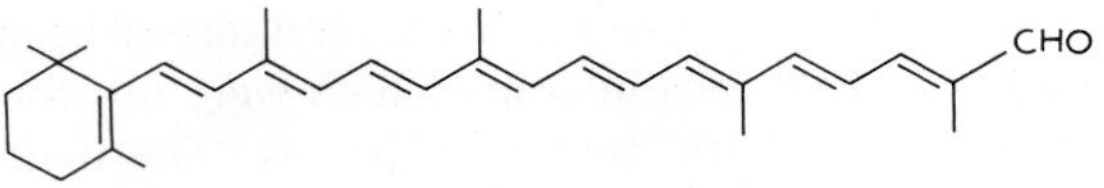

β-Apo-8'-carotenal
(8'-Apo-β-caroten-8'-al)

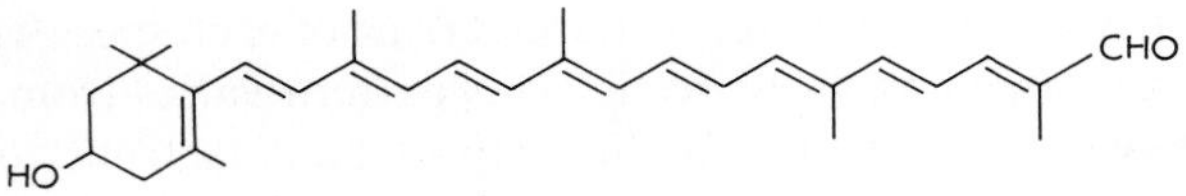

$C_{30}H_{40}O$

β-Citraurin
(3-Hydroxy-8'-apo-β-caroten-8'-al)

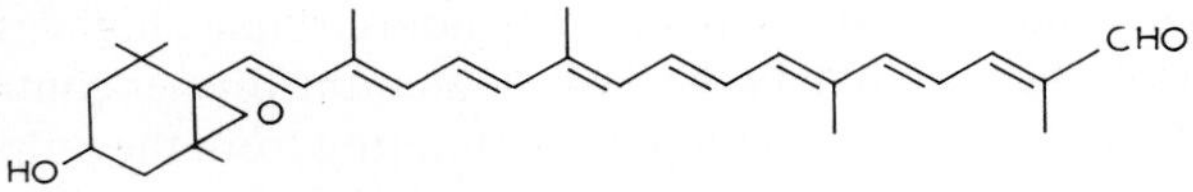

$C_{30}H_{40}O_2$

β-Citraurin-5,6-epoxide
(3-Hydroxy-5,6-epoxy-5,6-dihydro-8'-apo-β-caroten-8'-al)

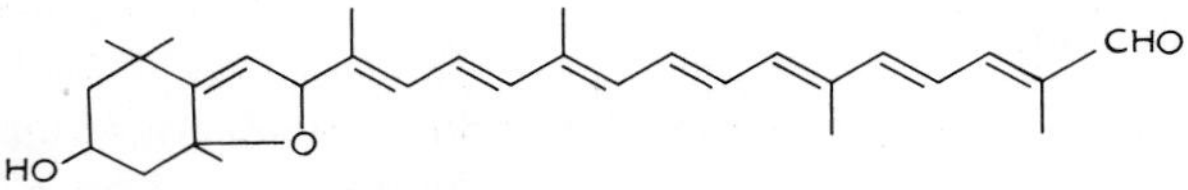

$C_{30}H_{40}O_3$

β-Citraurin-5,8-epoxide
(3-Hydroxy-5,8-epoxy-5,8-dihydro-8'-apo-β-caroten-8'-al)

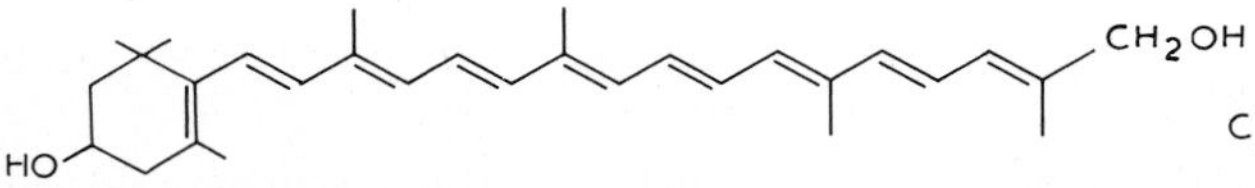

$C_{30}H_{40}O_3$

β-Citraurol
(8'-Apo-β-carotene-3,8'-diol)

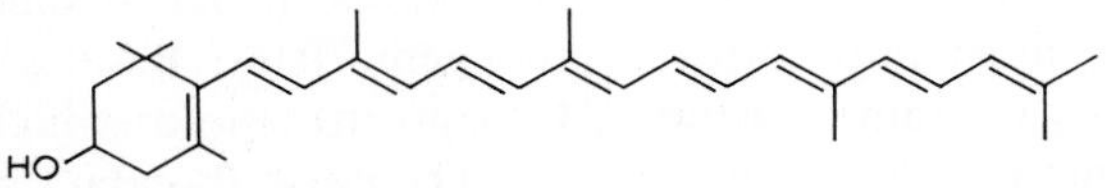

$C_{30}H_{42}O_2$

β-Citraurinene
(8'-Apo-β-caroten-3-ol)

$C_{30}H_{42}O$

Fig. 3.18 Structure of citrus C_{30}-apocarotenoids.

apo-8′-carotenal (8′-apo-β-caroten-8′-al) is generally found. It was isolated by Winterstein et al. (1960) from the peels of Clementines and sour oranges.

In the peel and pulp of Valencia oranges the 5,6- and 5,8-epoxides of citraurin (3-hydroxy-5,6-epoxy-5,6-dihydro-8′-apo-β-caroten-8′-al; 3-hydroxy-5,8-epoxy-5,8-dihydro-8′-apo-β-caroten-8′-al) were detected and their structures determined (Gross et al., 1975; Molnár and Szabolcs, 1980).

Two other C_{30}-apocarotenoids isolated from citrus hybrids were identified as β-citraurinene (8′-apo-β-carotene-3-ol) and β-citraurol (8′-apo-β-carotene-3,8′-diol) (Leuenberger and Stewart, 1976; Leuenberger et al., 1976). Citraurol can be obtained in vitro through the reduction of citraurin with sodium borohydride (see Fig. 3.11).

With regard to the biosynthesis of the C_{30}-apocarotenoids, citraurin was first thought to be a degradation product of zeaxanthin (Isler, 1971a), having the same absolute configuration (Bartlett et al., 1969). Recently the number of C_{30}-carotenoids found in citrus has increased to seven, suggesting that their biosynthesis may involve a special metabolic pathway (Leuenberger and Stewart, 1976; Leuenberger et al., 1976; Pfander, 1979).

Two different pathways for the biosynthesis of C_{30}-carotenoids are possible: through asymmetric degradation of a C_{10} fragment from one side of a C_{40}-carotenoid, or by the reaction to a C_{20} and a C_{10} compound, leading in both cases to C_{30}-apocarotenoids. The second possibility, a symmetric degradation of two C_5-fragments from both ends of a C_{40}-molecule or the biosynthesis via squalene, would lead to diapocarotenoids (Pfander, 1979). After comparing natural C_{30}-carotenoids with synthetic diapocarotenoids this possibility was rejected.

Another possible way via the asymmetric degradation of a C_{10}-fragment from one side of a C_{40}-carotenoid, the latter being cryptoxanthin has been confirmed (Farin et al., 1983). Comparing the flavedo carotenoids of a mandarin hybrid (Citrus reticulata) cv. Michal with that of its parents Dancy tangerine and Clementine it was found that there is a straightforward precursor–product relationship between cryptoxanthin and citraurin. Citraurin may undergo a further enzymatic reduction yielding as an intermediate citraurol and finally citraurinene. Treatment with ethylene induced the accumulation of cryptoxanthin, β-citraurin and citraurinene in the flavedo of Robinson tangerine, additional proof for the proposed pathway (Stewart and Wheaton, 1972).

(b) *Citrus apocarotenals.* Besides containing C_{30}-apocarotenals, citrus fruit are also a good source of other apocarotenals. Their structure is given in Fig. 3.19. β-apo-2′-carotenal and β-apo-10′-carotenal were isolated from peel of Clementines together with β-apo-8′-carotenal (Winterstein et al., 1960). β-apo-10′-carotenal was also found in the juice of different orange cultivars

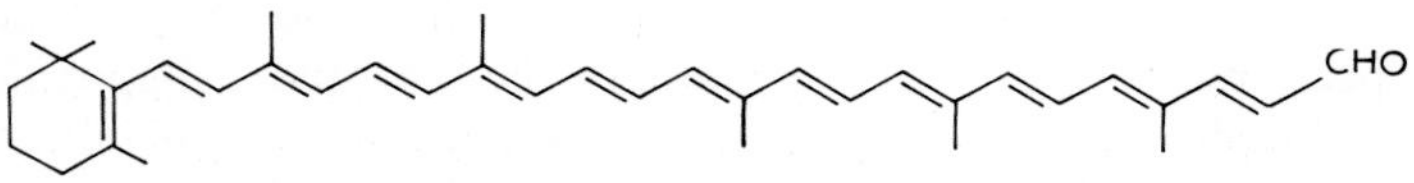

β-Apo-2'-carotenal
(3',4'-Didehydro-2'-apo-β-caroten-2'-al)

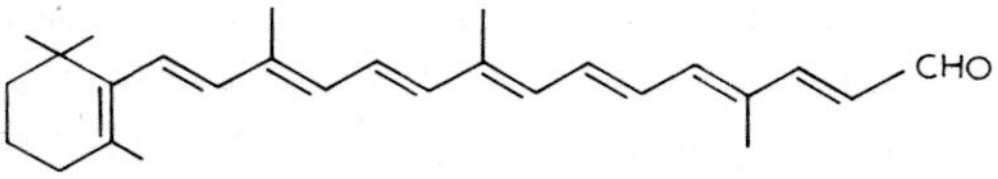

$C_{37}H_{48}O$

β-Apo-10'-carotenal
(10'-Apo-β-caroten-10'-al)

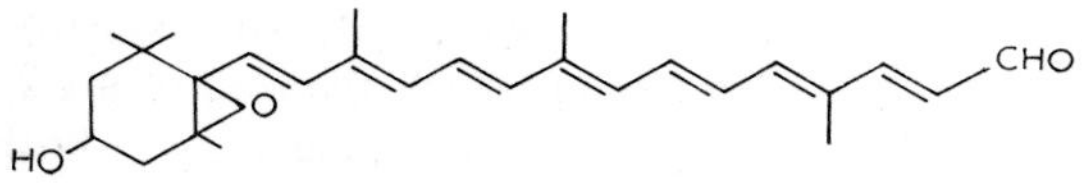

$C_{27}H_{36}O$

Apo-10'-violaxanthal
(5,6-Epoxy-3-hydroxy-5,6-dihydro-10'-apo-β-caroten-10'-al)

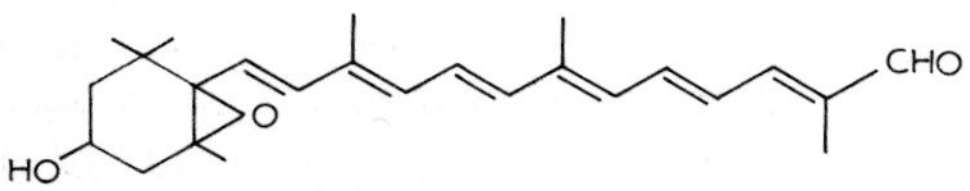

$C_{27}H_{36}O_3$

Apo-12'-violaxanthal
(5,6-Epoxy-3-hydroxy-5,6-dihydro-12'-apo-β-caroten-12'-al)

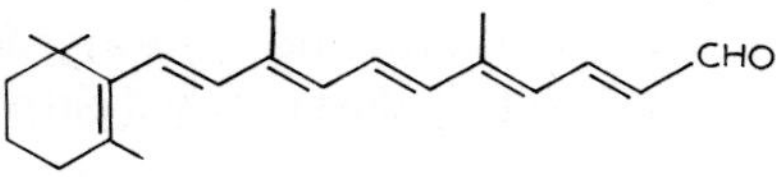

$C_{25}H_{34}O_3$

β-Apo-14'-carotenal
(14'-Apo-β-caroten-14'-al)

$C_{22}H_{30}O$

Fig. 3.19 Citrus apocarotenals.

(Gross et al., 1972a). In the flavedo of Washington Navel orange hydroxy-β-apo-10′-carotenal was detected, together with an even shorter carotenal β-apo-14′-carotenal (Malachi et al., 1974). Apo-10′-violaxanthal and apo-12′-violaxanthal, both probably metabolites of violaxanthin, were found by Curl in Valencia orange peels and tangerines respectively (Curl, 1965, 1967).

(c) *Other carotenoids of citrus and citrus relatives.* Recently the structures of some citrus carotenoids have been elucidated. From the most polar carotenoids of orange juice, two pigments were separated and characterized by Curl 30 years ago (Curl, 1956). Based on their similarity with a pigment isolated from the flower *Trollius europaeus* called trollixanthin (Straub, 1976), the polar citrus pigments were named by analogy "trollixanthin" and "trollichrome" by Curl.

A comparative study of the carotenoid pigments in juice of three orange (*Citrus sinensis*) varieties Shamouti, Valencia, and Washington, revealed that in the Shamouti orange "trollixanthin" was found at the highest level (Gross et al., 1972). The structure of two polar pigments found in orange juice was further elucidated by Gross et al., 1975. To avoid further confusion, the trivial names of these new pigments are adequately changed to "shamoutixanthin" and "shamoutichrome". Both pigments are rather uncommon tetrols (tetrahydroxy derivatives of β-carotene), containing a trihydroxylated ring. The following new structures have been assigned to "shamoutixanthin": 5,6-dihydro-β,β-carotene-3,3′,5,6-tetrol and to "shamoutichrome": 5,8-epoxy-5,8,5′,6′-tetrahydro-β,β-carotene-3,3′,5′,6′-tetrol.

Subsequently these two pigments were found also in other citrus species such as *Citrus reticulata*, Dancy tangerine (Gross, 1981, 1982) and in a mandarin hybrid (Farin et al., 1983), and in the endocarp of pummelo (*Citrus grandis*) (Gross et al., 1983c). With the use of new techniques it will probably be discovered that these pigments are widespread. Recently the absolute configuration of shamoutixanthin, which was found also in ripe hips of *Rosa pomifera* and of shamoutichrome found in flowers of *Rosa foetida*, was determined (Märki-Fischer et al., 1984; Märki-Fischer and Eugster, 1985).

From the red fruits of the citrus relatives *Murraya exotica* and *Triphasia trifolia* unique ketocarotenoids, semi-β-carotenone and β-carotenone, have been isolated and their structures elucidated (Yokoyama and White, 1968). These structures are shown in Fig. 3.20.

Another pigment occurring frequently in citrus is α-cryptoxanthin, the monohydroxy-derivative of α-carotene. It was first separated from orange juice by Curl (1956). The pigment was later found in many citrus species as well as in other plants such as corn, paprika, winter cherry fruit (*Physalis alkekengi*), flowers of *Calta palustris* (Petzold and Quackenbush, 1960; Szabolcs and Rónai, 1969; Bodea et al., 1978; Buchecker and Eugster, 1979).

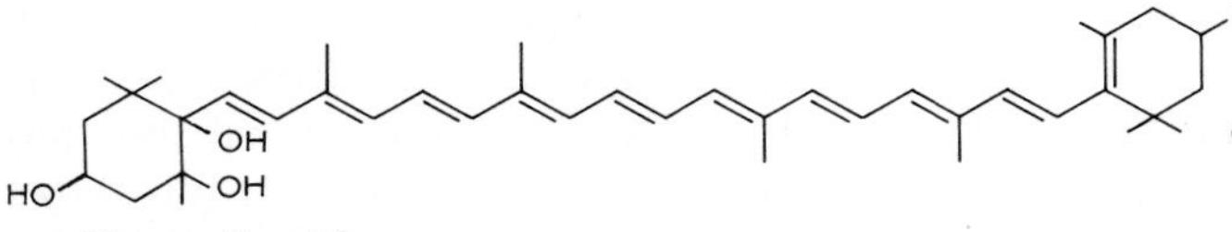

132 PIGMENTS IN FRUITS

Shamoutixanthin
(5,6-Dihydro-β,β-carotene-3,3',5,6-tetrol)

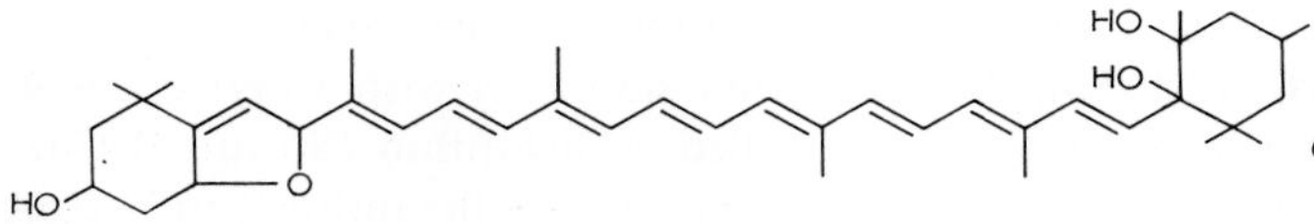

$C_{40}H_{58}O_4$

Shamoutichrome
(5,8-Epoxy-5,8,5',6'-tetrahydro-β,β-carotene-3,3,5',6'-tetrol)

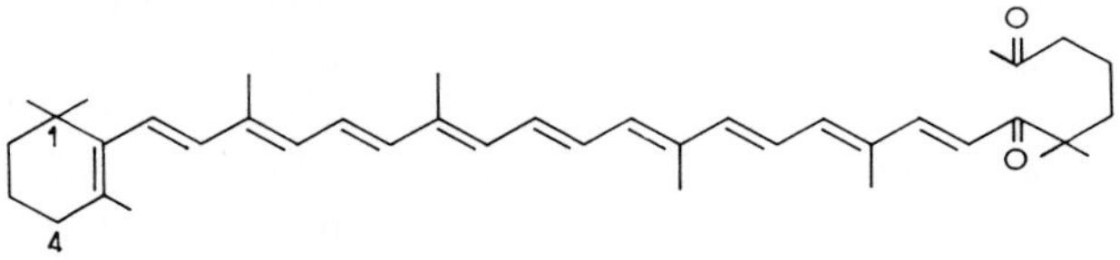

$C_{40}H_{58}O_5$

Semi-β-carotenone
(5',6'-Seco-β,β-carotene-5',6'-dione)

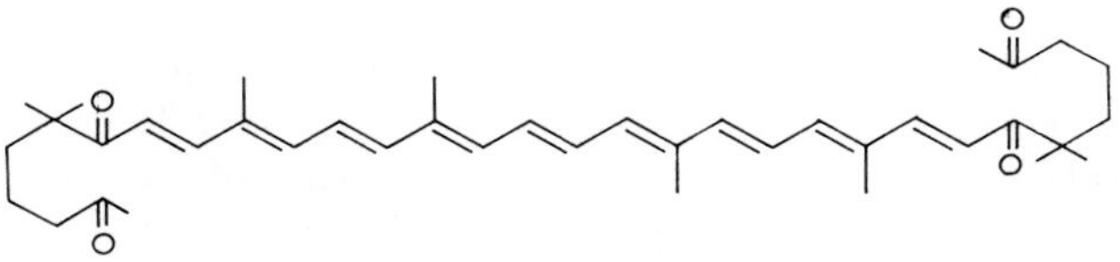

$C_{40}H_{56}O_2$

β-carotenone
(5,6,5',6'-Diseco-β,β-carotene,5,6,5',6'-tetrone)

$C_{40}H_{56}O_4$

Fig. 3.20 Unique pigments found in citrus and citrus relatives.

The structure of α-cryptoxanthin (also called "zeinoxanthin") that may have the hydroxyl group at the β- or ε-ring, was not unequivocally established for some time. Now it is known that the structure occurring in nature is (3R, 6'R)-β,ε-caroten-3-ol, (Fig. 3.4) as was first suggested by Curl, and therefore the pigment is devoid of provitamin A activity.

(d) *Capsicum carotenoids*. The fruit of the genus *Capsicum*, both the sweet bell pepper (*C. annum*) and the hot type, paprika (*C. frutescens*), contain unique ketocarotenoids with the keto group at the central chain and a cyclopentanol

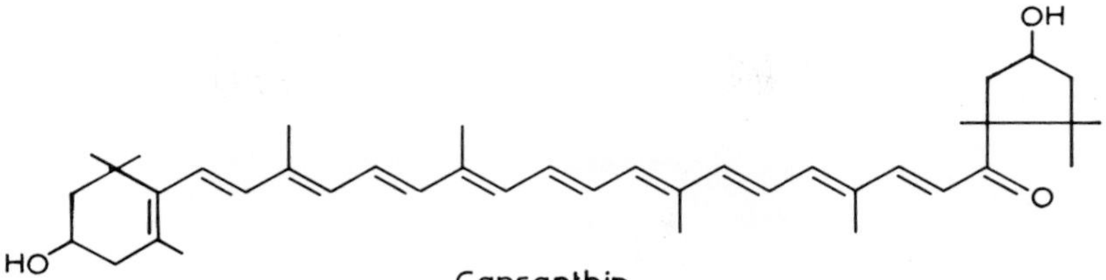

Capsanthin

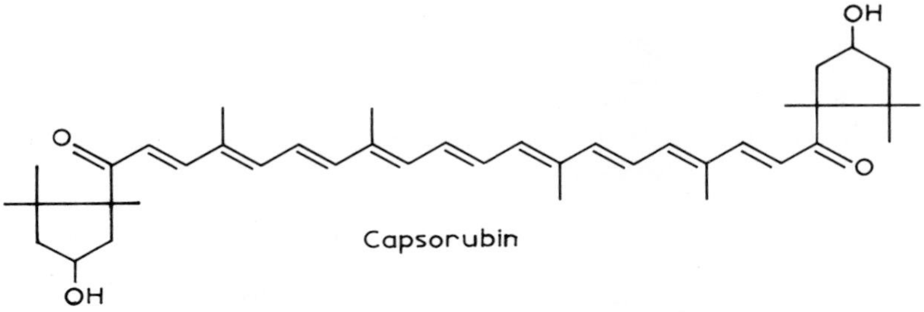

Capsorubin

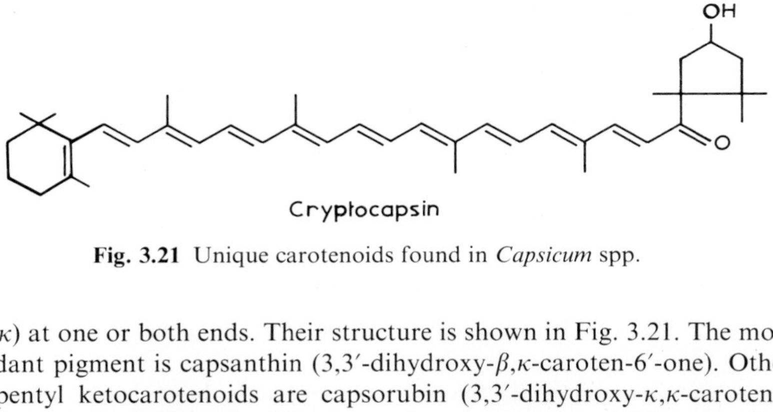

Cryptocapsin

Fig. 3.21 Unique carotenoids found in *Capsicum* spp.

ring (κ) at one or both ends. Their structure is shown in Fig. 3.21. The most abundant pigment is capsanthin (3,3′-dihydroxy-β,κ-caroten-6′-one). Other cyclopentyl ketocarotenoids are capsorubin (3,3′-dihydroxy-κ,κ-carotene-6,6′-dione) and cryptocapsin (3′-hydroxy, β,κ-caroten-6′-one). They were first isolated by Cholnoky *et al.* (1955). Capsanthin epoxide (5,6-epoxide-3,3′-dihydroxy-5,6-dihydro-β,κ-caroten-6-one) was subsequently also found in peppers (Curl, 1962a; Davies *et al.*, 1970; Camara and Monéger, 1978).

The biosynthesis of the ketoxanthophylls occurs via xanthophyll epoxides. The suggested mechanism involves a pinacolic rearrangement of the 3-hydroxy-5,6-epoxy end-group (Cholnoky and Szabolcs, 1960; Weedon, 1967). Thus cryptoxanthin 5,6-epoxide is the precursor of cryptocapsin, antheraxanthin the precursor of capsanthin and violaxanthin the precursor of capsorubin. This hypothesis has been corroborated recently through *in vitro* incubation of isolated *Capsicum* chromoplasts with labelled antheraxanthin and violaxanthin (Camara, 1980; Camara and Monéger, 1981).

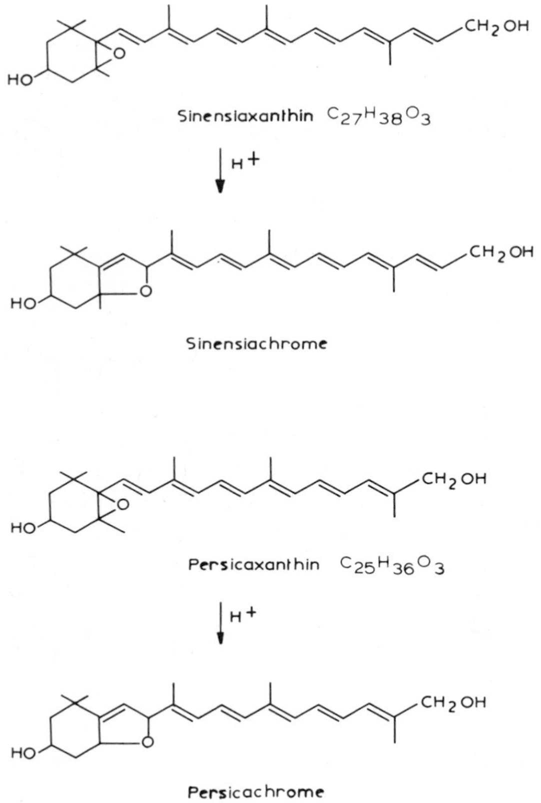

Fig. 3.22 Structures of various apocarotenols found in fruit.

(e) *Allylic apocarotenols.* Several apocarotenols have been detected in different fruits and their structures were recently elucidated by Gross and Eckhardt (1981). Their structure is shown in Fig. 3.22. They are polar UV fluorescent pigments, having absorption spectra with sharp fine structure, typical of a short in-chain chromophore consisting of 5–7 conjugated double bonds. All of them are 5,6- or 5,8-epoxides. They were first described by Curl, who isolated them from different fruits deriving their name from the fruit source. The C_{25}-apocarotenol persicaxanthin isolated from peach, apricot, plum, cranberry and fig has the structure 5,6-epoxy-5,6-dihydro-12'-apo-β-carotene-3,12'-diol (Curl, 1959, 1960, 1963, 1964). It is identical with "valenciaxanthin" isolated from Valencia oranges (Curl and Bailey, 1956). Persicachrome, the corresponding 5,8-furanoxide, accompanied persicaxanthin, the corresponding pigment of oranges being named valenciachrome.

Two C_{27}-apocarotenols sinensiaxanthin (5,6-epoxy-5,6-dihydro-10'-apo-β-carotene-3,10'-diol) and the corresponding 5,8-furanoxide were detected by Curl in Valencia and Navel oranges (Curl and Bailey, 1956, 1961) as well as in cranberries and figs (Curl, 1964). Recently the C_{27}-5,6-epoxide was detected in ripe Golden Delicious apples (Gross et al., 1978) and its structure determined (Gross and Eckhardt, 1978). The C_{27}-5,8 epoxide was isolated from avocado and its structure determined (Gross et al., 1974).

An additional apocarotenol is the C_{30}-apocarotenol citraurol, isolated from a citrus hybrid (Leuenberger et al., 1976). Its structure is shown in Fig. 3.18. The apocarotenols can be obtained in vitro through reduction of the natural apocarotenals. Through subsequent acidulation the furanoxides are formed. The apocarotenals can be obtained in vitro by alkaline permanganate oxidation of C_{40}-carotenoidepoxides and furanoxides (Molnár and Szabolcs, 1979). In fruits the metabolites may be formed, like vitamin A, through a Glover–Redfearn degradation of the most abundant C_{40}-pigment, followed by an enzymatic reduction. During in vitro oxidation it was observed that when the parent C_{40}-carotenoid is an epoxycarotenoid, the cleavage at carbons 9–10,9'–10' and 11–12, 11'–12' is favoured so that mainly C_{25}- and C_{27}-epoxyaldehydes are formed. This mechanism may also be valid for the enzymatic oxidation in the fruit. Evidence that subsequent enzymatic reduction occurs must still be produced. In the French plum cv. Saghiv the C_{25}-carotenal was present along with the 5,6- and 5,8-epoxyapocarotenols (Gross and Eckhardt, 1981).

C. Carotenoid Patterns

According to Goodwin (1982) there are eight general patterns of carotenoid distribution in fruits, although these patterns often merge into each other. The eight main patterns are:

(1) Insignificant amounts of carotenoids.
(2) Small amounts of carotenoids generally found in chloroplasts.
(3) Relatively large amounts of lycopene and the related more saturated acyclic polyenes, phytoene, phytofluene, ζ-carotene and neurosporene.
(4) Relatively large amounts of β-carotene and its hydroxy derivatives cryptoxanthin and zeaxanthin.
(5) Comparatively large amounts of epoxides, particularly furanoid epoxides.
(6) Unique carotenoids such as capsanthin.
(7) Poly-*cis* carotenoids such as prolycopene.
(8) Apocarotenoids such as β-citraurin.

XIII. Carotenoids of Various Fruits

A. Apple (*Malus domestica* L. Mill) *(Table A.1*)*

The carotenoids of the apple exocarp in which they accumulate have been extensively investigated. In one early report the carotenoids were determined both in peel and pulp and were found to be 37 μg(g fresh wt)$^{-1}$ in peel and 13 μg(g fresh wt)$^{-1}$ in pulp of Golden Delicious and 59 μg(g fresh wt)$^{-1}$ in peel and 16 μg(g fresh wt)$^{-1}$ in pulp of Grimes Golden (Workman, 1963). Subsequently it was found that apple peel has a lower carotenoid content containing pigments characteristic of the green leaf but in quite different proportions: violaxanthin 45–50%, neoxanthin 10–30%, β-carotene 10%, lutein 10% (Galler and Mackinney, 1965). The same chloroplast-carotenoid pattern was found to occur in various apple cultivars (Zelles, 1967; Knee, 1972; Gorski and Creasy, 1977).

The carotenoid pattern in apples grown in England was found to be different, with auroxanthin predominating in Golden Delicious and mutatochrome in Cox's Orange Pippin. The total carotenoid content (56 and 85 μg g^{-1}) was two- or three-fold higher in peel than in pulp (Valadon and Mummery, 1967). In a recent study by Gross *et al.* (1978), the total carotenoid content in the peel of Golden Delicious apples was *c.* 20 μg(g fresh wt)$^{-1}$. As well as the four chloroplast pigments in modified relative proportions, phytofluene, cryptoxanthin and its epoxide were also present. The ripe fruit contained a UV fluorescent C_{27}-apocarotenol. Approximately 80% of the total carotenoids were esterified. Lutein was present as a diester whereas violaxanthin and neoxanthin were found as both mono- and diesters. The major carotenoid pattern is that of chloroplast carotenoids with violaxanthin predominating (50%), that is (2), (5) from the list above. The vitamin A value is 200 μg retinol equivalents kg^{-1}.

B. Apricot (*Prunus armeniaca* L.) *(Table A.2)*

The carotenoids of the apricot were separated by Curl by countercurrent distribution and characterized individually (Curl, 1960a). The total carotenoid content was 35 μg(g fresh wt)$^{-1}$, of which 88% were carotenes. β-Carotene amounts to about 60% of the total carotenes. In the "polyol" fraction unknown unusual pigments like persicaxanthin, persicachrome, sinensiaxanthin, trollixanthin and trollichrome were separated and characterized by Curl for the first time. Their structure was elucidated recently: trollixanthin was proved to be identical with all-*trans*-neoxanthin (Buchecker

---

* See the Appendix.

et al. 1975), and thus different from citrus trollixanthin (Bickel and Schultz, 1976). The others were allylic apocarotenoids (Gross and Eckhardt, 1981).

The average carotene content is 20 µg g^{-1} in Italian apricots and in Hungarian apricots the total carotenoid content is 50 µg(g fresh wt)$^{-1}$, of which 70% consists of β-carotene (Mancini and Schianchi, 1963; Aczél, 1973).

The major carotenoid pattern is (4), although zeaxanthin is present only in trace amounts. The vitamin A value is 3770 µg equivalents kg^{-1}.

C. Avocado (*Persea americana* Mill) *(Table A.3)*

The avocado (Persea americana) is a fruit with an unusually high oil content (10–20%) and a relatively high chlorophyll content in the pulp layer adjacent to the peel p. 23). The carotenoid composition was determined by Gross *et al.* (1972) as 10–14 µg(g fresh wt)$^{-1}$. The predominant pigments were lutein (25%) and chrysanthemaxanthin (20%). The fruit contained all the chloroplast carotenoids and additionally cryptoxanthin (5%), isolutein (9%) and two uncommon polar pigments (10%). The major carotenoid patterns may be characterized as (2),(5). A typical chloroplast carotenoid pattern was found to occur in avocado peels and leaves (Gross *et al.*, 1973a). In the pulp, 60% of the xanthophylls were esterified, whereas in peel and leaves they were found entirely free.

The structure of one of the uncommon pigments was elucidated by Gross *et al.* (1974). It was found to be mimulaxanthin, that is a tetrol with only one allenic group (6′,7′-didehydro-5,6,5′,6′-tetrahydro-β,β-carotene-3,5,3′,5′-tetrol). Mimulaxanthin was first isolated from petals of *Mimulus guttatus* (Nitsche, 1972). One of the two present UV fluorescent pigments was the C_{27}-apocarotenol (Fig. 3.22) (Gross *et al.*, 1974). The structure of the second fluorescent pigment was only tentatively assessed. Although chromoplast carotenoids such as chrysanthemaxanthin, mimulaxanthin and the two apocarotenols were present in the avocado pulp, ultrastructural studies (Cran and Possingham, 1973) did not confirm the presence of the chromoplasts which were supposed by Gross *et al.* (1973a) to be present. This special occurrence of more oxidized xanthophylls may suggest that they are synthesized in the etioplasts, which are the plastids of the yellow flesh (see Fig. 1.13(c)). In etiolated seedlings too, chrysanthemaxanthin and auroxanthin are mainly synthesized, and instead of β-carotene, its diepoxide is found (Goodwin, 1982). Additional proof of this oxidative metabolism is the presence of two apocarotenols which may be formed through oxidative degradation of the most abundant pigment present, followed by an enzymatic reduction. The two main pigments which are degraded are chrysanthemaxanthin and probably mimulaxanthin, the latter being the parent of the apocarotenol with tentatively determined structure (Gross *et al.*, 1974).

Possibly the UV fluorescent pigments are also formed in etiolated plants but in undetectably small amounts.

The major carotenoid pattern may be characterized as an etioplast carotenoid pattern (2), (5). The vitamin A value is 150 μg retinol equivalents kg^{-1}.

D. Banana (*Musa* × *paradisiaca* L.) *(Table A.4)*

The carotenoids of banana of the Cavendish subgroup were only recently investigated by Gross *et al.* (1976) although the fruit is widely cultivated in tropical and subtropical areas and largely consumed throughout the world. The pigments were investigated both in peel and pulp, omitting saponification, so that the different xanthophyll esters could be separated. The major pattern of the pulp, which had a very low carotenoid content of 0.8 μg(g fresh wt)$^{-1}$, was α-carotene (31%), β-carotene (28%) and lutein (33%), the latter occurring in equal parts of monoester, diester and free form. This pattern, with three main carotenoids in the ratio (1:1:1) does not fit in any of the eight described fruit carotenoid patterns. The peel with a carotenoid content six times as high as in pulp had a pattern closer to that of chloroplasts but with all xanthophylls both fully and partially esterified or in the free form. Banana leaves, examined only for their main carotenes, had a ratio α/β-carotene = 1 which was similarly high in leaf chloroplasts. The pulp carotenoid pattern is (1) (uncommon). The vitamin A value is 60 μg retinol equivalents kg^{-1}.

E. Blackberry (*Rubus* spp.) *(Table A.5)*

Blackberries were characterized by Curl as "low carotenoid fruits" (1964). The total carotenoid content was 5.8 μg(g fresh wt)$^{-1}$. The carotenoids were separated into groups of different polarity by countercurrent distribution.

The major carotenoid pattern is that of chloroplast carotenoids (2). The vitamin A value is 100 μg retinol equivalents kg^{-1}.

F. Blueberry (*Vaccinium* spp.) *(Table A.6)*

The carotenoids of blueberries, also "low carotenoid fruits" were characterized by Curl in the same report (Curl, 1964) applying the same analytical technique. The total carotenoid content is 2.7 μg g^{-1}. They differ from the blackberry carotenoids in containing a relatively high percentage of monoepoxy-diols.

The carotenoid pattern is that of chloroplast carotenoids, (2). The vitamin A value is negligible, *c.* 50 μg retinol equivalents kg^{-1}.

G. Carambola (*Averrhoa carambola* L.) *(Table A.7)*

Carambola or starfuit is a tropical fruit which originates in Southeast Asia, where it is widely cultivated. The fruit is c. 12 cm long with a yellow waxy rind and yellow flesh and resembles a five-pointed star in cross-section. The carotenoids were investigated by Gross *et al.* (1983). The total carotenoid content was 22 μg(g fresh wt)$^{-1}$. The carotenoid pattern was uncommon, the main pigments being phytofluene (17%), ζ-carotene (25%), β-cryptoflavin (34%), and mutatoxanthin (14%). β-Carotene, β-apo-8'-carotenal, cryptoxanthin, cryptochrome and lutein were present in small amounts. Additionally, 10–15 pigments in trace amounts were characterized only by their electronic spectra. The predominant furanoxides are assumed to be naturally found as during analysis all precautions were taken to avoid a possible isomerization of the 5,6-epoxides. They may have been formed through isomerization *in situ* as oxalic acid found at high level in fruits may have penetrated the chromoplast envelope.

The structure of cryptoflavin as determined by MS, is 5,8-epoxy-5,8-dihydro-β,β-caroten-3-ol with the epoxy group at the hydroxylated ring. It follows that cryptoflavin is a provitamin A.

The major carotenoid pattern is (5). The vitamin A value is relatively high, 680 μg retinol equivalents kg^{-1}, deriving from a rather uncommon provitamin A.

H. Sour Cherry (*Prunus cerasus* L.) and Sweet Cherry (*Prunus avium* L.) *(Tables A.8 and A.9)*

The carotenoids of the sweet cherry were first investigated by Galler and Mackinney (1965), who determined a content of 5–11 μg(g fresh wt)$^{-1}$ in the Sweet Bing cultivar. Only the carotenes were identified, being predominantly β-carotene and in addition phytoene, phytofluene, β-zeacarotene and ζ-carotene.

The carotenoid content in the French Cherry cultivar Bigarreau Napoléon was c. 1 μg(g fresh wt)$^{-1}$. β-Carotene and xanthophylls (as a global group) were determined, as well as the ratio β-carotene/xanthophyll, which was 1.65 in the ripe fruit. The carotenoid pattern was that of chloroplast carotenoids, containing in addition trollixanthin and trollichrome (Okombi *et al.*, 1980). In the yellow cherry Dönissen's Gelbe, which contains c. 4 μg(g fresh wt)$^{-1}$ total carotenoids, the pattern was dominated by 70% epoxycarotenoids, mainly the 5,6- and 5,8-epoxides of zeaxanthin, along with β-carotene (9.5%), lutein (14.2%) and zeaxanthin (3.3%) (Gross, 1985). The major carotenoid patterns are (2), (5). The vitamin A value (calculable only in the yellow cherry) is c. 65 μg retinol equivalents kg^{-1}.

The pigments of the pulp of Montmorency sour cherry were analysed qualitatively and tentatively identified (Schaller and von Elbe, 1971).

I. Cranberry (*Vaccinium macrocarpon* Ait.) *(Table A.10)*

The carotenoids of cranberry were investigated within the report on "low-carotenoid fruits" by Curl (1964) in a more detailed manner than for the other fruits. The total carotenoid content was 5.8 μg g^{-1}. The carotene group contained β-carotene as major pigment and five other pigments of which phytofluene was relatively high. Lutein was the main xanthophyll (31.3%) followed by violaxanthin (20.7%). Uncommonly high was the level of isolutein (10.5%), exceeding that of neoxanthin. The fluorescent apocarotenols were also detectable. The major carotenoid pattern is that of chloroplast carotenoids, (2). The vitamin A value is 60 μg retinol equivalents kg^{-1}.

J. Red Currant (*Ribes rubrum* L.) *(Table A.12)*

The chlorophyll and carotenoid changes were investigated in red currant Rondom during ripening (Gross, 1982/83). The total carotenoid content in the ripe fruit was 3.6 μg(g fresh wt)$^{-1}$ as the fruit's colour is due to the anthocyanins. Lutein is the predominant pigment representing more than half of the total carotenoids (52.5%) followed by β-carotene (17.6%). Neochrome (18%) was found at higher levels than neoxanthin (5.4%), whereas violaxanthin was absent.

The major carotenoid pattern is a modified chloroplast carotenoid pattern, (2). The vitamin A value is 100 μg retinol equivalents kg^{-1}.

K. Date (*Phoenix dactylifera* L.) *(Table A.13)*

The carotenoids of the date were only recently investigated by Gross *et al.* (1983a). They were not investigated before due to the assumption that fruits of such dark colour do not contain carotenoids, since these pigments usually impart bright colours from yellow to red. The carotenoid content of two soft cultivars. Hayany and Barhee, and one semi-dry, Deglet-Noor, was 10–12 μg(g fresh wt)$^{-1}$. The carotenoid pattern in all three cultivars is that of the chloroplast type (2), with some variation in the proportions of the four main carotenoids. Thus Hayany has nearly double the β-carotene content of Deglet-Noor. In the latter, lutein constituted 50% of the total carotenoids. The violaxanthin level was low and the neoxanthin level remarkably high (25%). Deglet-Noor dates from commercial packages showed a much lower carotenoid content and a modified pattern, containing several degradation products with absorbance also in UV. This alteration may be due to heating during pre-processing.

The vitamin A value is 100–240 μg retinol equivalents kg^{-1}.

L. Fig (*Ficus carica* L.) *(Table A.14)*

The carotenoids of the black fig cv. Mission were investigated by Curl (1964). The total carotenoid content was 8.5 μg g^{-1}. The main carotenoid pattern was that of chloroplast carotenoids but with violaxanthin attaining the high level of 40%, so that the patterns may be characterized as (2), (5). In a more recent report, the pigments in the skin of Mission figs were similarly found to be chloroplast carotenoids, characterized only qualitatively (Peuch *et al.*, 1976b). The vitamin A value if 160 μg retinol equivalents kg^{-1}.

M. Gooseberry (*Ribes grossularia* L.) *(Table A.15)*

Within the same report on *Ribes* the chlorophyll and carotenoid pigments of four gooseberry cultivars were investigated (Gross, 1982/83). The total carotenoid content was low at 3–5 μg(g fresh wt)$^{-1}$. In three cultivars, two with ripe fruits of green colour and one blue–violet, the same chloroplast carotenoid pattern (2) was found. A fourth gooseberry cultivar with ripe fruits of yellow colour had different carotenoid patterns (2), (4), β-carotene being the major pigment (40%). The vitamin A values in the different cultivars were 150, 200 and 330 μg retinol equivalents kg^{-1} respectively.

N. Grapes (*Vitis vinifera* L.) *(Table A.16)*

The carotenoids of the grapes have been investigated very little considering the multitude of existing cultivars; this is in contrast to the grape anthocyanins which, being important for a wine's character, have been comprehensively investigated.

Within the report on "low carotenoid fruits", Curl (1964) investigated the carotenoids of the Thomson seedless cultivar. Recently the chlorophyll and carotenoid pigments of Dabouki, a dessert grape, and Riesling, a wine grape, have been investigated by Gross (1984). The carotenoid content was very low in Dabouki (1 μg(g fresh wt)$^{-1}$) and in Thomson (1.8 μg(g fresh wt)$^{-1}$), and higher in Riesling (3.5 μg(g fresh wt)$^{-1}$). In all three cultivars the carotenoid pattern was that of chloroplast carotenoids (2), showing variations in the proportions of both the main and the additional minor pigments. The relatively high percentage of β-carotene (30%) is noteworthy.

The vitamin A values were 60, 100 and 200 μg retinol equivalents kg^{-1} in the Dabouki, Thomson and Riesling cultivars respectively.

O. White Grapefruit (*Citrus paradisi* Macfad.) *(Table A.17)*

One of the earliest works on the carotenoids of grapefruit is that of Khan and Mackinney (1953). Although the pulp pigments could be resolved into 11 bands by chromatography, only the carotenes are mentioned in the white cultivar: 9 μg phytofluene and 6 μg ζ-carotene per 100 g pulp. Yokoyama and White (1967) investigated the carotenoids in the flavedo of Marsh seedless grapefruit in detail. The total carotenoid was very low, 1.5 μg(g fresh wt)$^{-1}$, corresponding to the light yellow colour of the peel. The colourless carotenoids, phytoene and phytofluene, accounted for 74% of the total carotenoids in the flavedo of mid-season fruit. The coloured carotenoids were a mixture of 17 pigments, less complex than in other citrus fruits, but nevertheless characteristic of citrus. The diol-diepoxides violaxanthin and luteoxanthin predominated, representing 42% of the total coloured carotenoids.

The carotenoids of both flavedo and pulp of the same cultivar were recently investigated (Romojaro *et al.*, 1979). In both flavedo and pulp the colourless carotenoids phytoene and phytofluene were more abundant, the coloured pigments amounting to 32% in flavedo and only 19% in pulp. The carotenoid pattern in flavedo was very similar to that found previously (Yokoyama and White, 1967), excepting citraurin which was absent. The major pigment is violaxanthin in flavedo and mutatoxanthin in the pulp. The accumulation of the colourless precursor is the result of a metabolic block, genetically determined, which hinders further dehydrogenation steps leading to coloured carotenoids.

The carotenoid pattern is uncommon (high amounts of colourless carotenes) along with pattern (5). The fruit has no vitamin A value.

P. Coloured Grapefruit (Pink and Red) *(Table A.18)*

The pink and red coloured grapefruit cultivars stimulated more interest than the white grapefruit and have therefore been more intensively investigated. They originated as mutants of the Marsh White grapefruit with lycopene and β-carotene as major pigments. This means that all the dehydrogenation and cyclization steps in carotenogenesis do occur normally, the genetic block being removed. Early investigations reported on lycopene and β-carotene as major pigments (Huffman *et al.*, 1953) and additional pigments were also detected (Khan and Mackinney, 1953).

A complete analysis of the carotenoids in Ruby Red grapefruit was carried out by Curl and Bailey (1957a). The total carotenoid content was 8.2 μg g^{-1} in pulp and 10.4 μg g^{-1} in flavedo. This equal level of total carotenoids in both pulp and peel is noteworthy as it is seldom found in fruits, especially in citrus. In accordance with previous investigations lycopene (40%) and β-carotene

(27%) were the principal pigments in the pulp, followed by phytoene (16%) and lower levels of phytofluene and ζ-carotene (4%). Various xanthophylls were detected in small amounts (less than 1%). In peel phytoene and phytofluene were the major pigments (60%) and lycopene (11%) and β-carotene (7%) were present at lower levels. The xanthophylls were similar to those found in pulp.

Observing that in some red cultivars the albedo and septa (segment walls) are also coloured, the carotenoids of all the fruit parts were recently investigated in our laboratory (Gross, unpubl. obs., 1983). Surprisingly in tissues such as septa and albedo, which usually do not contain chromoplasts, the total carotenoid content was very high at 84 and 44 μg g^{-1} respectively. In pulp and flavedo the content was nearly the same, c. 25 μg g^{-1}. The major pigment was lycopene, 70% in pulp and 60% in septa. The ratio lycopene/β-carotene fluctuated, but a precursor–product relationship can be observed. The second main pigment was phytofluene, exceeding the β-carotene level. Small amounts of xanthophylls were present.

In albedo the total carotenoid content was higher than in flavedo, 44 μg to 27 μg. The vascular bundles in the albedo were particularly highly coloured. Lycopene levels (43.3%) were twice as high as β-carotene levels (21.9%) and phytofluene attained its highest level in albedo (31.6%).

In flavedo the predominant pigment was phytofluene (24%). Lycopene, β-carotene and violaxanthin (two isomeric forms) were all three at the same level (17%). The xanthophylls formed a characteristic citrus carotenoid pattern containing a high level of violaxanthin, a typical citrus carotenoid citraurin and other carotenoids generally found in citrus, such as cryptoxanthin and antheraxanthin.

The major carotenoid patterns in flavedo are (3), (4), (5). The vitamin A value of the pulp is 400 μg retinol equivalents kg^{-1}.

Q. Guava (*Psidium guajava* L.) *(Table A.19)*

Guava is indigenous to tropical America and cultivated in most tropical and sub-tropical regions of the world. The highly aromatic fruit, round or pyriform in shape, has a rough yellow skin and a pulp coloured from white to yellow, pink and red–purple.

The single carotenoid in the peel of the white fleshed guava was found to be lutein 0.5 μg g^{-1} (Gross, unpubl. obs). Recently the carotenoids of the pink-fleshed Brazilian guava, a defined cultivar, and two other undefined varieties, were investigated by Padula and Rodriguez-Amaya (1984). The total carotenoid content in the whole fruit was found to be 60–70 μg g^{-1}. Lycopene at high levels (76–86%) is responsible for the fruits' colour. The second major pigment is β-carotene (6–17%). Additional pigments were α-cryptoxanthin (zeinoxanthin) and an unknown polar 5,8-furanoxide triol, tentatively identified. The major carotenoid pattern is (3) (modified). The vitamin A values are 620, 1980 and 915 µg retinol equivalents kg^{-1} respectively.

R. Kiwifruit or Chinese Gooseberry (*Actinidia chinensis* Planch.) *(Table A.20)*

The kiwifruit or Chinese gooseberry (*Actinidia chinensis*), indigenous to China, is a comparative newcomer to the world's markets. It was and is most successfully cultivated in New Zealand where it received its name "kiwifruit".

Pericarp pigments of fresh and dried fruit were analysed with separate determination of chlorophylls and total carotenoids within the darker outer and lighter inner pericarp tissues (Possingham *et al.*, 1980). The total carotenoid was higher in the outer pericarp at 7.5 µg g^{-1} compared with 5.9 µg g^{-1} in the inner pericarp.

Pigment changes during ripening in the pericarp of the kiwifruit cv. Bruno were recently investigated by Gross (1982). The total carotenoid content of the ripe fruit was 6.3 µg g^{-1}. The major carotenoid pattern was a modified chloroplast carotenoid pattern containing neochrome as additional fifth major pigment (6%). The ratio of the five main pigments differed from that of green leaves in lower β-carotene, very low violaxanthin level and the presence of neochrome at levels twice as high as violaxanthin. Besides these five main pigments, an additional 30 pigments in either small or trace amounts (characterized only by their electronic spectra) produced a carotenoid pattern of unusual complexity.

This complexity may be attributed both to the special type of chloroplasts found in the fruit (Possingham *et al.*, 1980, p. 21), which may have a special biosynthetic capacity, and to the high acidity of the fruit. In carambola too a similarly complex pattern may be attributed to acid penetration into plastids producing *in situ* isomerizations. The major carotenoid pattern is thus a modified (2), containing neochrome as a fifth main pigment.

The vitamin A value is 50 µg retinol equivalents kg^{-1}.

S. Kumquat (*Fortunella margarita* (Lour.) Swingle) *(Table A.21)*

The kumquat of the genus *Fortunella* is a relative of citrus, both belonging to the family of Rutaceae. The name of the genus is derived from Robert Fortune who introduced the fruits from China into Europe in 1846. The name kumquat corresponds to its Chinese name "Chin Kan"—golden orange.

According to Swingle the genus contains only four species of which *Fortunella margarita*, the oval kumquat Nagami, is the most important. The second in importance is the round kumquat Marumi (*F. japonica*). The fruits are oval, about 4 cm long and 2–3 cm in diameter, of yellow-orange to orange colour. They differ from true citrus in that their rind is thin and edible together with the mildly acid, juicy pulp which is contained in only 3–5 segments.

The carotenoid pigments of the Nagami kumquat grown in Italy were investigated (Calvarano et al., 1972). The total carotenoid content was 2.8 mg‰ of which the alcohol-ketone group represents approximately 70%. The percentages of three other carotenoid groups were also determined.

In another study of carotenoids of the trigeneric hybrid Sinton citrangequat *Fortunella margarita* × (*Poncirus trifoliata* × *Citrus sinensis*) the carotenoids of the parent genera are only mentioned as containing or being totally devoid of low methyl ketone carotenoids (Yokoyama and White, 1966). These unusual apocarotenones found in the hybrid were later proved to be artefacts of the carotenoids which are actually present in the citrus (Stewart and Wheaton, 1973).

The pigments of the oval kumquat *Fortunella margarita* Nagami were investigated and the carotenoids characterized at four ripening stages (Huyskens et al., 1985). Since the fruit is consumed whole, the carotenoids of the entire fruit were analysed. Only in the ripe fruit were the carotenoids of the peel and pulp analysed separately. The total carotenoid content in the pulp is c. 8 μg g^{-1}, whereas the peel contains a level as high as 172 μg(g fresh wt)$^{-1}$. Two major pigments are present in the pulp, violaxanthin representing about 25% of the total carotenoid, followed by cryptoxanthin (13.2%). Zeaxanthin together with its epoxides antheraxanthin and mutatoxanthin reaches a level of 18.5%. Noteworthy is the presence of citraurin.

In peel the pattern is dominated by violaxanthin, mainly in the *cis*-form. Together with the isomer luteoxanthin, the whole group amounts to about 60% of the total carotenoids. Cryptoxanthin is accompanied by almost all its possible epoxides. An unusual feature is the level of citraurin, which is lower in the peel than in the pulp.

In both pulp and peel the complex carotenoid pattern dominated by epoxides is typical of the *Citrus* genus. The pattern of the pulp is similar to the orange (*Citrus sinensis*) (Gross et al., 1972), whereas that of the flavedo is very similar to *Citrus reticulata* (Farin et al., 1983). The *Fortunella* genus resembles the *Citrus* genus as it contains citraurin, as yet found only in citrus and thus considered genus specific. Uncommon and characteristic of *Fortunella* is its capacity to synthesize citraurin both in pulp and peel. Through hybridization this capability to biosynthesize C_{30}-apocarotenals is transmitted and considerably enhanced. In Sinton citrangequat citraurin represents 50% of total carotenoids (Yokoyama and White, 1966).

The major carotenoid patterns in both pulp and peel are (5), (8). The vitamin A value of the whole fruit is 480 μg retinol equivalents kg^{-1}.

T. Lemon (*Citrus limon* L. Burm.) *(Table A.22)*

The carotenoids of lemons, both mature-green and yellow-ripe, were investigated by Yokoyama and Vandercook (1967) in pulp and peel. The light colour of the fruit is determined by a low content of total carotenoids, 0.6 μg g^{-1} in pulp and 1.4 μg g^{-1} in peel. The colourless carotene content was about 25% both in pulp and peel. The coloured carotenoid mixture was less complex and also different from other citrus species. In the pulp β-carotene is accompanied by its 5,6-epoxide (14%), and cryptoxanthin is most abundant (30%). The peel has a high content of ζ-carotene (17%) and also a high content of all epoxides of cryptoxanthin. The carotenoids of Meyer lemons (*C. limon* × *C. sinensis*) were investigated by Curl (1962b). The main carotenoid in the pulp was found to be cryptoxanthin, whereas in the peel all cryptoxanthin epoxides were present. The general pattern is predominantly that of lemons.

The major carotenoid pattern in pulp is (4) (cryptoxanthin predominating) and in peel (4), (5). The vitamin A value is minimal.

U. Loquat (*Eriobotrya japonica* (Thunb.) Lindl.) *(Table A.24)*

The loquat, also known as the Japanese plum, is a sub-tropical fruit believed to have originated in China. It has been cultivated in Japan since antiquity and is an important crop both there and in Israel. It is cultivated in various other sub-tropical regions (Shaw, 1980). The plant belongs to the family Rosaceae, sub-family Maloideae.

The fruits, which grow in loose clusters, are ovoid or pyriform and about 4 cm long. The fruit has a golden orange colour with an agreeable acid flavour. It contains a few large ovoid brown seeds that constitute about 25% of the total fruit weight.

The carotenoids of the Golden Nugget cultivar grown in Israel were investigated both in pulp and peel by Gross *et al.* (1973b). The total carotenoid content in the pulp was 20–25 μg(g fresh wt)$^{-1}$ in the pulp and about 100–120 μg(g fresh wt)$^{-1}$ in the peel. The major carotenoids of the pulp were β-carotene (38%), cryptoxanthin (22%), γ-carotene (6%). Additional pigments were lutein, violaxanthin and neoxanthin each about 3–4%. In the peel β-carotene represented 55% of the total carotenoids; the second main pigment was lutein (13%). Cryptoxanthin and γ-carotene attained both the same level of *c.* 5%. The carotenoids of the loquat (sub-family Maloideae) were very similar to those of the apricot (sub-family Prunoideae) (Curl, 1960).

Seven Japanese loquat cultivars, recently investigated, had carotenoid contents between 16 and 27 μg g^{-1} (Kobayashi *et al.*, 1978). The major carotenoids are cryptoxanthin (30–40%) and β-carotene (23–40%) accompanied by relatively high levels of cryptoxanthin 5,6,5′,6′-diepoxide (7–14%).

The Japanese cultivars differ from the cultivar Golden Nugget in having higher levels of cryptoxanthin and cryptoxanthin diepoxide.

The major carotenoid pattern in pulp is (4), and in peel a modified (4). The vitamin A value in pulp is 2100 μg retinol equivalents kg^{-1}, in peel 11 200 μg retinol equivalents kg^{-1}, and in the whole fruit of Japanese cultivars 1800 μg retinol equivalents kg^{-1}.

V. Mandarin (*Citrus reticulata* Blanco) *(Table A.25)*

Mandarins (called tangerines in the U.S.A.) of orange–reddish colour have the deepest colour of all citrus fruits. Zechmeister and Tuzson (1933) investigated the carotenoids of Italian mandarin both in pulp and peel and reported a complex mixture from which β-carotene, cryptoxanthin, lutein, zeaxanthin and β-citraurin were identified.

Huggard and Wenzel (1955), measuring the external colour with the Hunter colour difference meter, confirmed that tangerine colour is more intense than the colour of the orange. Curl and Bailey (1957b) investigated the tangerine carotenoids in pulp and peel. Twenty-eight pigments could be separated and the deeper coloration was attributed to a higher concentration of cryptoxanthin and β-citraurin, the latter pigment being isolated after some years (Winterstein *et al.*, 1960; Curl, 1965). In fact the reddish tint is due to the presence of pigments absorbing at longer wavelengths, such as β-citraurin and cryptoxanthin, even if they do not exceed the level of the predominating yellow pigments.

In 1971, Stewart and Wheaton tried the new method of HPLC on a Dancy tangerine peel extract which represents a highly complex carotenoid mixture. From a total of 23 chromatographic peaks six were predominant. Listed in decreasing order these were: violaxanthin (*cis* and *trans*) cryptoxanthin, antheraxanthin (*cis* and *trans*), β-citraurin, carotenes and zeaxanthin (Stewart and Wheaton, 1971). The carotenoids of the Japanese Satsuma mandarins (*C. unshiu*) were investigated by Umeda *et al.* (1971a, b). The carotenoid pattern of *C. unshiu* flesh was characterized by 70% or more monols and monol-monoepoxides. In the peel the amount of β-citraurin was about 20% of the total.

Recently the carotenoid content in the juice of seven mandarin cultivars was followed during the fruit maturation period, the analysis being performed with HPLC. Violaxanthin, antheraxanthin and cryptoxanthin were major pigments. The latter was the main source of provitamin A, being highest in Murcott and intermediate in Robinson and Dancy tangerine. (Murcott and Robinson are hybrids of *C. reticulata*) (Stewart, 1977b).

Pigment changes in both flavedo and juice of Dancy tangerine during ripening were investigated by Gross (1981, 1982b). The total carotenoid content in the ripe Dancy tangerine was 12.0 µg ml^{-1} in juice and 300 µg g^{-1} in flavedo. In juice 20 carotenoids were separated, the major pigment being cryptoxanthin (40%), followed by violaxanthin and antheraxanthin (c. 11%), phytofluene (8%) and ζ-carotene (6.3%).

The flavedo carotenoid pattern was dominated by violaxanthin, which represented more than 50% of total carotenoids. The second major pigment was cryptoxanthin (14%) accompanied by its various epoxides. Citraurin as third major pigment was found at a level of 8%, lower than is found in the major *C. reticulata* carotenoid pattern characterized by Gross (1977). On the other hand, the carotenoid pattern of juice corresponds to the general *C. reticulata* pattern. Cryptoxanthin predominates in pulp. In peel, although the amount of cryptoxanthin is high, the violaxanthin fraction is considerably greater and citraurin is found as a major pigment. The total carotenoid content in the juice of four Italian cultivars grown in southern Italy, had a medium value of 15.0 µg ml^{-1} (Schachter, 1977).

Mandarins are a good source of apocarotenoids, as has been shown by all the investigations carried out and as was previously discussed (see Section XII.B.1(a). β-Apo-8′-carotenal however is found only in very low amounts.

Citraurinene, which was isolated from the hybrid mandarin Robinson, was recently found to be present also in Clementine, thus not being a pigment produced only in hybrids (Farin *et al.*, 1983). It seems that the biosynthetic potential of the Clementine cultivar to form citraurinene is inherited by its hybrids Robinson (Orlando tangelo × Clementine) and Michal (Dancy tangerine × Clementine). This latter is an accidental hybrid and its putative parents were identified by comparing the respective carotenoid patterns of potential parents with that of the hybrid. Besides the numerous similarities found, the main clue was precisely the presence of citraurinene both in Clementine and Michal.

The major carotenoid patterns are: in juice (4), (5), in flavedo (5), (8). The vitamin A value of juice is 500 µg retinol equivalents l^{-1}.

W. Mango (*Mangifera indica* L.) *(Table A.26)*

The mango is the most popular tropical fruit, cultivated in India for more than 4000 years. It is cultivated in different parts of the world, with India's production being the largest. The fruit has a smooth skin of green, yellow or violet and yellow colour. The flesh is sweet, juicy, of yellow–orange colour.

In 1932 Yamamoto *et al.* reported that α- and β-carotenes are the pigments found in mango. The carotenoids of mango pulp (the skin is tough and not eaten) were investigated in detail by Jungalwala and Cama (1963). The best cultivar Alphonso was analysed; its total carotenoid content was 125 µg g^{-1} pulp. β-Carotene is the most abundant pigment, making up 60% of the total carotenoids. An additional 15 pigments were separated. The diol diepoxides violaxanthin (mainly in *cis*-form), and its furanoxide isomers, luteoxanthin and auroxanthin, amounted to 24%. Changes in carotenoid pigments during ripening were followed in the cultivar Badami (John *et al.*, 1970). The ripe fruit contained a similar carotenoid pattern to Alphonso, but the total carotenoid content was lower at c. 90 μg g^{-1} and β-carotene 50%. The total carotenoid content varies among cultivars between 24 μg in the cultivars Baneshan and Safeda to 90 μg in Dashehari (Roy, 1973; Adsule and Roy, 1974).

The major carotenoid patterns are (4), (5). The vitamin A value is 12 500 μg retinol equivalents kg^{-1}.

X. Melon (*Cucumis melon* L.)—Orange-fleshed and Green-fleshed cultivars (*Tables A.27 and A.28*)

Muskmelon cultivars are very numerous, the colour of the flesh ranging from pale green to deep orange. The orange-fleshed muskmelons, commonly called cantaloupes, are a good source of carotenes. The carotenes were first investigated by Vavich and Kemmerer (1950), and β-carotene was found to amount to 94% of the total carotenoids. A complete analysis of the cantaloupe carotenoids was carried out by Curl (1966). The total carotenoid content was 20.2 μg(g fresh wt)$^{-1}$, β-carotene representing the major carotenoid at a level of 85%. The second main pigment was ζ-carotene (6.5%), followed by smaller amounts of the colourless precursors phytoene and phytofluene. The usual xanthophylls were present in very small or trace amounts. The total carotenoid content of three other orange-fleshed cultivars varied between 8 and 25 μg g^{-1} (Reid *et al.*, 1970).

Pigments of green-fleshed cantaloupes were little investigated. In the cultivars Honeydew and Casba the total carotenoid content was very low, 0.2–0.3 μg(g fresh wt)$^{-1}$ (Pratt, 1971). Pigment and plastid changes in mesocarp and exocarp of the ripening muskmelon Galia were investigated by Flügel and Gross (1982). The carotenoid content in the fully ripe flesh was 1.9 μg g^{-1} and in the rind 32.0 μg(g fresh wt)$^{-1}$. The major carotenoid pattern was that of chloroplast carotenoids (2) in both pulp and skin, but the relative proportions of the four major pigments differed markedly in the two fruit parts. In pulp the respective percentages were β-carotene 20%, lutein and violaxanthin c. 28% and neoxanthin 1.5%, whereas in the skin they were β-carotene 7.5%, lutein 32%, violaxanthin 14% and neoxanthin 3.4%. The colour of the pale green flesh was due to the small quantities of chlorophyll present which could mask the carotenoid pigments present in even smaller amounts. The ultrastructural studies (Chapter 1, Section X.B) revealed the presence of chromoplasts called gerontoplasts, which are unable to carry out a *de novo* carotenoid biosynthesis, a feature characteristic of this genotype. The major carotenoid pattern in the orange-fleshed melon is (4), without cryptoxanthin and zeaxanthin. The vitamin A value is 2800 μg retinol equivalents kg^{-1}.

Y. Orange (*Citrus sinensis* (L.) Osbeck) *(Table A.30)*

The carotenoids of the orange are the most investigated pigments in citrus; only tomato carotenoids have been studied to the same extent. Earlier, the orange aroused interest as representing a rich carotenoid source. Later colour became a major quality parameter, especially in citrus fruits and their processed products. The colour of pulp, particularly of juice, is also used as a criterion for determining grade; quality standards for orange juice as set by the U.S. Department of Agriculture assign 40% of the total value score to colour (Stewart and Leuenberger, 1976).

In the earliest investigations of orange fruit carotenoids several major or newly discovered pigments were characterized (Zechmeister and Tuzson, 1937; Karrer and Jucker, 1948; Winterstein *et al.*, 1960), some of which were found to be carotenoids unique to the citrus genus (Thommen, 1962; Curl, 1965, 1967).

Curl was the first to succeed in the difficult separation of complex citrus carotenoid mixtures, investigating in detail the carotenoids of the major orange cultivars Valencia and Washington Navel, both in pulp and peel (Curl and Bailey, 1956, 1961). The total carotenoid content was 15 and 23 μg ml^{-1} in the pulp and 120 and 67 μg(g fresh wt)$^{-1}$ in the peel of the Valencia and Washington Navel respectively. The carotenoid pattern was a complex citrus pattern dominated by the diol-diepoxides (mainly violaxanthin), which in pulp reached 36% and in peel up to 62% of total carotenoids. The monol fraction dominated by cryptoxanthin is much higher in pulp (7–18%) than in peel (1–8%) (Gross, 1977).

The carotenoids of the Shamouti orange, the principal cultivar grown in Israel, were investigated by Gross *et al.* (1971) in order to find out whether the light pigmentation of the juice is due to a low concentration of total carotenoids or to qualitative differences among the carotenoids. More than 50 carotenoids could be separated. In addition to the carotenoids usually found in citrus, others not previous detected in common orange cultivars were identified. These include γ-carotene, rubixanthin, sinthaxanthin and its hydroxy-derivative as well as lutein 5,6-monoepoxide. The light colour of Shamouti orange juice was found to be due to its low carotenoid content rather than to the absence of any specific coloured pigments.

The juices of two other cultivars grown in Israel, Washington Navel and Valencia, were also investigated (Gross *et al.*, 1972). The pigment pattern was similar in all three cultivars, and the total carotenoid content decreased from Valencia (10–15 μg ml^{-1}), through Shamouti (6–10 μg ml^{-1}) to Washington

(4–7 μg ml^{-1}). Cryptoxanthin was found at levels of c. 10% and in the diol fraction zeaxanthin and its monoepoxides antheraxanthin (7%) and mutatoxanthin (14%) predominated. Shamoutixanthin was found at levels up to 10%. The ratio of the orange pigments (λ_{max} 445 nm) to the yellow pigments (λ_{max} 420 nm) which may influence the colour intensity, decreased from Valencia through Shamouti to Washington.

The carotenes in genuine orange juices represent 3–10% of the total carotenoids. The ratio carotenes over total carotenoids is used in order to establish if a product is genuine or adulterated.

The major carotenoid patterns in oranges, represented graphically in Gross (1977), are (4), (5), (6) in pulp and (5), (6) in peel.

The vitamin A values in juice are 90, 140 and 200 μg retinol equivalents l^{-1} increasing from Washington Navel through Shamouti to Valencia.

Z. Papaya (*Carica papaya* L.) *(Table A.31)*

Papaya is a tropical fruit, round or pear-shaped, with green or yellow–orange smooth skin. The abundant flesh is yellow–orange to salmon–pink in colour. The colour of ripe papaya flesh is due to the presence of carotenoids; cryptoxanthin, violaxanthin, β-carotene and ζ-carotene have been identified by Karrer and Jucker (1948). The carotenoids of the yellow-fleshed and red-fleshed fruits of Hawaii were compared to Yamamoto (1964). The total carotenoid contents were 37 and 40 μg g^{-1} in the two cultivars. Only the major carotenoids were determined. The yellow cultivar contained cryptoxanthin (38.9%), ζ-carotene (24.8%), cryptoxanthin monoepoxide (15.6%) and β-carotene (4.8%). In the red-fleshed fruit lycopene was the most abundant pigment (63.5%). Cryptoxanthin (19.2%), cryptoxanthin epoxide (4.4%) and ζ-carotene (5.9%) were found at lower levels than in the yellow-fleshed fruit, whereas β-carotene was found at the same level (4.8%).

The carotenoids of the Indian cultivar Bangalore were investigated in detail by Subbarayan and Cama (1964). The total carotenoid content was 13.8 μg(g fresh wt)$^{-1}$. Nineteen different carotenoids have been identified. The main carotenoid was cryptoxanthin (48%), followed by β-carotene (30%) and cryptoflavin (13%). ζ-Carotene (3%) and violaxanthin (3.5%) were present at lower levels, while additional pigments were found at very low levels or as trace amounts; three of these were not identified. A reinvestigation with new methods may be required to identify the unknown pigments and to clarify if cryptoflavin is genuine or an artefact, as the level of cryptoxanthin 5,6-epoxides varies markedly in the different fruits.

The major carotenoid patterns are (4) in yellow-fleshed fruit and (3), (4) in red-fleshed fruit.

The vitamin A values are 1100 and 2000 retinol equivalents kg^{-1} respectively.

A'. Passion Fruit (*Passiflora edulis* Sims) *(Table A.32)*

The passion fruit is a tropical fruit. The purple passion fruit (*P. edulis*) and the yellow fruit *P. edulis* f. *flavicarpa*) are the most important species. The earliest investigation of the carotenoids of *P. edulis* was carried out by Pruthi and Lal (1958). The main pigments discovered were three carotenoids: phytofluene, β-carotene and ζ-carotene. Three unidentified pigments with λ_{max} at 425 nm were detected. Leuenberger and Thommen (1972) found the same main carotenes, phytofluene, β-carotene and ζ-carotene, in all parts of the fruits. In addition, two carotenals β-apo-12'-carotenal and β-apo-8'-carotenal, as well as cryptoxanthin, auroxanthin and mutatoxanthin were also identified.

The carotenoids in processed passion fruit juice have been investigated (Cecchi and Rodriguez-Amaya, 1981). Although the carotenoid pattern may be modified during processing, the main carotenoids were again the carotenes ζ-carotene, β-carotene and neurosporene, the latter having previously been undetected. The major carotenoid pattern is uncommon—dominated by ζ-carotene.

B'. Peach (*Prunus persica* L.) *(Table A.33)*

Peach carotenoids were first investigated by Mackinney (1937) in the cultivar Muir. Five carotenoids were identified: α- and β-carotene, cryptoxanthin, lutein and zeaxanthin. The European yellow peaches were reported to contain a carotenoid mixture of lutein, carotene and lycopene, the ratio between them being 8:2:1 (Thale and Schulte, 1940). The first detailed investigation of the carotenoids of peaches was carried out by Curl (1959) using his method of counter current distribution and chromatography.

The cling peach Halford was found to have a total carotenoid content of 27 $\mu g\, g^{-1}$. The complex carotenoid mixture consisted of 45 pigments, dominated by xanthophylls in esterified form. The main carotenoids were violaxanthin (26%), cryptoxanthin (12%), β-carotene (10%), phytoene (10%) and persicaxanthin (12%). The latter was described for the first time. Less abundant were phytofluene, ζ-carotene, lutein, zeaxanthin and its epoxides, trollixanthin-like polyols and "persicachromes" *a* and *b*.

The pigment changes during ripening of Halford peach were followed by Katayama *et al.* (1971). The major pigments of the ripe fruit were chloroplast carotenoids dominated by β-carotene and violaxanthin and containing in addition cryptoxanthin at higher levels than neoxanthin. More recently pigment changes in ripening peaches were followed in the cultivar Earliglo by Lessertois and Monéger (1978) and in the Redhaven cultivar by Gross (1979a).

The ripe Earliglo peach was found to contain carotenoids of the β-ionone series: β-carotene, cryptoxanthin and its various epoxides, zeaxanthin, antheraxanthin, auroxanthin and trollichrome. The total carotenoid content of the ripe Redhaven peach was 4.7 μg(g fresh wt)$^{-1}$. The carotenoid pattern was similar to that found by Curl in the Halford cultivar. The predominant carotenoid was violaxanthin (34.6%) followed by cryptoxanthin (15%), phytofluene (10%), isolutein (8%) and zeaxanthin (7%). β-Carotene and neoxanthin were found at lower levels of c. 5%. The UV fluorescent pigments persicaxanthin and persicachrome were not detected. The major carotenoid patterns are (4), (5) (modified).

The vitamin A value if 100 μg retinol equivalents kg^{-1}.

C'. Pear (*Pyrus communis* L.) *(Table A.34)*

The carotenoids of pears were first investigated by Galler and Mackinney (1965). The values for the mesocarp of the Anjou and Bartlett cultivars were very low: 0.3–1.3 μg g^{-1}, but the peel of the Comice cultivar had a higher content of 5.6 μg g^{-1}. The major carotenoids were found to be lutein and neoxanthin, with phytoene, phytofluene, β-carotene and cryptoxanthin also being detected.

The chlorophyll and carotenoid pigments of the peel of two pear cultivars, Spadona and Super Trévoux were investigated by Gross (1984). The cultivar Spadona, which retains its green colour when ripe, contained 9.3 μg g^{-1} total carotenoids and the other cultivar Super Trévoux, which becomes yellow when ripe, contained 10.2 μg g^{-1} total carotenoids.

In the green cultivar the carotenoid pattern was that of the chloroplast type. In the yellow cultivar the proportions of the residual chloroplast carotenoids were appreciably modified, with lutein reaching a level as high as 70% of the total carotenoids and β-carotene as low as 6%. Shamoutixanthin was detected in small amounts. The major carotenoid patterns are thus (2) and modified (2) respectively.

The vitamin A value in peel is 50–100 μg retinol equivalents kg^{-1}.

D'. Pepper (*Capsicum annuum* L.) *(Table A.35)*

The pepper carotenoids were and continue to be intensively investigated as the fruit produces unique carotenoids containing acylcyclopentanol rings, capsanthin and capsorubin, which are of theoretical interest. Pepper carotenoids also have a practical importance, paprika being one of the oldest and most important natural food colourant.

The carotenoids of red pepper were investigated by Cholnoky *et al.* (1955), who also determined their structure (Cholnoky *et al.*, 1955; Cholnoky and Szabolcs, 1960). A complex carotenoid pattern was found, containing in addition to capsanthin, capsorubin and cryptocapsin, β-carotene, cryptoxanthin, zeaxanthin and its epoxides. Curl investigated the carotenoids of the Red Bell pepper cultivar, both green and red (Curl, 1962a, 1964b). In the Red Bell pepper, the total carotenoid content was 127–284 μg g^{-1}. More than 30 pigments were separated. Capsanthin accounted for 35%, β-carotene and violaxanthin about 10% each, cryptoxanthin and capsorubin about 6% each and cryptocapsin about 4%. Numerous other carotenoids were present in amounts of 2% or less. In the green pepper the carotenoids were chloroplast carotenoids and minor pigments included phytoene, phytofluene, α-carotene and ζ-carotene.

Kirk and Juniper (1967) investigated the chloroplast ultrastructure of different colour cultivars of the ornamental pepper, correlating it with their various total carotenoid levels. Davies *et al.* (1970) investigated in detail the carotenoids of the white, yellow, orange and red cultivars of the ornamental pepper, allowing him to outline the biosynthetic steps leading to the formation of the ketocarotenoids.

Pepper carotenoids have been reviewed by de La Torr-Boronet and Farré-Rovira (1975). Carotenoids of pepper *C. annuum* var. *grossum* at different stages of maturity were investigated by Valadon and Mummery (1977), while free and esterified carotenoids in green and red fruits of the Yolo Wonder A cultivar were investigated by Camara and Monéger, (1978). Capsanthin, the main carotenoid in the ripe fruit appeared in the diester, monoester and free form, while capsorubin occurred as a diester only. Cryptocapsin was not esterified. Other carotenoids present were β-carotene, cryptoxanthin, zeaxanthin, antheraxanthin, violaxanthin and neoxanthin, of which only zeaxanthin and antheraxanthin were found as monoesters. The pepper fruit was largely used by Camara and Monéger (1982) for enzymatic studies on carotenogenesis.

Camara *et al.* (1983) succeeded in isolating and purifying *Capsicum* chromoplasts. Their biochemical characteristics were investigated including the carotenoid composition, which is very similar to that found in the fruit.

Generally the major carotenoid patterns (4), (6) are found in all red cultivars. The vitamin A value is 6000 μg retinol equivalents kg^{-1}.

E'. Persimmon (*Diospyros kaki* Thunb.) *(Table A.36)*

The Japanese persimmon (*Diospyros kaki*), considered to be native to Japan, is the most important of the known persimmon species. It has at least 1000 varieties which are divided into groups according to colour and astringency (Itoo, 1980). The colour of the ripe fruit, varying from yellow–orange to red, is due to the carotenoid pigments of which the fruit is a rich source. They were first analysed in 1932 by Karrer and co-workers who identified lycopene, zeaxanthin and "various pigments" (Karrer *et al.*, 1932). A Japanese investigation of two persimmon varieties attributed the marked colour intensification as ripening is completed to a large increase in lycopene, up to 30–40% of the total carotenoids (Tsumaki et al., 1954).

In a detailed investigation of the Hachiya cultivar using the counter current distribution method, Curl (1960b) separated 23 carotenoids, excluding cis-isomers. The total carotenoid content was 54 μg(g fresh wt)$^{-1}$. About 84% were xanthophylls, approximately half of which were monols. Nearly all the xanthophylls were esterified. The major pigments were cryptoxanthin (38%), zeaxanthin (18%) and antheraxanthin (10%) of the total carotenoids. Other pigments in order of decreasing levels were lycopene (8%), β-carotene (7%), lutein (4%), ζ-carotene and violaxanthin (3% each) and neoxanthin (2%).

A comprehensive survey of carotenoid distribution in 40 varieties of Japanese persimmons revealed that cryptoxanthin was the major constituent, amounting to 30–35% of the total carotenoids. The hydrocarbon fraction was found to be highly variable; in particular, lycopene content varied from 0–30%. The levels of lycopene were not related to total carotenoid content. The varieties investigated were classified into three categories with respect to their lycopene content: high lycopene (20–40%), intermediate, and low lycopene groups (Brossard and Mackinney, 1963).

The carotenoids of the Triumph cultivar grown in Israel have been investigated by Ebert and Gross (1985). The carotenoid changes in the peel of persimmons were followed for an entire season. In addition the carotenoid pattern changes during post-harvest ripening in both pulp and peel were characterized at three ripening stages from harvest ripe to fully ripe (Ebert, 1984; Ebert and Gross, 1984). The total carotenoid content in the peel of fruit ripened on the tree was 310 μg g^{-1}. Cryptoxanthin and lycopene were both major pigments contributing about 20% each. Zeaxanthin, antheraxanthin (cis and trans), violaxanthin (cis and trans) amounted to c. 10% each. The total carotenoid content in post-harvest ripened fruit was 68 μg(g fresh wt)$^{-1}$ in pulp and 490 μg(g fresh wt)$^{-1}$ in peel, of the fully ripe fruit. In pulp cryptoxanthin amounted to c. 50% of total pigments, zeaxanthin and antheraxanthin to c. 10% each and violaxanthin to 4%. Lycopene was found at a low level of 4.5%. In peel cryptoxanthin represented half of the total carotenoids, followed in decreasing order by lycopene (8.2%), β-carotene, zeaxanthin, antheraxanthin and violaxanthin at c. 6% each.

The major carotenoid patterns are (3), (4). The vitamin A value in peel is (Triumph) 9000–27 000 μg retinol equivalents kg^{-1} and in pulp 3350 μg retinol equivalents kg^{-1}; in the whole fruit (Japanese cultivars) 2200–3500 μg retinol equivalents kg^{-1}.

F'. Pineapple (*Ananas comosus* (L.) Mill) *(Table A.37)*

The carotenoids of the fresh pineapple have been analysed by Morgan (1966) using the counter current distribution method and column chromatography. The total carotenoid content was very low at *c.* 1 µg(g fresh wt)$^{-1}$. Twenty-one pigments were separated including isomers. The major constituents were violaxanthins (52%), luteoxanthins (13%), neoxanthin (*c.* 10%) and β-carotene (7%). Less abundant carotenoids included ζ-carotene, hydroxy-α-carotene (α-cryptoxanthin), lutein, trollixanthin, auroxanthin and neochrome. Flavoxanthin, the epimer of chrysanthemaxanthin (lutein 5,8-epoxide), was found in amounts less than 1%.

The major carotenoid patterns are (1) and (5) as the epoxides constitute 90% of the total carotenoids. The fruit has no vitamin A value.

G'. Plum (*Prunus domestica* L.) and other Subspp. *(Table A.38)*

Plums are the most widely distributed stone fruits. Despite the very large number of plum cultivars there have been very few reports on plum carotenoids. A cultivar of Italian prunes (*P. domestica*) was investigated by Curl over twenty years ago (Curl, 1963). The fruits were deep purple outside and greenish to brownish inside. The total carotenoid content was 21.0 µg(g fresh wt)$^{-1}$. The major constituents were violaxanthin (35%), β-carotene (19%) and lutein (16%). Less abundant constituents included cryptoxanthin (7%) and its various epoxides (6%), persicaxanthin (3%), luteoxanthin, antheraxanthin, phytoene and phytofluene. About 20 other carotenoids were found as very minor constituents. In a more recent study the principal carotenoids of the plum Prune d'Ente were identified and estimated globally as carotenes and xanthophylls (Moutounet, 1976). The main carotenoids were β-carotene, lutein, violaxanthin and its isomers and neochrome.

Recently the carotenoids of three cultivars belonging to three different subspecies were investigated by Gross (1984). The three cultivars were: Saghiv (*P. domestica*), a French plum ovoid in shape with a blue-black skin and yellow-brown flesh; Golden King (*P. salicilina*), a Japanese plum of cordate shape and golden yellow colour; and Mirabelle (*P. insititia*), a round golden yellow fruit. The first two cultivars are grown in Israel and the third in Germany. Their total carotenoid contents were 7.5, 24.4 and 12.0 µg(g fresh wt)$^{-1}$ respectively. The carotenoid pattern comprised five main pigments: β-carotene, cryptoxanthin, lutein, violaxanthin and neoxanthin. A similar pattern was found in Italian prunes (Curl, 1963).

Noticeable differences existed between the cultivars. Thus only Saghiv contained the UV fluorescent C_{25}-apocarotenoids persicaxanthin and persicachrome, which were also found in the Italian prunes. Additionally apo-12'-

violaxanthal was also detected. Golden King had the highest total carotenoid content and the highest β-carotene level. Violaxanthin was absent and neochrome, not found in the other cultivars, was present in two isomeric forms. Neochrome has also been found in Prune d'Ente (Moutounet, 1976). Whereas in Golden King the cryptoxanthin epoxides were totally absent, in Mirabelle an almost complete set of all the epoxy-derivatives of cryptoxanthin, including *cis*- and *trans*-isomers, were present. Thus this cultivar had the most complex pattern.

The carotenoid pattern is a modified chloroplast pattern as cryptoxanthin is also present at higher levels and the ratios between the four chloroplast pigments are totally modified. The major carotenoid pattern is (2) (modified).

Vitamin A values are 340, 1730 and 410 μg retinol equivalents kg^{-1} in Saghiv, Golden King and Mirabelle, respectively.

H'. Pomegranate (*Punica granatum* L.) *(Table A.39)*

The carotenoids of the pomegranate have been investigated by Curl (1964) in his report on "low carotenoid fruits". As the main pigments of the fruit are anthocyanins, the carotenoids are found at the very low levels of 0.2 μg g^{-1}, representing residual chloroplast pigments. The percentages of the three main groups were: hydrocarbons (15%), monols (5%) and diols and polyols (80%). They were not analysed further.

The major carotenoid patterns are (1), (2). The fruit has no vitamin A value.

I'. White Pummelo (*Citrus grandis* (L.) Osbeck) *(Table A.40)*

The pummelo or shaddock (*Citrus grandis* Osbeck) is considered to be one of the three true species among *Eucitrus* (Barrett and Rhodes, 1976). The fruit surpasses all other citrus fruits in size, having a diameter of up to 25 cm and weighing 2–6 kg. The species includes numerous varieties: seedy, seedless, yellow, red and colourless (white).

Pigment and ultrastructural changes in the developing white pummelo cultivar Goliath grown in Israel were investigated by Gross *et al.* (1983). The total carotenoid content in the ripe fruit was 0.4 μg g^{-1} in pulp and 5 μg g^{-1} in flavedo, corresponding to the greenish pale-yellow tint of the pulp and the lemon yellow colour of flavedo. In pulp the levels of the major pigments in decreasing order were lutein (23%), phytofluene (19%), β-carotene (12.5%), and violaxanthin together with its isomer luteoxanthin (11%). Pigments of the β-ionone series were also found, such as mutatochrome, cryptoxanthin, cryptoflavin, zeaxanthin and its epoxides. Additionally the rarer pigments shamoutixanthin and valenciachrome were also present. In flavedo phytofluene was the major pigment at levels as high as 67% of the total carotenoids.

The coloured carotenoids were dominated by violaxanthin and lutein, albeit at low levels up to 6% followed by ζ-carotene and cryptoflavin, c. 3% each, and other pigments at even lower levels.

The accumulation of the colourless precursor in pummelo occurs because further dehydrogenation which would lead to other carotenoids does not occur. This metabolic block is genetically determined. The occurrence of the membraneous chromoplast type which predominated in the ripe fruit (Fig. 1.26d) may be related to this fruit's special carotenogenesis.

According to Barrett and Rhodes (1976), *C. grandis* is thought to be the progenitor of various hybrids such as the grapefruit (*C. paradisi* = *C. grandis* × *C. sinensis*), and the lemon (*C. limon* = *C. medica* × *C. grandis* × *Microcitrus*), in which the first parent is the predominating genotype. The similarity between the carotenoid pattern and metabolism in the flavedo and endocarp of the grapefruit with the pummelo carotenoids is striking (Yokoyama and White, 1967; Romojaro *et al.*, 1979). The only difference between the two fruits is the absence of phytoene, which could not be identified in pummelo although all available methods were tried. A carotenoid pattern similar to that of *C. grandis* is also found in lemon (Curl, 1962; Yokoyama and Vandercook, 1967). This is remarkable because *C. grandis* is not the predominating genotype of the progenitors of the lemon. The major carotenoid patterns are (2) (modified) in pulp and uncommon in peel, containing 67% colourless carotenes (2), (5). The fruit has no vitamin A value.

J'. Red Pummelo (*Citrus grandis* (L.) Osbeck *(Table A.41)*

The carotenoids of the red pummelo cultivar Chandler have been investigated in our laboratory by Gross (unpubl. obs. 1983). As reported on the red grapefruit, the different parts of fruit pulp and septa, albedo and flavedo, which all were coloured from pink to red, were separately investigated. The total carotenoid content was highest in septa at 45–60 μg g^{-1}, in albedo at 24.7 μg g^{-1}, in flavedo at 19.3 μg g^{-1} and lowest in the pulp at 13.2 μg(g fresh wt)$^{-1}$. Lycopene was the pigment responsible for the pink and red tints of all the parts, excluding flavedo, attaining levels of 90–95%. The flavedo contained 25 μg g^{-1} chlorophyll and a relatively high level of lutein (32%), indicating that the fruit was not fully ripe. The major discernible pattern was a citrus pattern dominated by epoxides. In the other parts of the fruit, pulp, septa and albedo, the pattern was much simpler. The second major pigment was β-carotene presented at between 1 and 6%. Some xanthophylls, such as rubixanthin, were detectable in very small amounts.

As there is between white pummelo and white grapefruit, a similarity exists between the two red fruits. But noticeable differences can also be observed, such as the absence of phytofluene, a higher level of lycopene and lower level of β-carotene in the red pummelo. Its flavedo carotenoid pattern had more characteristics of a chloroplast carotenoid than a citrus pattern.

One factor responsible for these differences may be the different ripening stages of the two fruits, the grapefruit being at a more advanced stage, as obvious in flavedo. The carotenoid pattern of the pulp of red grapefruit varies too during ripening. These changes have been followed by Lime et al. (1954). As the fruit reaches maturity the lycopene content decreases and the β-carotene content increases. The absence of phytofluene is more difficult to explain; it may accumulate in the fully ripe fruit. The major carotenoid patterns are in pulp, septa and albedo (3) and in flavedo (2), (5).

The vitamin A value is 100 μg retinol equivalents kg^{-1} in the pulp.

K'. Quince (*Cydonia oblonga* Mill.) *(Table A.42)*

The carotenoids of the peel of the quince cultivar Portugal grown in Israel have been investigated by Gross (unpubl. data. 1980). The total carotenoid content was c. 7 μg(g fresh wt)$^{-1}$. The carotenoid pattern was dominated by phytofluene (43.5%). The coloured carotenoids included chloroplast carotenoids as major pigments, violaxanthin (*trans* and *cis*) (15.3%), β-carotene (9.2%), lutein and neoxanthin c. 7% each. Chromoplast pigments, such as ζ-carotene (2.1%), cryptoxanthin (4.3%) and cryptoflavin (3.6%) and zeaxanthin (5%) were minor pigments.

The carotenoid pattern is a modified chloroplast carotenoid pattern (2), but is uncommon as it contains high levels of phytofluene.

The vitamin A value is 150 μg retinol equivalents kg^{-1}.

L'. Strawberry (*Fragaria* × *ananassa* Duch.) *(Table A.44)*

The red colour of strawberries is due to anthocyanins. Carotenoids are also present in very small amounts, being masked by the anthocyanins. Curl (1964), investigating the "low-carotenoid" fruits, found that the total carotenoid content of strawberries was 0.64 μg g^{-1}, of which 14% were hydrocarbons and 85% xanthophylls. Galler and Mackinney (1965) re-examined the carotenoids of the strawberry varieties Marshall and Shasta, and found a similar content, 1.5 μg g^{-1} and a similar distribution of carotenes and xanthophylls. They identified two carotenes, β- and ζ-carotene, and five xanthophylls: cryptoxanthin, lutein, antheraxanthin, luteoxanthin and auroxanthin.

Changes of chlorophylls and carotenoids in developing strawberry fruits of the cultivar Tenira were investigated by Gross (1982c). In the fully ripe, red fruit, the total carotenoid content was very low at 0.4 μg g^{-1}. The carotenoid pattern was that of the chloroplast type (2) with lutein as major pigment

(51.5%), violaxanthin (21.5%) and β-carotene and neoxanthin c. 8% each; in fact residual chloroplast carotenoids with modified relative percentages.

M′. Tomato (*Lycopersicon esculentum* L.) *(Table A.45)*

The major pigment of the tomato is lycopene, which is found at high levels up to 90% of the total carotenoids and is thus responsible for the characteristic colour of the fruit. As early as in 1932 Kuhn and Grundmann found that tomato carotenoids are carotenes with lycopene as the most abundant pigment. The carotenes predominate, constituting 90–95% of total carotenoids (Trombly and Porter, 1953; Kargl *et al.*, 1960; Curl, 1961; Lee and Robinson, 1980).

The total carotenoid content varies between 90 and 190 μg(g fresh wt)$^{-1}$. Besides lycopene, the common red tomato contains the colourless precursors phytoene and phytofluene (15–30%) and minor pigments such as β-carotene, ζ-carotene, γ-carotene and neurosporene (Koskitalo and Ormrod, 1972; Raymundo *et al.*, 1976; Johjima and Oguro, 1983).

Curl (1961) separated the tomato xanthophylls into 20 pigments of which 15% were monols, 49% diols, 4% monoepoxide diols, 22% diepoxide diols and 11% polyols. Besides the common xanthophylls lycoxanthin, 16-hydroxylycopene (ψ,ψ-caroten-16-ol) was also identified. A more recent reinvestigation of tomato carotenoids revealed the presence of various carotene epoxides of phytoene, phytofluene and lycopene in very small amounts. Two apocarotenals which may be metabolites of lycopene, apo-6′-lycopenal and apo-8′-lycopenal, were identified as well as lycoxanthin (Ben-Aziz *et al.*, 1973).

Besides the common red tomatoes there are numerous tomato mutants of various colours ranging from white to deep orange. The characterization of their carotenoid pattern together with genetic data has permitted the elucidation of carotene biogenesis (Porter and Lincoln, 1950). An interesting mutant is the Tangerine tomato which synthesizes poly-*cis*-carotenes. The predominant carotene is prolycopene which is 7,9,7′,9′-tetra-*cis*-ψ ψ-carotene. Its structure was elucidated by Clough and Pattenden (1979) and Englert *et al.* (1979). The same fruit also produces poly-*cis* precursors of lycopene, a fact which may change the current views on carotene biogenesis.

The plastid ultrastructure changes in red tomatoes during ripening have been investigated and compared with the plastids of mutants. The differences in carotenoid patterns markedly affect the ultrastructure (Harris and Spurr, 1969). In the "cherry" tomatoes plastid changes and pigment evolution were followed during growth and ripening by Laval-Martin (1974) and Laval-Martin *et al.* (1975).

The major carotenoid pattern is (3). The vitamin A value as calculated by Zakaria *et al.* (1979) is 200 μg retinol equivalents kg^{-1}.

N'. Watermelon (*Citrullus lanatus* (Thunb.), Red-fleshed) *(Table A.46)*

According to Palmer the flesh pigment of the watermelon was first isolated in 1879. Because of its red colour it was called rubidin. Later the resemblance to the tomato pigment was noted and it was confirmed as lycopene (Palmer, 1922). Zechmeister and Tuzson (1930) isolated lycopene and β-carotene from watermelon. The carotenoids of watermelon pulp were found by Zechmeister and Polgar (1941) to be predominantly hydrocarbons, the main component of which was lycopene. It was accompanied by a *cis*-isomer. Similarly, β- and γ-carotene were found in two isomeric forms.

Tomes *et al.* (1963) investigated the carotenoids of nine red-fleshed watermelon cultivars, varying in flesh colour from deep red to pale pink. The total carotenoid content varied from 20–62 μg(g fresh wt)$^{-1}$. Lycopene, the main pigment, was found at levels between 60 and 80%. The four additional carotenes were present in decreasing order as: β-carotene (6–20%), phytofluene (5–8%), ζ-carotene (4–10%), and γ-carotene (*c.* 1%). The colour of all cultivars correlated closely with the lycopene contents of the flesh.

The carotenes of a Queensland watermelon cultivar Candy Red were investigated by Morgan (1967). The total carotenoid content was 25 μg(g fresh wt)$^{-1}$. The hydrocarbon fraction contained twenty-one pigments. Lycopene and its *cis*-isomers were the major pigments, forming 73.7% and 7.6% respectively of total carotenoids. Other minor pigments were phytoene (2.1%), phytofluene (1.4%), β-carotene (4.1%), ζ-carotene (1.6%) and γ-carotene (0.4%). α-Carotene, poly-*cis*-lycopene and an unknown pigment were also present in trace amounts.

Carotenes of an orange-fleshed watermelon were investigated by Tomes and Johnson (1965). The major pigment responsible for the flesh colour was polycopene (46.8%), accounting for nearly half of the total carotenoids. Other carotenes present were phytoene (15.5%), phytofluene (5.9%), β-carotene (4.1%), ζ-carotene (16.8%), proneurosporene (6.5%) and lycopene (4.4%) of total carotenoids.

There exists an "amazing" parallelism between the carotenoid patterns of normal red-fleshed tomatoes and watermelons and the patterns of tangerine tomatoes and orange-fleshed watermelons. As in tomatoes, it is postulated that the orange-fleshed fruits should differ by a single recessive gene. In order to verify this suggestion a reinvestigation of the carotenoids of the orange-fleshed watermelon is necessary, as well as genetic studies.

The major carotenoid pattern in the red watermelon is (3), in the yellow (7). The vitamin A value is 80–370 μg retinol equivalents kg^{-1}.

XIV. General Considerations on Carotenoid Distribution

After the detailed description and characterization of each carotenoid pattern found in the edible fruits, some general considerations may be drawn. Fruit colour imparted by carotenoids is directly correlated with the level of the pigments, this constituting the basis of the non-destructive external colour assessment by physical methods. However, this is not always valid; the same pigment level may exhibit different colour effects depending on internal factors such as the pattern localization of plastids within the tissue. This has been observed in different citrus species and cultivars. For fruits with increasing carotenoid amounts in flavedo the Hunter a/b values did not increase correspondingly but were lower or higher than expected (Gross, 1977).

The total carotenoid content is decisive but the colour is not equally affected by the presence of each pigment. The effect is more pronounced for pigments absorbing in the region of longer wavelengths, such as β-carotene, cryptoxanthin and especially citraurin, lycopene and the ketopentylcarotenoids. They may impart deeper colours from orange to red even when present at lower levels. The examples are numerous: citrus pulp containing more cryptoxanthin is deep orange; the flavedo of mandarins is orange-red due to the presence of both cryptoxanthin and more particularly citraurin; apricots are orange because of their high β-carotene level; Japanese persimmons are orange-red when containing only cryptoxanthin and redder when lycopene is also present; tomatoes are red due to lycopene as are red peppers due to the unique paprika ketones.

Pattern (1) is described as insignificant amounts of carotenoids. As modern chromatographic methods permit analysis at the μg level, the carotenoids of many fruits with a low carotenoid content can be investigated and reveal interesting results. For example in banana pulp with a total carotenoid content of 1 μg(g fresh wt)$^{-1}$, the main pattern was uncommon consisting of nearly equal levels of α-carotene, β-carotene and lutein. The pulp of pineapple with an equal total carotenoid content contained 23 pigments dominated by violaxanthin, 50% of total carotenoids.

Pattern (2), the chloroplast carotenoids, in spite of its uniformity in consisting of the same four basic carotenoids, is extremely diverse due to the variations in the relative proportions of these pigments and to the nature of the additional minor pigments. It is one of the most widespread patterns, found generally in fruits whose colour is imparted by anthocyanins and in fruits in which ripening does not involve the chloroplast–chromoplast transformation.

In blackberries, blueberries, strawberries and some pear cultivars which probably contain some chloroplast remnants, lutein is predominant, as it is in leaves. In grapes and plums, β-carotene is found at almost the same level as lutein. In dates, neoxanthin is the second major pigment. In some fruits in which ripening involves the chloroplast–chromoplast transformation, one of the chloroplast carotenoids is increasingly synthesized during ripening. In various apple cultivars violaxanthin rises to levels as high as 50% of the total carotenoids. On the other hand, in one plum cultivar violaxanthin is absent. In some fruits a fifth pigment, at high levels, is found along the chloroplast carotenoids. In avocado, chrysanthemaxanthin is the second major pigment, while in kiwifruit neochrome is the fifth main chloroplast pigment.

Pattern (3), as characterized by Goodwin (1982), is found as such in red tomatoes and red watermelons, but more saturated precursors are not always detectable, as for example in red pummelo and red guava.

Pattern (4), which is relatively widespread, includes three pigments: β-carotene, cryptoxanthin and zeaxanthin, but generally all the three pigments are not always present. Zeaxanthin is usually absent or present at low levels, as in apricot, loquat and orange-fleshed melons. In persimmon pulp β-carotene is found at low levels and cryptoxanthin is the major pigment, followed by zeaxanthin.

Pattern (5) is characterized as a mixture of epoxides, particularly furanoxides, which is found in almost all citrus species—but the predominant epoxide is always violaxanthin, a $5,6,5',6'$-diepoxide.

A characteristic of some fruits of yellow-lemon colour, as they are not able to synthesize large amounts of carotenoids, is the presence of large amounts of colourless carotenes—up to 70% of the total carotenoids. Such fruits are the lemon, the white pummelo, the white grapefruit; this is not confined only to citrus species but also occurs in quince.

Tropical fruits appear to have carotenoid patterns that are harder to characterize. Carambola contains as major carotenoids ζ-carotene and cryptoflavin; passion fruit contains three major carotenoids at equal levels: phytofluene, ζ-carotene and β-carotene.

Fruits with medium vitamin A value (up to 500 μg retinol equivalents kg^{-1}) are carambola, guava, kumquat and mandarins. Fruits with high vitamin A value (up to 4000 μg retinol equivalents kg^{-1}) are apricot, loquat, orange-fleshed melon, papaya and persimmon. Red pepper contains 6000 μg retinol equivalents kg^{-1} and the highest value is attained by mango at 12 500 μg retinol equivalents kg^{-1}.

XV. Importance of Carotenoids in Taxonomy

The usefulness of carotenoids, which are secondary metabolites, as chemotaxonomic markers depends on a discontinuous distribution. Properties which promote their use in taxonomy were discussed by Liaaen-Jensen (1979). These are their structural diversity (c. 450 structures), their distribution in all photosynthetic organisms and occasional occurrence in other bacteria, yeasts and fungi, the knowledge of their biosynthetic pathway and function, and finally, the possibility of ready analysis on a μg scale.

In lower organisms the structure, distribution and biosynthetic pathways of carotenoids have given useful taxonomic information, improving the existing classification and revealing unknown phylogenetic relationships. In fruits insufficient information on carotenoid distribution hinders their use as taxonomic markers, a problem which is likely to be overcome with the improvement of analytical techniques and the application of computerization.

As previously discussed, according to Goodwin (1982), eight carotenoid patterns of distribution can be distinguished in fruits. Different genera are characterized by specific carotenoid patterns, correlating with the existing plant classification; e.g. in kiwifruit (*Actinidia chinensis*) the major carotenoid pattern is that of the chloroplast carotenoids; in tomato (*Lycopersicon esculentum*) large amounts of lycopene are accompanied by the related more saturated polyenes; in citrus the complex carotenoid mixture is predominated by epoxides, etc. Of interest is the striking similarity of the tomato carotenoid pattern with that found in watermelon (*Citrullus lanatus*), fruits of two different families Solanaceae and Cucurbitaceae.

Even when the same major carotenoid pattern is discernible, for example pattern (2), chloroplast carotenoids, the patterns differ markedly from one another in avocado (*Persea americana*) and in grapes (*Vitis vinifera*). The same is valid for genera belonging to the same family. The patterns (2), (5) chloroplast carotenoids and large amounts of epoxides are characteristic of apple (*Malus domestica*) and the sweet yellow cherry (*Prunus avium*), both of the Rosaceae family, but differ from one another by the amount and nature of the major epoxides.

In other genera of the Rosaceae family the major carotenoid pattern (4), large amounts of β-carotene and its hydroxy derivatives, is found in both the genus *Prunus*, apricot and peach, and in *Eriobotrya*, loquat. It is of interest that this pattern is more similar between these two genera of different subfamilies Maloideae and Prunoideae than it is among the other genera of Maloideae (Valadon and Mummery, 1967) or even among the Prunoideae species (Katayama *et al.*, 1971).

The citrus genus is remarkable for producing the largest number of carotenoids found in fruit. About 115 different pigments have been reported, including a large number of isomers (Stewart, 1973). Still it has not been possible to determine which of the isomers occur in nature and which are formed during isolation (Stewart, 1980).

The numerous citrus species are characterized by well distinguished carotenoid patterns, the variation in type and amount of the individual pigments producing the various tints from yellow to deep orange and red of the fruits. Some species such as *Citrus reticulata* produce specific C_{30}-apocarotenoids, especially citraurin. Only recently, citraurin was detected at lower levels in the *Fortunella* genus of the same Rutaceae family (Huyskens *et al.*, 1985).

Unique carotenoids such as capsanthin and capsorubin are synthesized in pepper fruits. They are rare in nature, but have also been reported in other fruits such as barberry (*Berberis* spp. (Bubicz, 1965) and asparagus fruit (*Asparagus officinalis* (Simpson *et al.*, 1977)).

Three plum cultivars belonging to three different sub-species exhibited noticeable differences in the carotenoid pattern although the main carotenoids were common in all three sub-species (Section XIII.G').

As previously discussed for anthocyanins, the most important taxonomic use is in the differentiation of fruit cultivars. Even though the cultivars may have the same major carotenoid pattern, the presence or absence of one pigment must be considered as significant. Unfortunately there are few detailed surveys with comparisons of more than 2–4 cultivars (see Appendix) excepting the survey of nine strains of watermelons carried out by Tomes *et al.* (1963). Even so, the intercultivar differences are significant, as is shown in Section XIII.

Of equal practical importance is the use of carotenoids in genetic studies. Studies of different constituents such as flavonoids of both natural and artificial hybrids have shown that inheritance is normally additive, although occasionally either some parental constituents are missing or some additional hybrid compounds are present. This also applies to carotenoids. The trigeneric hybrid Sinton citrangequat produces citraurin at levels as high as 50% of the total carotenoids (Yokoyama and White, 1968), whereas in one of the parents, *Fortunella*, citraurin is found at much lower levels. Also in other citrus hybrids Minneola tangor and Temple orange, citraurin is found at higher levels than in *Citrus reticulata*, one of its parents (Yokoyama *et al.*, 1972).

Two other C_{30}-apocarotenoids, citraurinene and citraurol, were first isolated from a citrus hybrid Robinson (Orlando tangelo × Clementine) (Leuenberger and Stewart, 1976; Leuenberger *et al.*, 1976). Recently citraurinene has been found to be present also in Clementine, thus not being produced exclusively in hybrids (Farin *et al.*, 1983). This biosynthetic potential of the Clementine is inherited by both its hybrids Robinson and Michal and was decisive in ascertaining the putative parents (Section XIII.V).

Equally edifying is the comparison of the pummelo (*C. grandis*) carotenoid pattern with that of its hybrids the grapefruit and lemon. The similarities found between the hybrids and *C. grandis*, one of the parents, are striking and also provide the confirmation of the citrus classification according to Barrett and Rhodes (1976) (Section XIII.I′).

XVI. Carotenoid Changes During Fruit Ripening

During ripening carotenogenic fruits turn gradually yellow, orange or red as chlorophylls decompose and carotenogenesis takes over. This change is correlated with the conversion of chloroplasts into chromoplasts which is induced by the interaction of different plant hormones, ethylene being the "ripening hormone".

The chloroplast carotenoid pattern which is the same in both leaves and green fruits changes into a chromoplast carotenoid pattern. The chromoplasts are characterized by their rapid and high biosynthetic ability which is reflected in the appearance of multiple carotenoids of different types. In early reports it was observed that the total carotenoid content increases during ripening. An increase in total carotenoid content in citrus fruit endocarp during ripening was first reported by Miller and Winston (1939). Higby (1963) analysed Valencia and Navel oranges and reported on variations due to the effect of maturity. Seasonal variations in different varieties of Italian oranges showing a marked increase in carotenoid content were reported by Pennisi *et al.* (1955) and Di Giacomo and Rispoli (1960). In Italian mandarins a similar increase was reported by Di Giacomo *et al.* (1968a, b).

In tomatoes the studies were more advanced as changes of individual carotenoids were followed during maturation (Goodwin and Jamikorn, 1952; Mackinney and Jenkins, 1952; Edwards and Reuter, 1967).

The chloroplast–chromoplast transformation has been followed in different fruits including pepper (Kirk and Juniper, 1967; Spurr and Harris, 1968; Simpson and Lee, 1976), tomato (Rosso, 1967; Harris and Spurr, 1969; Mohr, 1979) and orange (Thomson, 1966), revealing the high diversity of the process in different fruits. However, in these earlier studies the ultrastructural changes were not correlated with pigment changes although it has been suggested the chromoplast ultrastructure is related to the nature of the carotenoid present (Harris and Spurr, 1969b).

Systematic investigations of pigment changes in ripening fruits have recently been carried out, providing a basis for detailed studies on carotenogenesis as a function of ripening. Consecutive accumulation, decrease and disappearance, precursor–product relationships of the pigments as well as structural consideration of the carotenoid formed in the process, all permit a better interpretation of carotenogenesis. A most advanced step involves the correlation of ultrastructural changes with pigment changes. These recent

A. Ripening Patterns

By following the variation of the total carotenoid content during an entire season, a "ripening curve" can be established. It has been observed that the total carotenoid level does not always increase continuously during development as was thought earlier, but follows different curves.

1. Ripening Curve Displaying a Mininum at Mid-season

In fruits which synthesize large amounts of carotenoids during ripening, the total carotenoid content follows a ripening curve which displays a minimum at mid-season. This was observed for the first time by Laval-Martin (1975) in "cherry" tomato. The development involves three distinct steps and shows that the growth period is separated from that of maturation by a lag phase during which the new regulatory phenomena which induce pigment transformation occur. This lag phase corresponds to the disintegration of the granal system as chlorophyll disappears (Laval-Martin, 1974).

A similar ripening curve was observed in the peel of Golden Delicious apple (Gross et al., 1975; Gorski and Creasy, 1977). During the early period from July to mid-September the carotenoid content decreased from 18 $\mu g\ g^{-1}$ to a minimum of 7 $\mu g\ g^{-1}$ and then increased again to 13 $\mu g\ g^{-1}$, a level lower than in the unripe fruit. The absorption spectra of the total carotenoid extract showed an increasing hypsochromic shift up to 10 nm, an indication that a substantial change had occurred.

This kind of carotenoid metabolism was detected only by a few investigators at first, but with the intensification of the studies on pigment changes during ripening, it seems to have become general. It is very pronounced in citrus fruits, particularly in flavedo of orange (Fig. 1.11) (Eilati et al., 1975), of Satsuma mandarin (Fig. 1.9) (Noga, 1981), of Clementine (Farin et al., 1983), Dancy tangerine (Gross, 1981) and a mandarin hybrid (Farin et al., 1983). It was observed also in other fruits such as persimmon (Ebert and Gross, 1984), kumquat (Huyskens et al., 1985), peach (Fig. 1.12), (Lessertois and Monéger, 1978) and banana (Gross and Flügel, 1982).

The ripening curve of oranges shown in Fig. 1.11 presented a minimum level; at this stage there was a hypsochromic shift of 6–8 nm in the absorption spectrum of total carotenoids indicating not only the completion of the chloroplast–chromoplast conversion but also the appearance of carotenoids of a different type. In Satsuma mandarins the minimum occurring in the ripening curve was even more pronounced (Fig. 1.9). In flavedo of Dancy tangerine, in which the pigment changes were followed at two-week intervals (Gross, 1981), the ripening diagram does not show the expected sharp minimum. The total carotenoid content decreased by half immediately at the beginning of the season and remained constant over a period of 14 weeks. Afterwards the level increased at an accelerated rate, reaching a maximum of c. 300 μg(g fresh wt)$^{-1}$. The maintenance at a low total carotenoid content level of about 50 μg(g fresh wt)$^{-1}$—a plateau—reflects a low rate of development from June to October.

In banana peel the level of total carotenoids was reported to remain unchanged during ripening by Von Loeseke (1929). This uncommon ripening curve not encountered in any other fruit is still quoted today, although the report is outdated (Goodwin, 1982). A reinvestigation carried out by Gross and Flügel (1982) revealed that the level of total carotenoids decreased to half at colour break and subsequently increased until total carotenoids in the ripe fruit reached virtually the initial level of the green fruit.

2. Continuous Increase of the Ripening Curve

In the carambola (Gross et al., 1983) and a yellow gooseberry cultivar (Gross, 1982/83) only two ripening stages were studied and an increase of the total carotenoid level was observed. Presumably systematic studies beginning at the green unripe stage would show the expected minimum in the ripening curve. In orange-fleshed muskmelons the ripening curve shows a small minimum very early in the season, which may not exactly reflect the general colour development (Reid et al., 1970). A continuous increase of the total carotenoid content was observed to occur in the pulp of ripening Shamouti orange (Rotstein et al., 1972), Dancy tangerine (Gross, 1982) and seven other citrus cultivars (Stewart, 1977b).

3. Continuous Decrease of the Ripening Curve

(a) *Fruits containing pigments other than carotenoids when ripe.* In fruits in which the final colour is imparted by anthocyanins such as the sweet cherry (Okombi et al., 1986), red currant (Gross, 1982/83) and strawberry (Gross, 1982b), the total carotenoid content decreases continuously, paralleling the decrease in the chlorophylls. A low level of total carotenoids is reached at the onset of anthocyanin biosynthesis, as was observed in strawberry (Fig. 1.10 (Hager and Bertenrath, 1962)) and continues decreasing in red currant (Gross, 1982/83). A decreasing carotenoid content was observed in ripening dates; the loss of the carotenoids is attributable more to loss of moisture than to the gradual darkening of the fruit (Gross et al., 1983). In kiwifruit, which remains green at the ripe stage, a decreasing ripening curve which parallels a partial decrease of chlorophylls is evidence that maximum carotenoid synthesis occurs when the chlorophyll content is highest (Gross, 1982).

(b) *Fruits with a low carotenoid content.* Some fruits become yellow during ripening as their carotenoids become unmasked as the chlorophylls gradually disintegrate. However the total carotenoid content decreases continuously. This was observed in the peel of a pear cultivar (Gross, 1984), in the exocarp of muskmelons (Flügel and Gross, 1982), in a yellow cherry cultivar (Gross, 1984), in the flavedo of grapefruit (Banet *et al.*, 1981), pummelo (Gross *et al.*, 1983) and lemon (Yokoyama and Vandercook, 1967).

(c) *Special ripening curves.* A special colour development was observed in the pulp of red grapefruit during ripening (Lime *et al.*, 1954; Cruse *et al.*, 1979). At the beginning of the season the total carotenoid content increased, reaching a maximum very early in the season and then it decreased continuously. A decrease of the carotenoid content at the end of the season occurred in peach mesocarp (Gross, 1979), a total disappearance occurring in another peach cultivar (Lessertois and Monéger, 1978).

XVII. Changes of the Carotenoid Pattern during Ripening

High amounts of chlorophyll are present in all unripe fruits, together with the typical basic chloroplast carotenoids β-carotene, lutein, violaxanthin and neoxanthin (Chapter III Section XI.A, p. 122). During fruit development the chlorophylls gradually disintegrate and the chloroplast carotenoids decrease or totally disappear. A *de novo* enhanced carotenogenesis follows and the chloroplast carotenoid pattern changes gradually into a complex chromoplast carotenoid pattern. The predominance of the carotenoids of the β,ε-series is replaced by carotenoids of the β,β-series. Generally at the stage corresponding to the mid-season minimum of the ripening curve the synthesis of new chromoplast carotenoids, which become major pigments, begins. The epoxides accumulate at the last ripening stage. At these stages the early precursors of the carotenoids, the colourless phytoene and phytofluene and the more saturated carotenes, ζ-carotene and neurosporene, accumulate as the turnover rate slows down. In the process the xanthophylls are gradually esterified.

In some fruits the ripening process does not involve the synthesis of new chromoplast carotenoids but an increased synthesis of one or two chloroplast pigments, usually β-carotene or violaxanthin.

In the peel of Golden Delicious apple at the beginning of the season the chloroplast carotenoid pattern consisted of lutein (47%), β-carotene (34%), neoxanthin (5%) and violaxanthin (<1%). The seasonal changes of this

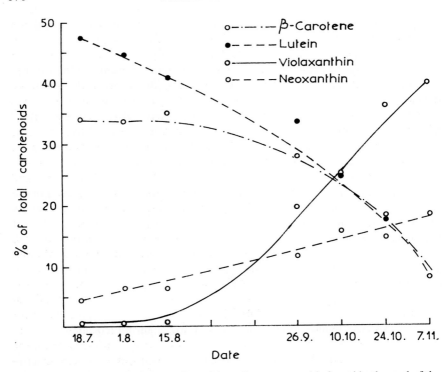

Fig. 3.23 Seasonal changes in the four chloroplast carotenoids found in the peel of the Golden Delicious apple. From Gross *et al.* (1978).

pattern are shown in Fig. 3.23. During ripening a spectular *de novo* synthesis of violaxanthin occurs (a 50-fold increase) paralleled by a 5-fold increase of neoxanthin, even though the total carotenoid level rises only slowly. The pattern becomes a complex chromoplast pattern as cryptoxanthin and its epoxide, isolutein and a C_{27}-apocarotenol are detectable in harvest ripe fruit.

The ripening of peach mesocrap involves also the biosynthesis of violaxanthin, which becomes the major pigment in the ripe fruit (Gross, 1979a). In contrast with the apple, violaxanthin follows a different development. Found already at high levels in the unripe fruit (32%), it continues to increase, reaching a level of 50%, and then decreases again to a level lower than the initial level. The second major pigment is cryptoxanthin, the synthesis of which begins at the minimum of the ripening curve. Its monoepoxide is gradually formed and in the ripe fruit the number of cryptoxanthin epoxides increases. Phytofluene and ζ-carotene accumulate later in the season.

In fruits such as apricot and mango, β-carotene increased rapidly to become the major pigment, making up 50% of the total carotenoids (John et al., 1970; Katayama et al., 1971). Numerous other pigments of the β,β-series are also formed. In other fruits the principal event during ripening involves the massive synthesis of one single chromoplast pigment such as lycopene in tomato. Along with lycopene other carotenoids, namely its more saturated precursors, are also detectable, their levels increasing gradually during the process (Meredith and Purcell, 1966). Recently it was shown that the synthesis of the pigments in the inner pulp and the outer region of the pericarp of the "cherry" tomato is different during growth and maturation (Laval-Martin et al., 1975). The foliar pigments accumulate in both parts of the green fruit and then remain constant excepting neoxanthin, which decreases. Phytofluene and lycopene appear only at the end of the ripening and almost exclusively in the external part of the fruit; the hydroxy derivatives of lycopene, lycoxanthin (ψ,ψ-caroten-16-ol) and lycophyll (ψ,ψ-carotene-16,16'-diol) can be detected during the entire growth and maturation period, whereas in another tomato cultivar they were detectable at a more advanced stage (Edwards and Reuter, 1967).

In other fruits the changes are so complicated that a schematization is hardly possible, this being especially true in citrus fruits. In the flavedo of Satsuma mandarin (Noga, 1981), Dancy tangerine (Gross, 1981), Clementine and a mandarin hybrid (Farin et al., 1983), all chloroplast pigments decrease during ripening except violaxanthin, which rises to a level of 50%. With the onset of colour break the synthesis of new pigments begins, in particular of β-citraurin and cryptoxanthin (Gross, 1981). Phytofluene, ζ-carotene and the epoxides of cryptoxanthin are detectable at low levels at the last maturity stages.

In the kumquat Nagami a similar metabolism occurs, a relatively simple chloroplast carotenoid pattern gradually changing into a complex chromoplast pattern comprising 25 pigments. As well as violaxanthin, cryptoxanthin and β-citraurin are also synthesized. The complexity of the pattern is due to the accumulation of phytofluene, ζ-carotene, and especially to numerous cryptoxanthin epoxides and zeaxanthin epoxides (Huyskens et al., 1985).

In the peel of persimmon ripened on the tree a chromoplast pattern dominated by pigments of β,β-series is formed through the gradual synthesis of cryptoxanthin, zeaxanthin and antheraxanthin (Ebert and Gross, 1985). A noteworthy feature is the continuous increase of cryptoxanthin from 1.7–42% at the harvest stage. With the completion of ripening the carotenoid pattern changes, lycopene appearing as an additional major pigment. This is a very peculiar development not encountered in other fruit. A possible explanation for the unusual carotenoid pattern change may be that the turnover rate slows down at the final ripening stage. As a consequence early precursors accumulate, usually phytofluene and ζ-carotene and not lycopene as in this fruit. In persimmon pulp a similar pattern is formed in the ripe fruit but the different relative proportions of the pigments are the result of a different biosynthetic rate. Cryptoxanthin reaches approximately the same level as in the peel (48%), being the major pigment, followed by zeaxanthin and antheraxanthin. The carotenes which accumulate in pulp are phytofluene, ζ-carotene and lycopene.

In ripening pepper, as the chloroplast pigments decrease gradually, chromoplast pigments, predominated by the unique pigments capsanthin and capsorubin, are synthesized (Valadon and Mummery, 1977). Their synthesis begins at the yellow stage and increases continually. Violaxanthin, the precursor of capsorubin, is still found at the ripe stage but antheraxanthin, the immediate precursor of capsanthin, is absent. In another pepper cultivar, Yolo Wonder, both precursors of the ketocarotenoids are found (Camara and Monéger, 1978).

In citrus pulp the carotenoid metabolism does not involve the usual chloroplast–chromoplast pattern transformation. In Dancy tangerine juice the carotenoid pattern of the unique fruit at the earlier stages is already predominated by carotenoids of the β,β-series, namely cryptoxanthin (20%), antheraxanthin (29%) and violaxanthin (39%) (Gross, 1982). The principal event during the ripening process is the biosynthesis of cryptoxanthin, which doubles in level from 20–40%. The changes in the level of the main carotenoids of the Dancy tangerine juice during ripening are shown in Fig. 3.24. The synthesis of both β-carotene and lutein during the process is noteworthy. The carotenoid pattern at the beginning of the season is as complex as that of the fully ripe fruit.

A similar development appeared to have occurred in the juice of Shamouti orange, the immature fruit and the ripe fruit showing the same complex carotenoid pattern (Rotstein et al., 1982). The changes occurring in this species result in the synthesis of lower amounts of cryptoxanthin and higher amounts of zeaxanthin and its epoxides.

In fruits in which the final colour is due to anthocyanins or chlorophyll, the chloroplast pattern does not change during ripening. As the level of each pigment decreases during the process, the ripe fruit contains residual chloroplast carotenoids showing changed relative proportions than at the beginning. This occurs in strawberry (Gross, 1982c), red currant (Gross, 1982/83), date (Gross et al., 1983), kiwifruit (Gross, 1982a) and the pulp of green-fleshed cantaloup (Flügel and Gross, 1982).

The qualitative distribution of the chloroplast pattern as well as the turnover rate show intergeneric differences determined by the genotype. In the unripe red currant berry the carotenoid pattern included 18 pigments; in kiwifruit five major pigments and additionally thirty minor pigments. The metabolism may in some cases resemble that of carotenogenic fruits in that

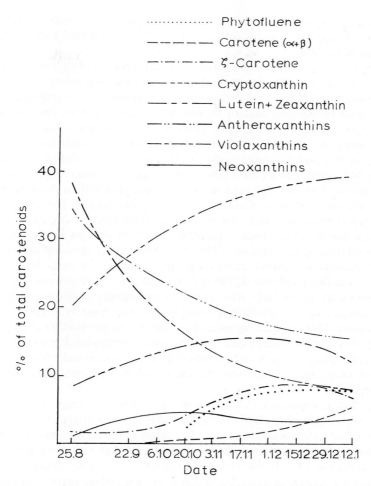

Fig. 3.24 Seasonal changes in the levels of the main carotenoids found in Dancy tangerine juice. From Goss (1982).

cryptoxanthin appear in the ripe fruit or the level of one pigment may increase slighly as violaxanthin in kiwifruit or isolutein in the green-fleshed muskmelon.

Generally, β-carotene is more rapidly destroyed than lutein which then represents more than half of the residual chloroplast carotenoids. This metabolism is reversed in the sweet cherry in which β-carotene predominates in the ripe fruit (Okombi et al., 1980).

In fruits in which the yellowing is due to the unmasking of carotenoids by the degradating chlorophylls the carotenoid metabolism may or may not involve changes from a chloroplast carotenoid pattern into a chromoplast carotenoid pattern.

In many fruits such as melon (exocarp), pear (peel) and yellow cherry, the individual carotenoids decrease during ripening, following the general decreasing trend of the total carotenoids. As a result the ripe fruit contains a chloroplast carotenoid pattern in which the ratio between the individual pigments is considerably changed. Thus in the ripe melon exocarp the lutein level decreased from 50 to 40%, whereas in the ripe pear lutein reached a level as high as 70% and β-carotene as low as 6%. In the yellow cherry cultivar, the unripe fruit contained an unusually high violaxanthin level and additionally other epoxides. As the epoxides decreased more slowly than β-carotene and lutein, the final pattern is dominated by epoxy carotenoids (70%).

In other fruits, although the total carotenoid content decreases during ripening, the initial chloroplast carotenoid pattern changes gradually into a chromoplast carotenoid pattern. This was observed in lemon, the carotenoids of which were compared at two ripening stages, both in pulp and peel (Yokoyama and Vandercook, 1967). In fact the carotenoid pattern of the pulp of the mature green fruit was already a chromoplast carotenoid pattern containing as major pigment cryptoxanthin accompanied by its epoxides and phytofluene. During ripening α-carotene, lutein and neoxanthin disappeared totally, whereas β-carotene and more especially cryptoxanthin were increasingly synthesized. In peel, while the chloroplast carotenoids decreased, the level of cryptoflavin increased markedly and phytofluene and ζ-carotene accumulated.

In pummelo endocarp, although there are almost no changes in the level of total carotenoids, the initial chloroplast carotenoid pattern changes into a chromoplast carotenoid pattern as two chloroplast carotenoids, β-carotene and lutein, decrease and phytofluene accumulates as the major pigment at levels as high as 19%. In the flavedo the changes are more drastic, phytofluene reaching a level as high as 67% and other carotenes of the earlier biosynthetic steps, such as ζ-carotene and neurosporene are detectable. A similar metabolism has also been described in the developing white grapefruit (Yokoyama and White, 1967; Banet et al., 1981). In peel the colourless carotenes appeared at the colour break stage and increased continually, becoming the major pigments in the ripe fruit. A chromoplast pattern resulted, predominated by carotenoids of the β,β-series such as β-cryptoxanthin, cryptoflavin, zeaxanthin and its epoxides. In the unripe pulp a carotenoid pattern nearly as complex as in the ripe pulp was found, the ripening involving a marked increase of the colourless precursors phytoene and phytofluene.

The accumulation of the colourless polyenes in all these citrus fruits suggests the existence of a genetic block which inhibits the further dehydrogenation steps in the carotene pathway. However it is not total as in parallel carotenoid biosynthesis, albeit at low levels, occurs in the normal way.

In the red grapefruit pulp the seasonal changes of only two major pigments, lycopene and β-carotene, have been followed. The lycopene level reached a maximum early in the season and then decreased continuously, whereas the β-carotene levels followed a similar curve but later in the season. This would explain the unusual development of the ripening curve (Lime et al., 1954).

XVIII. Plastid Changes during Ripening

Ultrastructural changes in ripening fruits have been studied both independently and simultaneously following the pigment changes as was done in more recent studies. A strict correlation exists between ultrastructural and pigment changes during ripening. In carotenogenic fruits which synthesize large amounts of carotenoids during ripening, such as citrus fruits, tomato and pepper, chloroplasts changes into chromoplasts, determining the parallel change of the carotenoid pattern. The gradual disintegration of the chloroplasts involves the disappearance of the chlorophylls and the decrease or disappearance of the chloroplast carotenoids. The minimum in the ripening curve corresponds to the grana disaggregation and marks the beginning of the enhanced carotenogenesis that occurs in the chromoplasts. At this stage the biosynthesis of new chromoplast pigments begins, for example violaxanthin in apple, β-citraurin and cryptoxanthin in citrus, capsanthin and capsorubin in pepper, and lycopene in tomato. Plastoglobules appear and increase gradually in size and number until they become the predominant structural elements in which the lipophilic carotenoids accumulate. In chromoplasts of other types they may crystallize out or accumulate in the respective structural elements.

Although the beginning of the chromoplast carotenoid synthesis follows the breakdown of the thylakoid membranes, this does not mean that the carotenoid-carrying structures of chromoplasts are derived from decomposing or transforming thylakoids (Sitte et al., 1980). Camara and Brangeon (1981) observed that the beginning of an active biogenesis of ketocarotenoids in pepper fruit plastids, coincides with the development of achlorophyllous membranes derived in part from the inner envelope membranes. These membranes are considered to be a framework for enzyme binding in carotenoid synthesis. Such membranes were also observed in flavedo of oranges by Thomson (1966), who rightly assumed that they might play a role in carotenogenesis.

The ultrastructural changes in the pulp of various citrus fruits during ripening were followed by Shomer (1975). In juice-sac primordia and young juice sacs, chloroplast-containing grana and relatively large starch granules were present. The plastid differentiation showed unusual features. As the grana gradually disappeared a lamellar body began to appear in the centre of the plastid, developing continuously and being present in chromoplasts. It differs from the prolamellar body of etioplasts by its lamellar organization, consisting of short and twisted lamellae. At a later stage osmiophilic globules of various dimensions appear around the lamellar body. The plastids found in cells of the central region of the juice sac differ from those of the surrounding cells in that they contain large amounts of plastoglobules and either a relatively small lamellar body or only remnants of it.

The possible role of the lamellar body in carotenogenesis should be investigated since a connection seems to exist between its formation and carotenoid accumulation. The fact that in all carotenoid investigations of citrus fruit pulp a chromoplast carotenoid pattern was discernible may be due to the earlier completion of the ripening process in the pulp than in the peel so that the very early stages were skipped over. Moreover, the very young fruits containing chloroplasts are not analysable because they provide only a minimal quantity of liquid. On the other hand, in fruit pulp immature plastids without chlorophyll may develop very early into active chromoplasts. This would explain the increased biosynthesis of β-carotene and lutein in the juice of ripening Dancy tangerine (Gross, 1982b) or the absence of the common chloroplast carotenoid pattern in various unripe fruit (Gross, 1979, 1985).

In fruits in which the final colour is due to chlorophyll or anthocyanins the carotenoid pattern is characterized by chloroplast carotenoids or by residual chloroplast carotenoids. The fine structure of plastid in such fruits was followed at various depths in young and mature avocado (Cran and Possingham, 1973), in fresh and dried kiwifruit (Possingham *et al.*, 1980) and during ripening in the mesocarp and exocarp of green-fleshed muskmelons (Flügel and Gross, 1982, see also Figs 1.13–1.17). The presence, in avocado, of a special "etioplast" carotenoid pattern (Section XIII.C), a pattern dominated by lutein (25%) and its epoxides (isolutein + chrysanthemaxanthin, 29%) may be the result of an oxidative metabolism occurring in etioplasts. In kiwifruit two types of chloroplasts present in the inner and outer pericap may account for its special chloroplast carotenoid pattern (Gross, 1982a).

An unchanged chloroplast carotenoid pattern was found both in the mesocarp and the exocarp of the green-fleshed muskmelon. The plastid of the inner mesocarp did not change markedly during ripening, whereas those of the outer pericarp and the exocarp with unusual highly stacked grana changed into plastids called gerontoplasts by Sitte (1977). These plastids are the site of catabolic processes exclusively, containing the same chloroplast carotenoids as at the green stage, but at modified levels.

From the category of fruits in which yellowing is due to the unmasking of the carotenoids, the only ultrastructure described during ripening is that of flavedo plastids in pummelo (Fig. 1.26(a)–(d); Gross et al., 1983). The chloroplast carotenoid pattern changing into a chromoplast carotenoid pattern reflects the plastid changes. Two types of chromoplasts are formed: the common globulous type and an uncommon membranous type. As this latter type predominates it may be possible that these membranes are the site of the dehydrogenase responsible for the first step in carotenogenesis that leads to phytofluene, which accumulates in large amounts.

XIX. Xanthophyll Esterification during Ripening

Mature fruits contain esterified xanthophylls (Goodwin, 1952), the esterification process occurring progressively during maturation. The phenomenon has a physiological significance, as acylation increases the lypophilic character of the xanthophylls, facilitating their accumulation within the globules of the chromoplasts. There appears to be little information about the rate and extent of esterification throughout the season and even less about the nature of the acids.

The esterification of xanthophylls during ripening has been followed in the flavedo of various citrus fruits (Eilati et al., 1972). The gradual accumulation of total xanthophylls is accompanied by a parallel increase in xanthophyll esters, the latter constituting about 60% of the total xanthophylls. The correlation between total and esterified xanthophylls for ripening Shamouti appears to hold also in different citrus cultivars and species such as Washington Navel, Valencia, sour orange, Dancy tangerine and Clementine. On the other hand in species with low carotenoid content, such as lemon, only 5% of the xanthophylls were esterified (Yokoyama and Vandercook, 1967).

Generally the earlier analytical data concerning the percentage of esters of the total carotenoids are very different. In orange juices of Mediterranean origin ester values ranging from 10–16% have been reported (Koch and Haase-Sajak, 1965; Di Giacomo et al., 1968a). In our laboratory it was found that 86% of total carotenoids in the Shamouti orange juice are esterified, 34% of them being monoesters (J. Gross and M. Carmon, unpubl. data). Similar data were reported by Philip (1973a, b, c) who found, when examining different citrus species and cultivars that most if not all of the xanthophylls are esterified. The main acid was identified as lauric acid.

In some recent studies xanthophyll esterification has been followed over an entire season. The carotenoids of the kiwifruit appeared not to be subject to esterification, all the xanthophylls in the ripe fruit being free (Gross, 1982). In apple, esterification began at a more advanced ripening stage (Gross et al.,

1978). The esters increased gradually from 13–23% of total carotenoids in the month before harvest and reached 54% in the tree-ripened fruit. Lutein appeared generally as diester, violaxanthin as both diester and monoester, the ratio diester monoester gradually increasing. The same was true for neoxanthin esters.

A gradual esterification also occurred in the ripening Redhaven peach (Gross, 1979). In the ripe fruit 93% of the xanthophylls were present as both diester and monoesters, only 7% being in the free form. In the Earliglo peach a progressive xanthophyll esterification was also observed (Lessertois and Monéger, 1978).

In the ripe green-fleshed melon, 28% of the xanthophylls in mesocarp and 36% of the xanthophylls in the exocarp were esterified (Flügel and Gross, 1982). The carotenoids of the banana fruits were investigated omitting the saponification step (Gross *et al.*, 1976). A major pigment in pulp, lutein, occurred as diester (11.0%), as monoester (8.6%) and in the free form (12.6%). All other xanthophylls present as minor pigments occurred as esters: both monols α-cryptoxanthin, β-cryptoxanthin; violaxanthin and luteoxanthin were found as monoesters. In the peel 17% of total carotenoids were free xanthophylls. Both monols were esterified. Lutein appeared in all three forms, violaxanthin as mono- and diester and other minor xanthophylls such as isolutein, chrysanthemaxanthin, auroxanthin and neochrome both as monoesters and in the free form. The ester of neoxanthin was not definitively characterized. According to Kleinig and Nitsche (1968), who investigated xanthophyll esters in the flower petals, neoxanthin never occurs as triester as the tertiary hydroxyl never becomes esterified.

In pepper at the fully green stage all xanthophylls were unesterified, while in the red ripe fruit they were mainly esterified: monoesters (26%), diesters (54%) and free (20%) (Camara and Monéger, 1978). Like lutein in banana, the major carotenoid capsanthin appeared in all three forms. The fact that cryptocapsin, violaxanthin and neoxanthin were not esterified was probably due to a selectivity of the esterification reactions. The presence of different monoesters suggests that esterification is a late step, affecting only newly synthesized xanthophylls.

XX. Carotenoids as an Index of Maturity

The external fruit colour which changes during ripening is considered as a maturity index. In many cases the colour is due mainly to the predominating pigment which is the determinant for the fruit's colour, such as lycopene in tomato, red grapefruit and red watermelon. The systematic studies on carotenoid changes during ripening have revealed that often in fruits which at

Table 3.2 *Violaxanthin variation in the peel of Golden Delicious apple during ripening*

Harvest Date	Violaxanthin (% of Total Carotenoids)	Ripening Stages
Mid-July	1	
End-July	1	
Mid-August	1	I. Cell multiplication
End-August	5	
Mid-September	10	
End-September	20	II. Pre-climacteric
Mid-October	25	Harvest ripe
End-October	36	III. Climacteric
November	40	Edible
December	52	Senescence

From Gross and Lenz (1979).

ripeness contain a complex carotenoid pattern, the major occurring event is the biosynthesis of a single pigment which becomes the main pigment, such as violaxanthin in the peel of Golden Delicious apple (Fig. 3.23) cryptoxanthin in persimmon and in pulp of *Citrus reticulata*, and β-citraurin in flavedo of *Citrus reticulata*.

In Golden Delicious apples distinct values of the violaxanthin percentage of total carotenoids for each two-week interval are in reasonable correlation with the maturity stages (Table 3.2). Consequently, the variation of the violaxanthin content in fruit peel may be used as an index of maturity to determine optimal harvest time (Gross and Lenz, 1979). This is of considerable practical use because establishing the optimal harvest time is a most difficult task. Subsequently the method was tried also on other apple cultivars including Cox Orange, James Grieve and Boskop, and was found to be suitable and highly recommended (Henze, 1983).

In the peel of ripening persimmon the cryptoxanthin level increased at a uniform rate of 8–10% in each two-week interval. This constant increase of the cryptoxanthin level paralleling the ripening process suggests its use as an index of maturity. This was applied by Gross *et al.* (1984) and was found to permit the exact assessment of the delaying effect of gibberellin on colour development induced in persimmons.

For tomatoes lycopene content has been proposed as the maturity index as it changes significantly from one maturity stage to another (Cabibel and Ferry, 1979).

XXI. Other Carotenoid Changes during Ripening

A. Cis-isomers

Numerous *cis*-isomers occur in higher plants and should not be considered as artefacts but as genuine natural products. This was recently proved by Tóth and Szabolcs (1981) for some mono-*cis*-isomers of asymmetric C_{40}-carotenoids. They may be formed by stereoselective or non-stereoselective biosynthesis or stereomutation from the all-*trans* forms.

In most reports on fruit carotenoids the configuration of the *cis*-isomers present has not been thoroughly investigated. Only the widespread *cis*-isomers of violaxanthin, occurring in fruits as major pigment, have been studied. It is known that it is the 9-*cis*-isomer of violaxanthin called violeoxanthin. In Valencia orange peel this is the principal carotenoid, constituting 61% of total carotenoids (Molnár and Szabolcs, 1980). A feature of interest is the fact that *cis*-violaxanthin is increasingly synthesized in ripening fruits, becoming the predominant isomer. This has been observed especially in citrus fruits, but also in others such as plums, persimmons, kumquat, yellow cherry and mango. In some fruits, such as kumquat and citrus fruits, this phenomenon has also been observed for *cis*-antheraxanthin.

B. Allylic Apocarotenols

In various fruits a series of allylic apocarotenols (Section XII.B.1(e) was detected and their structure elucidated (Gross and Eckhardt, 1981). The supposed metabolism is an oxidative degradation of the parent C_{40}-carotenoid, followed by a reduction of the formed apocarotenol. So far these C_{25}- and C_{27}-epoxyapocarotenoids have been detected only in ripe fruits. In apple peel, in which the carotenoid changes have been followed during ripening, the C_{27}-apocarotenol appeared only at the ripe stage (Gross and Eckhardt, 1978; Gross *et al.*, 1978).

As the fruit approaches senescence, oxidation reactions are intensified and probably the degradation of the epoxycarotenoids, which are generally the most abundant, occurs more rapidly. These metabolites may be considered as characteristic pigments of a fully ripe fruit. Additional evidence is still to be produced.

C. Ripening after Harvest

In many fruits the ripening process, including carotenogenesis, continues after harvest. A classical example is the tomato, and additionally various citrus fruits, pepper, apple, persimmon, mango, etc. Whether fruits ripened off the tree synthesize more carotenoids than when ripened on the tree depends on the species. Vine-ripened tomatoes produce more carotenoids than those ripened in storage (Goodwin, 1982). Similarly, oranges harvested while still green and stored for an extended period at 20°C had a lower total carotenoid content than fruits ripened on the tree, but the carotenoid changes occurring during ripening of attached and detached fruits were essentially similar (Eilati et al., 1975).

For apples and persimmons the results were the reverse. The carotenoid content in the peel of Golden Delicious apple ripened during storage was higher (24 μg g^{-1}) than that of fruit ripened on the tree (13.0 μg g^{-1}) (Gross et al., 1978). The apple ripened on the tree did not attain the maximum degree of ripeness as did the stored fruit. But the carotenoid metabolism was the same, violaxanthin being increasingly biosynthesized up to a level as high as 58%, phytofluene accumulating and cryptoxanthin being accompanied by more epoxides. Xanthophylls were increasingly esterified, reaching a level of 81–86% of total carotenoids.

In the peel of detached persimmon a very active carotenogenesis occurred, the ripe fruit containing 0.5 mg(g fresh wt)$^{-1}$ (Ebert and Gross, 1984). Although the carotenoid pattern was the same as in fruits ripened on the tree, the metabolism was different. Cryptoxanthin continued to increase, reaching a higher level of 50% as opposed to 40%, the maximum in attached fruit; lycopene, the second major pigment, accumulated at lower levels, evidence of an enhanced turnover rate.

In stored apricots, peaches and pepper the carotenoid content continued to increase but at the beginning of storage a decrease of the total carotenoid occurred (Strmiska and Holčiková, 1970). No explanation has been given for this initial decrease, which may possibly correspond to the minimum of the total carotenoid content which occurs during the ripening process.

In kiwifruit ripened after storage, the total carotenoid content continued to decrease, involving a similar metabolism as in fruits ripened on the vine (Ben-Arie et al., 1982/83; Gross, 1983).

XXII. Factors Affecting Carotenoid Biosynthesis in Fruits

A. Hormones

1. *Ethylene*

Ethylene is widely used to promote ripening in fruits (Chapter 1, Section XI A.) as it hastens both chlorophyll breakdown and carotenoid biosynthesis. A visible colour modification results which is generally measured by non-destructive colorimetric methods. There is dearth of data concerning detailed investigation of the modification of the carotenoid pattern. One of the first detailed reports of this kind, that of Stewart and Wheaton (1972), revealed that the enhanced colour of certain ethylene-treated citrus cultivars is due to the accumulation of specific carotenoids, such as β-citraurin and citraurinene, the former contributing especially to the red tint. The ethylene treatment was more effective for fruits at a more advanced ripening stage. Other degreening assays with Shamouti oranges confirmed that the ethylene effectiveness is dependent on the maturity stage of the treated fruits. The qualitative carotenoid changes were the same as observed in normal ripened fruits (Eilati et al., 1975).

2. *Gibberellin*

The delaying effect on maturation of gibberellin has been discussed in Chapter 1, Section XII. As for ethylene, the colour measurements are generally limited to non-destructive methods. Only recently the changes in the carotenoid pattern in the peel of Triumph persimmon induced by pre-harvest gibberellin treatment were followed in detail (Gross et al., 1984). The gibberellin (GA_3) treatment did not alter either the carotenoid pattern or the carotenoid metabolism during ripening. The sole effect was the delaying effect of the fruit ripening. The different maturity stages could be assessed by comparing the respective carotenoid pattern which characterizes each ripening stage, or simply by using the cryptoxanthin level as a maturity index.

B. Light

Light is not essential for inducing carotenogenesis, which occurs within the fruit where only low light intensities can penetrate. The effect of light on carotenoid biosynthesis in detached green tomatoes has been thoroughly investigated. It has been shown that it augments carotenoid synthesis, red light being more effective than white or green light (Jen, 1974). In fact red light

duced carotenoid synthesis and far red light inhibited carotenoid synthesis in ripening tomatoes, suggesting a phytochrome-mediated response (Thomas and Jen, 1975a, b). Comparative effects of light and ethephon on the ripening of detached tomatoes indicated that carotenoid biosynthesis is light-dependent and that ethephon has no effect on the total carotenoid content (Paynter and Jen, 1976).

On the other hand, tomatoes ripened in darkness and in light both contained the same carotenoid pattern, which indicates that carotenogenesis is light-independent; it may be stimulated by light but light is not required for its induction (Raymundo et al., 1976).

C. Oxygen

Although ethylene has a marked effect on the colour of citrus flavedo, it does not affect the colour of the endocarp. But endocarp and juice colour of Shamouti oranges during storage could be improved by oxygen-enriched atmospheres (Aharoni, 1968). The results were confirmed on Navel oranges grown in California (Houck et al., 1978) and three other Florida orange cultivars (Aharoni and Houck, 1980). The optimal endocarp colour was obtained at 15°C in an atmosphere containing 80% O_2 + 20% N_2. The different cultivars reacted differently to the treatment. The colour of the peel was also affected, being improved at lower oxygen of 40% and partially destroyed at the higher level. Carotenoids were not analysed. Additional work is necessary in order to interpret the phenomenon.

D. Temperature

Generally the optimal temperature for carotenogenesis in plants is relatively low. The synthesis of certain pigments has been shown to be temperature-sensitive; the temperature sensitivity varies from plant to plant.

Temperature sensitivity was studied thoroughly in connection with lycopene formation in tomato. As early as 1937, Vogele studied the effect of environmental factors upon the colour of the tomato and the watermelon (Vogele, 1937). The optimum temperature for lycopene formation in the tomato is 24°C; above 30°C it is not formed. Fruits ripened at temperatures between 32 and 38°C became yellow; after being returned to lower temperatures (20–24°C) they developed normally. The results were subsequently confirmed and additionally it was observed that the synthesis of β-carotene is not affected by temperature (Goodwin and Jamikorn, 1952; Tomes et al., 1956; Tomes, 1963; Laval-Martin et al., 1975).

Electron microscopic studies indicated that accumulation of β-carotene precedes that of lycopene (Ben-Shaul and Naftali, 1969). In more recent studies on carotene synthesis in tomato it was also noted that β-carotene was synthesized first before the massive lycopene accumulation (Raymundo et al., 1970; Laval-Martin et al., 1975). To explain this phenomenon it was suggested that a physically separated pathway of carotenoid biosynthesis is superimposed on the earlier chloroplast pathway (Goodwin and Jamikorn, 1952) or even a compartmentalization of similar enzymes (Raymundo et al., 1967).

The influence of sub-optimal temperatures on the colour and pigment composition of ripening tomatoes was investigated by Koskitalo and Ormrod (1972). Fruits exposed to the 18/26°C regime had the highest pigment content with lycopene accounting for 90% of the carotenes. Lower temperature regimes were less effective.

In other fruits in which lycopene is also the main carotenoid, such as red watermelons and red grapefruits, its synthesis is not temperature sensitive. In watermelon a shift in temperature from 20 to 37°C did not affect lycopene biosynthesis (Vogele, 1937). In red grapefruits a higher temperature of 28–35°C produced more lycopene than in fruits grown at 4–16°C, when lycopene disappeared rapidly and β-carotene increased (Purcell et al., 1968). A study of the effect of high temperature showed that lowering the temperature from 42–36°C to 32–21°C produces a high level of lycopene. The β-carotene concentration changed only at lower temperature, which caused a larger increase (Meredith and Young, 1971). The special temperature requirement of citrus fruits, i.e. low temperature to induce chlorophyll breakdown and thus a good fruit coloration has been discussed in Chapter 1, Section XIII.B.

The effect of two different root temperatures, 14°C and 30°C, on carotenoid formation in peel of Satsuma mandarins and calamondins was investigated in detail by Sonnen et al. (1979). In both cultivars, total carotenoid content was twice as high at low temperature. Satsuma mandarins ripened at low temperature had a more complex carotenoid pattern, although they appeared to be at a less advanced ripening stage than the fruits ripened at 30°C. In calamondins the situation was reversed. In other harvested citrus, significant amounts of colour developed at 15°C, without addition of ethylene (Wheaton and Stewart, 1973). The pigment changes in stored green oranges were markedly inhibited at 8°C in comparison with fruits ripened at 20°C (Eilati et al., 1975).

Colour improvement in post-harvest ethylene-treated fruits is dependent on temperature due to the temperature sensitivity of β-citraurin synthesis (Stewart and Wheaton, 1971). In fruit stored at 30°C virtually no β-citraurin was formed, but at 25, 20 and 15°C it increased progressively. Optimal ethylene concentration decreased as temperature decreased. This makes it possible to control carotenoid synthesis effectively by changing temperature and ethylene concentration.

In stored Golden Delicious apples the ripening process continued at a different biosynthetic rate determined by different temperatures, but producing the same final carotenoid pattern. At 18°C fruits were ripe after 2 weeks and at 4°C after 8 weeks (Gross *et al.*, 1978). During storage for a period of five months at 4°C, yellowing of fruit was achieved at different times, depending on the physiological ripening status of fruit at the beginning of storage, but the carotenoid content and pattern were almost identical (Gross, 1979b).

In stored apricots, peaches and peppers at two temperatures 0°C and 20°C, a higher carotenoid content accumulated at a higher rate at 20°C (Strmiska and Holčiková, 1970). On the other hand, the carotenoid content of green bell peppers was reduced following 14 days storage at 3°C and 7 days at 21°C (Mathews *et al.*, 1975). As in other fruits, the differences in carotenoid biosynthetic rates in pepper depend on the ripening status of fruits at the beginning of storage, on storage temperatures, and are different in various cultivars. As peppers to not respond to ethylene treatment, it was found that storing Israeli-grown California-type bell peppers of all ripening stages except the deep green, for 48 hours at 25°C and relative humidity of 90–95%, induced and accelerated fruit reddening (Temkin-Gorodeiski *et al.*, 1983).

In papaya fruits stored at both 10 and 15°C, the higher storage temperature accelerated colour development in peel and pulp (Abou Aziz *et al.*, 1975).

The mango, being a tropical fruit, is very susceptible to low temperatures, so that lengthy storage at temperatures low enough to delay ripening is out of the question. Accumulation of carotenoids was maximal in Alphonso mangoes stored at tropical ambient temperatures of 28–30°C. Storage of pre-climacteric fruits at 7–20°C for 16–43 days caused a substantial reduction of carotenoid formation even when these fruits were subsequently ripened under optimal conditions (Thomas and Janave, 1975). Similarly, freshly harvested physiologically mature Kent mangoes, stored at 8, 10 and 13°C for 10, 16 and 22 days and subsequently transferred to ripening chambers, synthesized less than half of the total carotenoids produced in fruits ripened at 25°C (Saucedo *et al.*, 1977).

E. Fertilizers

Increasing the amount of potassium supplied to tomatoes increased the total carotenoid content. Lycopene followed a similar trend whereas it had the opposite effect on the β-carotene levels. The latter effect is due more to the precursor–product relationship existing between the two pigments than to the fertilizer (Trudel and Uzbun, 1971). The effect of different supplies of nitrogen on flavedo carotenoids of calamondin fruit was investigated by Gross *et al.* (1979c). The total carotenoid content was enhanced by increasing nitrogen levels up to 8 meq N l^{-1} and decreased at higher doses. At the optimal nitrogen level a typical mandarin carotenoid pattern was formed, whereas the other nitrogen treatments induced a carotenoid pattern in which 5,8-epoxides predominated.

In apples the yellow colour is enhanced by a low supply of nitrogen (Kvåle, 1971; Hansen, 1980). The carotenoids were investigated in detail in four apple cultivars receiving low and high N levels (Lenz and Gross, unpublished data). As the total carotenoid content is not relevant to the ripening stage due to the special development of the ripening curve (Section XVI.A.1), the violaxanthin percentage of total carotenoids was used as an index of maturity. Fruits receiving the low N treatment were at a more advanced ripening stage.

F. Miscellaneous

A tree's fruit load affects fruit colour, as was found in Golden Delicious apples from thinned trees which were bigger, heavier and had a higher carotenoid content than fruits from loaded trees (Lenz and Gross, 1979).

Different cultural methods such as glass and plastic greenhouses or field, as well as the place of the fruit on the plant, affect the carotenoid content of tomatoes. The highest carotenoid content was found in the vine grown fruit (Cabibel and Ferry, 1980).

Appendix: Pigment Distribution in Fruits

I. Introduction

The following tables give details of the pigment distribution in fruits. Listed below are the table numbers and the fruits to which they refer.

- A. 1 Apple
- A. 2 Apricot
- A. 3 Avocado
- A. 4 Banana
- A. 5 Blackberry
- A. 6 Blueberry
- A. 7 Carambola
- A. 8 Cherry sour
- A. 9 Cherry sweet
- A. 10 Cranberry
- A. 11 Currant black
- A. 12 Currant red
- A. 13 Date
- A. 14 Fig
- A. 15 Gooseberry
- A. 16 Grape
- A. 17 Grapefruit white
- A. 18 Grapefruit red
- A. 19 Guava
- A. 20 Kiwi
- A. 21 Kumquat
- A. 22 Lemon
- A. 23 Litchi
- A. 24 Loquat
- A. 25 Mandarin
- A. 26 Mango
- A. 27 Melon (orange-fleshed)
- A. 28 Melon (green-fleshed)
- A. 29 Olive
- A. 30 Orange
- A. 31 Papaya
- A. 32 Passion fruit
- A. 33 Peach
- A. 34 Pear
- A. 35 Pepper
- A. 36 Persimmon
- A. 37 Pineapple
- A. 38 Plum
- A. 39 Pomegranate
- A. 40 Pummelo white
- A. 41 Pummelo red
- A. 42 Quince
- A. 43 Raspberry
- A. 44 Strawberry
- A. 45 Tomato
- A. 46 Watermelon (red- and orange-fleshed)

II. Classification of Anthocyanin Patterns

Pattern	Anthocyanidin	Glycoside Type
I	One	One, various monosides
II	Various	One monoside
III	Various	Various monosides
IV	One	Various biosides
V	Various	One, various biosides
VI	Various	3,5-Diglucosides
VII	One, various	Various triosides
VIII	Various	Acylated monosides and diosides

III. Classification of Carotenoid Patterns

(1) Insignificant amount of carotenoids.
(2) Small amounts of chloroplast carotenoids: β-carotene, lutein, violaxanthin and neoxanthin.
(3) Large amounts of lycopene and the related partly saturated acyclic polyenes, phytoene, phytofluene, ζ-carotene and neurosporene.
(4) Large amounts of β-carotene and its hydroxy-derivatives, cryptoxanthin and zeaxanthin.
(5) Large amount of epoxides.
(6) Unique carotenoids.
(7) Poly-*cis*-carotenoids.
(8) Apocarotenoids.

IV. Abbreviations Used in the Tables

Pg = pelargonidin
Cy = cyanidin
Dp = delphinidin
Pn = peonidin
Pt = petunidin
Mv = malvidin
RE = retinol equivalents
fresh wt = fresh weight

Table A. 1 *Rosaceae APPLE* (*Malus domestica* (L.) Mill)

Anthocyanins

Anthocyanins	Cultivars						(%)
	Red Delicious	*Stoke Red*	*Jonathan*	*Tremletts Bitter*	*Cox's Orange Pippin*	*Ingrid Marie*	
Cy 3-galactoside	+++	+++	+++	+++	+++	+++	80–90
Cy 3-arabinoside	+	+	+	+	+	+	
Cy 3-glucoside	+	+	+	+	+	+	
Cy 3-xyloside	+	+	+	+	+	+	
Their acylated derivatives	+	+	+	+	+	+	

From Timberlake and Bridle (1971).

Cy 3-galactoside				+++			
Cy 3-arabinoside				++			
Cy 7-arabinoside				+			

From Sun and Francis (1968).
Major anthocyanin pattern: (I).

Carotenoids

Quantitative changes of Carotenoid pattern in the peel of Golden Delicious apples during ripening

Carotenoid Pattern	% of Total Carotenoids						
	July 18	Aug. 1	Aug. 15	Sept. 26	Oct. 10	Oct. 24	Nov. 7
β-Carotene	34.0	33.8	35.5	27.6	23.0	18.2	7.9
Cryptoxanthin	2.0	—	—	—	—	2.0	2.4
Cryptoxanthin 5,6-epoxide	—	—	—	—	—	—	0.6
Lutein	47.5	44.4	40.6	33.3	24.0	17.3	9.2
Isolutein	0.2	0.3	0.2	2.7	3.7	3.8	6.8
Violaxanthin	0.8	0.6	0.7	19.5	24.8	35.9	39.6
Zeaxanthin	1.7	3.9	1.2	0.9	2.5	2.0	3.1
Mutatoxanthin	2.8	3.4	—	—	—	—	1.1
cis-Luteoxanthin	2.5	1.1	—	1.1	2.5	0.9	7.6
Apo-carotenol	—	—	—	—	—	3.2	—
Neochrome	3.7	4.4	5.6	—	—	0.4	1.7
trans-Neoxanthin	4.2	6.2	6.2	11.7	15.6	14.7	18.2
cis-Neoxanthin	—	—	—	—	—	7.9	1.2
Auroxanthin	—	—	6.8	—	—	—	—
Mixture of unidentified traces of carotenoids	0.6	1.9	4.4	3.2	3.9	1.6	0.6
Total carotenoids (μg(g fresh wt)$^{-1}$)	17.8	16.0	14.2	7.6	13.4	12.0	13.4

From Gross *et al.* (1978).

Quantitative changes of carotenoid pattern in the peel of Golden Delicious apples ripened on the tree and during storage

Carotenoid Pattern	% of Total Carotenoids		
	Nov. 7[a]	Nov. 4–17[b]	Oct. 25–Dec. 20[c]
Phytofluene	—	1.8	3.1
β-Carotene	7.9	3.6	3.5
Cryptoxanthin	2.4	2.3	2.5
Cryptoxanthin 5,6-epoxide	0.6	1.3	2.8
Cryptoxanthin 5,6,5′,6′-diepoxide	—	2.0	2.0
Lutein	9.2	8.6	9.0
Isolutein	6.8	—	3.8
trans-Luteoxanthin	—	3.6	—
Violaxanthin	39.6	58.0	52.0
Zeaxanthin	3.1	0.3	0.6
Mutatoxanthin	1.1	0.1	0.3
cis-Luteoxanthin	7.6	1.6	1.3
Apo-carotenol	—	1.9	2.6
Neochrome	1.7	0.8	0.6
trans-Neoxanthin	18.2	13.3	15.2
cis-Neoxanthin	1.2	0.6	0.7
Mixture of unidentified traces of carotenoids	0.6	0.2	—
Total carotenoids (μg(g fresh wt)$^{-1}$)	13.4	23.8	22.0

[a] Ripened on the tree.
[b] Stored at 18°C.
[c] Stored at 4°C.
From Gross *et al.* (1978).
Major carotenoid patterns: (2), (5).
Vitamin A value: 200 μg RE kg^{-1}.

Table A. 2 Rosaceae APRICOT (*Prunus armeniaca* L.)
Carotenoids

Carotenoid Pattern	% of Total Carotenoids
Phytoene	10.0
Phytofluene	6.1
α-Carotene	0.2
β-Carotene	54.0
Cis-β-carotene	5.9
Phytofluene-like	1.4
Zeta-carotene	0.6
Pro-γ-carotene	0.5
Cis-γ-carotene	0.3
γ-Carotene	3.9
Polycis lycopene a	0.19
Polycis lycopene b	2.3
Polycis lycopene c	0.4
Neo-lycopene B	0.9
Lycopene	1.0
Hydroxy-α-carotene	0.14
Cis-cryptoxanthin	0.5
Cryptoxanthin	3.1
Cis-cryptoxanthin	0.4
Hydroxy-zeta-carotene	0.3
Cis-hydroxy-zeta-carotene	0.2
Cryptoflavin	0.4
Rubixanthin	0.16
Lutein	2.1
Zeaxanthin	0.7
Carbonyl	0.01
Lutein 5,6-epoxide	0.02
Antheraxanthin isomer	0.06
Cis-antheraxanthin	0.07
Flavoxanthin	0.03
Mutatoxanthin a	0.14
Mutatoxanthin b	0.09
Violaxanthin isomer	0.09
Cis-violaxanthin	0.19
Luteoxanthins	0.09
Cis-Luteoxanthin a	0.2
Cis-Luteoxanthin b	0.2
Auroxanthins	0.3
Persicaxanthin	0.5
Sinensixanthin	0.19
Trollixanthin-like	0.06
Trollichrome-like	0.15
Persicachrome	0.09
Cis-trollichrome-like	0.05
Total carotenoids (μg (g fresh wt)$^{-1}$)	35.0

From Curl (1960).
Major carotenoid pattern: (4).
Vitamin A value: 3770 μg RE kg^{-1}.

APPENDIX

Table A. 3 *Lauraceae AVOCADO (Persea americana* Mill.)
Anthocyanins

Anthocyanins	Cultivar Purple	(%)
Cy 3-galactoside	+++	90
Cy 3-(*p*-coumarylglucoside)-5-glucoside	+	10

From Prabha *et al.* (1980).
Major anthocyanin pattern: (I), (IV).

Carotenoids

Carotenoid distribution in pulp of avocado, cv. Nabal

Carotenoid Pattern	% of Total Carotenoids
α-Carotene	0.9
β-Carotene	4.0
ζ-Carotene	0.5
α-Cryptoxanthin	1.2
β-Cryptoxanthin	5.2
α-Citraurin	0.7
Lutein	25.0
Isolutein	9.0
Violaxanthin	4.0
Chrysanthemaxanthin	20.4
Luteoxanthin	1.1
C_{27}-apocarotenol (5,8-epoxide)	1.0
Neochrome	9.2
trans-Neoxanthin	7.3
Mimulaxanthin	8.6
cis-Neoxanthin	0.5
Total carotenoids (μg(g fresh wt)$^{-1}$)	12.0

From Gross *et al.* (1972, 1973, 1974).
Major carotenoid pattern: (2) (modified).
Vitamin A value: 150 μg RE kg^{-1}.

Table A.4 *Musaceae BANANA (Musa × paradisiaca L.)*

Carotenoids

Qualitative and quantitative distribution of carotenoids in Banana pulp Cavendish

Identification	Semi-systematic Name	Absorption Maxima (nm) in EtOH	Colour	Epoxide Test (HCl Treatment) Hypsochromic Shift (nm)	Carbonyl Test Reduced pigment Shift (nm)	Saponi-fication Test	Acetyla-tion Test	% of Total Carotenoids
Phytofluene	15-cis-7,8,11,12,7'	327,348,368						1.2
α-Carotene	8'-hexahydro-ψ,ψ-carotene	420,442,472						31.5
β-Carotene	β,ε-carotene	425,450,480						22.5
Neo-β-carotene	β,β-carotene	422,449,478						5.3
3-Hydroxy-α-carotene ester	β,β-carotene	420,442,472				+		0.7
Cryptoxanthin ester	β,ε-carotene-3-ol	424,449,476				+		1.5
Lutein diester	β,β-carotene-3-ol	420,443,472				+		11.0
Mutatochrome	β,ε-carotene-3,3'-diol	398,422,448	blue	—				2.2
	5,8-epoxy-5,8-dihydro-β,β-carotene							
Luteochrome	5,6,5',8'-diepoxy-5,6,5', 8'-tetrahydro-β,β-carotene	400,420,444	green	20				0.4
Aurochrome	5,8,5',8'-diepoxy-5,8,5', 8'-tetrahydro-β,β-carotene	375,395,423	blue					traces
Carbonyl 440		440			392,415,440			traces
Lutein monoester	β,ε-carotene-3,3'-diol	420,443,472				+	+	8.6
Violaxanthin monoester	5,6,5',6'-diepoxy-5,6,5', 6'-tetrahydro-β,β-carotene-3,3'-diol	416,440,463	green–blue	40		+		1.0
Luteoxanthin monoester	5,6,5',8'-diepoxy-5,6,5', 8'-tetrahydro-β,β-carotene-3,3'-diol	400,420,442	green	20		+		1.5
Lutein free	β,ε-carotene-3-ol	420,443,472						12.6
Total carotenoids (μg(g fresh wt)$^{-1}$)								1.0

Major carotenoid pattern: (1) (uncommon). Vitamin A value: 60 μg RE kg^{-1}.

Carotenoids

Qualitative and quantitative distribution of carotenoids in Banana peel Cavendish

Identification	Semi-systematic Name	Absorption Maxima (nm) in EtOH	Epoxide Test (HCl Treatment) Colour	Hypsochromic Shift (nm)	Saponification Test	Acetylation Test	% of Total Carotenoids
Hydrocarbons							
α-Carotene	β,ε-carotene	420,442,472					6.7
β-Carotene	β,β-carotene	425,450,480					9.5
Neo-β-carotene	β,β-carotene	422,448,476					4.0
Monoesters							
cis-Hydroxy-α-carotene	β,ε-carotene-3-ol	330,418,441,468			+		2.0
Hydroxy-α-carotene	β,ε-carotene-3-ol	418,443,473			+		
Hydroxy-α-carotene-5,6-mono-epoxide	5,6-epoxy-5,6-dihydro-β,ε-carotene-3-ol	416,440,469	green	20	+		0.1
Hydroxy-α-carotene-5,8-mono-epoxide	5,8-epoxy-5,8-dihydro-β,ε-caroten-3-ol	398,420,440	blue				traces
Cryptoxanthin isomer	β,β-caroten-3-ol	420,444,473					1.0
Cryptoxanthin	β,β-caroten-3-ol	425,449,475			+		4.2
Cryptoflavin	5,8-epoxy-5,8-dihydro-β,ε-caroten-3-ol	395,423,450	blue				0.3
Fluorescent F₁		332,349,368			+		traces
Fluorescent F₂	5,8-epoxy-5,8-dihydro-10′-apo-β-carotene-3,10′-diol	353,371,396	blue				traces
Diol and triol esters							
Lutein diester	β,ε-carotene-3,3′-diol	420,443,472			+		13.5
Neoxanthin triester?	5′,6′-epoxy-6,7-didehydro-5,6,5′,6′-tetrahydro-β,β-carotene-3,5,3′-triol				+		
Violaxanthin diester		420,438,465	green	18	+		1.8
	5,6,5′,6′-diepoxy-5,6,5′,6′-tetrahydro-β,β-carotene-3,3′-diol	414,437,465	green–blue	37	+		1.4
cis-Lutein monoester	β,ε-carotene-3,3′-diol	330,415,440,467			+		1.6

Identification	Semi-systematic Name	Absorption Maxima (nm) in EtOH	Epoxide Test (HCl Treatment)		Saponi-fication Test	Acetyla-tion Test	% of Total Carotenoids
			Colour	Hypsochromic Shift (nm)			
Lutein monoester	β,ε-carotene-3,3'-diol	420,443,472			+	+	28.0
Isolutein monoester	5,6-epoxy-5,6-dihydro-β,ε-carotene-3,3'-diol	418,440,465	green	18	+	+	+0.5
Chrysanthemaxanthin monoester	5,8-epoxy-5,8-dihydro-β,ε-carotene-3,3'-diol	400,422,448	blue		+	+	1.5
Violaxanthin monoester	5,6,5',6'-diepoxy-5,6,5',6'-tetrahydro-β,β-carotene-3,3'-diol	414,438,462	green–blue	38	+	+	1.5
Auroxanthin monoester	5,8,5',8'-diepoxy-5,8,5',8'-tetrahydro-β,β-carotene-3,3'-diol	381,400,424	blue		+	+	0.5
Luteoxanthin monoester	5,6,5',8'-diepoxy-5,6,5',8'-tetrahydro-β,β-carotene-3,3'-diol	398,420,447	green	20	+	+	3.7
Neochrome monoester	5',8'-epoxy-6,7-didehydro-5,6,5',8'-tetrahydro-β,β-carotene-3,5,3'-triol	396,420,448	blue		+	+	0.8
Free xanthophylls							
Lutein	β,ε-carotene-3,3'-diol	422,443,472					14.0
Isolutein	5,6-epoxy-5,6-dihydro-β,ε-carotene-3,3'-diol	418,440,465	green	17			2.4
Chrysanthemaxanthin	5,8-epoxy-5,8-dihydro-β,ε-carotene-3,3'-diol	398,420,447	blue				0.2
Auroxanthin	5,8,5',8'-diepoxy-5,8,5',8'-tetrahydro-β,β-carotene-3,3'-diol	382,400,424	blue				traces
Neochrome	5',8'-epoxy-6,7-didehydro-5,6,5',8'-tetrahydro-β,β-carotene-3,5,3'-triol	396,420,448	blue				0.8
Total carotenoid content (μg(g fresh wt)$^{-1}$)							6.0

From Gross *et al.* (1976).
Major carotenoid pattern: (2).

Table A. 5 *Rosaceae BLACKBERRY* (*Rubus* spp.)
Anthocyanins

	Cultivar	
Anthocyanins	*Merton Early*	*John Innes*
Cy 3-glucoside	+++	+++
Cy 3-rutinoside	+	+

From Harborne (1967).
Major anthocyanin patterns: (I), (IV).

Carotenoids

Carotenoid Pattern	*% of Total Carotenoids*
Phytoene	1.5
Phytofluene	0.2
α-Carotene	1.1
β-Carotene	9.5
Monols	3.2
Lutein	44.3
Zeaxanthin	2.3
Monoepoxide diols	1.9
Diepoxide diols	22.0
Polyols	12.2
Total carotenoids (μg(g fresh wt)$^{-1}$)	5.8

From Curl (1964).
Major carotenoid pattern: (2).
Vitamin A value: 100 μg RE kg^{-1}.

Table A. 6 *Ericacaea BLUEBERRY* (*Vaccinium* spp.)

Anthocyanins

Species: Lowbush (*V. angustifolium*)
Highbush (*V. corymbosum*)

Anthocyanins

Mv 3-galactoside	+++[a]
Dp 3-galactoside	+++
Pt 3-galactoside	+++
Cy 3-galactoside	+
Pn 3-galactoside	+
Mv 3-arabinoside	+++
Dp 3-arabinoside	+++
Pt 3-arabinoside	+++
Cy 3-arabinoside	+
Pn 3-arabinoside	+
Mv 3-glucoside	+
Dp 3-glucoside	+
Pt 3-glucoside	+
Cy 3-glucoside	+
Pn 3-glucoside	+

[a] +, +++: relative quantities.
From Francis *et al.* (1966); Ballinger *et al.* (1970).
Other species: Same aglycone, glucose absent, see Ballinger *et al.* (1979).
Major anthocyanin pattern: (III).

Carotenoids

	Carotenoid Pattern	*% of Total Carotenoids*
Carotenes		11.0
Monols		2.2
Lutein		39.2
Zeaxanthin		2.1
Monoepoxide diols		13.7
Diepoxide diols		19.4
Polyols		11.4
Total carotenoids ($\mu g\,(g\,fresh\,wt)^{-1}$)		2.7

From Curl (1964).
Major carotenoid pattern: (2).
Vitamin A value: 50 μg RE kg^{-1}.

Table A. 7 *Oxalidaceae CARAMBOLA (STAR FRUIT)* (*Averrhoa carambola* L.)

Carotenoids

Characterization and quantitative distribution of carotenoids of the carambola fruit (Averrhoa carambola) cv. Golden Star

Carotenoid Pattern	λ_{max}^{EtOH}(nm)	% of Total Carotenoids	
		Unripe	Ripe
Phytofluene	331,348,367	8.0	16.7
β-Carotene	425,450,478	0.8	0.6
ζ-Carotene	380,400,424	14.9	25.3
Neurosporene	413,438,465	—	0.2
β-Apo-8'-carotenol	454	trace	1.0
β-Apo-8'-carotenol	402,425,450	—	—
β-Cryptoxanthin	425,450,478	1.5	1.3
β-Cryptoflavin	404,427,453	37.0	34.2
β-Cryptochrome	380,402,428	2.7	2.8
Mixture (unidentified)[a]	—	5.3	—
Lutein	420,445,473	2.4	1.3
Mutatoxanthin	404,427,453	21.1	13.9
Mixture (unidentified)[a]	—	6.3	2.7
Total carotenoids (μg(g fresh wt)$^{-1}$)	—	15.0	22.0

[a] Mixture of 10–15 pigments in trace amounts characterized only by electronic spectra.
From Gross *et al.* (1983).
Major carotenoid pattern: (5).
Vitamin A value: 680 μg RE kg^{-1}.

Table A. 8 Rosaceae CHERRY SOUR (*Prunus cerasus* L.)

Anthocyanins

(a)

Anthocyanins	Cultivars					
	Wye Morello	Olivet	Empress Eugenie	Late Duke	Archduke	Ostheimer Weichsel
Cy 3-glucoside	+	+	+	+	+	+
Cy 3-rutinoside	+	+	+	+	+	+
Cy 3-sophoroside	+	+	+	+	+	+
Cy 3-(2^G-glucosyl-rutinoside)	+	+	+	+	+	+

From Harborne and Hall (1964).

(b)

Anthocyanins	Cultivars		
	Montmorency Marasca Moscata Königliche Amarelle Triaux	Wezesna Z Prinn Del Nord Schattenmorelle Marasca di Ostheim	Flemish Red
Cy 3-gentiobioside	+	+	+
Cy 3-rhamnoglucoside	+	+	+
Cy 3-glucoside	+	+	−
Cy 3-diglycoside	−	+	−

From von Elbe *et al.* (1969).

Anthocyanin pigment changes during maturation and ripening of red cherries Montmorency

(c)	Maturity stages			
	Very immature	Immature	Partially mature	Mature
Anthocyanins	% of Total Anthocyanins			
Cy	11.8	5.1	6.8	7.0
Pn	4.3	5.9	4.7	3.2
Cy 3-glucoside	16.5	19.5	16.6	18.7
Cy 3-rutinoside	28.0	21.2	27.1	26.6
Pn 3-rutinoside	6.6	4.0	4.0	3.0
Cy 3-sophoroside	11.5	18.3	14.6	16.2
Cy 3-(2^G-glucosyl-rutinoside)	21.2	26.0	26.2	25.2
Total anthocyanins (mg(100 g))	2.0	10.3	29.0	43.6

Recalculated from Dekazos (1970).
Additional pigment found in Montmorency cv: Cy 3(2^G-xyloxylrutinoside), see Shrikhande and Francis (1973).
Major patterns: (IV), (VII).

Table A. 9 *Rosaceae CHERRY SWEET (Prunus avium L.)*

Anthocyanins

Anthocyanins	Cultivar Bing
Cy 3-glucoside	+ +
Cy 3-rutinoside	+ +
Pn 3-glucoside	+
Pn 3-rutinoside	+

From Lynn and Luh (1964).
Major anthocyanin patterns: (II), (V).

Carotenoids

Chlorophyll and carotenoid pigment changes in the ripening yellow cherry (Prunus avium) cv. Dönissen's Yellow

	Content (μg (g fresh wt)$^{-1}$)		
Chlorophyll *a*	10.0	—	—
Chlorophyll *b*	2.6	—	—
Total carotenoids	9.4	5.7	3.8

Carotenoid Pattern	% of Total Carotenoids		
β-Carotene	14.0	11.2	9.5
β-Carotene 5,6-epoxide	—	0.4	—
β-Carotene 5,6,5'6'-diepoxide	—	2.4	—
Mutatochrome	—	—	3.2
α-Cryptoxanthin	—	1.5	—
Cryptoxanthin	—	2.5	—
Lutein	17.5	16.8	14.2
Zeaxanthin	2.6	3.4	3.3
Mutatoxanthin	—	1.1	0.8
Isolutein	2.1	—	—
trans-Antheraxanthin	2.7	5.3	5.7
cis-Antheraxanthin	2.0	6.8	5.8
trans-Luteoxanthin	5.5	8.8	7.1
trans-Violaxanthin	34.9	10.6	9.7
cis-Violaxanthin	—	14.7	15.7
Auroxanthin	3.4	—	8.2
C_{27}-Apocarotenol	1.2	1.3	1.7
cis-Luteoxanthin	—	1.8	4.4
Neoxanthin	14.1	11.4	10.8

From Gross (1985).
Major carotenoid patterns: (2), (5).
Vitamin A value: 60 μg RE kg^{-1}.

Table A. 10 *Ericaceae CRAN-BERRY (Vaccinium macrocarpon* Ait.)

Anthocyanins

Anthocyanins	% of Total
Cy 3-galactoside	36.4
Pn 3-galactoside	35.9
Cy 3-arabinoside	18.9
Pn 3-arabinoside	8.8

From Fuleki and Francis (1968).

Cy 3-glucoside	+
Cy 3-arabinoside	+
Cy 3-galactoside	+
Pn 3-glucoside	+
Pn 3-arabinoside	+
Pn 3-galactoside	+

From Fuleki and Francis (1967).

Cy 3-galactoside
Cy 3-arabinoside
Pn 3-galactoside
Pn 3-arabinoside

From Sapers *et al.* (1983).
Major anthocyanin pattern: (III).

Carotenoids

Carotenoid Pattern	% of Total Carotenoids
Phytoene	4.7
Phytofluene	0.9
α-Carotene	0.14
β-Carotene	5.3
ζ-Carotene	0.6
Mutatochrome	0.6
α-Cryptoxanthin	1.5
Cryptoxanthin	1.8
Lutein	31.3
Zeaxanthin	2.8
Carbonyl 417	0.2
Isolutein	10.5
cis-Antheroxanthin	4.6
Chrysanthemaxanthin	0.7
Mutatoxanthins	0.33
Carbonyl 406	0.37
Violaxanthin	20.7
Luteoxanthins	1.8
Valenciaxanthin	1.1
Sinensiaxanthin	2.0
Neoxanthin	6.9
Total carotenoids (μg(g fresh wt)$^{-1}$)	5.8

From Curl (1964).
Major carotenoid pattern: (2).
Vitamin A value: 60 μg RE kg^{-1}.

Table A. 11 *Saxifragaceae CURRANT BLACK (Ribes nigrum L.)*
Anthocyanins

Anthocyanins
Dp 3-glucoside
Dp 3-rutinoside
Cy 3-glucoside
Cy 3-rutinoside

From Harborne *et al.* (1964).

Anthocyanins	% of Total
Cy 3-rutinoside	35
Dp 3-rutinoside	30
Cy 3-glucoside	17
Dp 3-glucoside	13
Pg 3-rutinoside	3
Cy 3-sophoroside	1
Dp 3-sophoroside	1

From LeLous *et al.* (1975).
Major anthocyanin patterns: (II), (V).

Table A. 12 *Saxifragaceae CURRANT RED (Ribes rubrum L.)*
Anthocyanins

(a)

Anthocyanins	Cultivars		
	Earliest of the Fourlands	Fay's Prolific	Red Lake
Cy 3-glucoside	+	+	+
Cy 3-rutinoside	+	+	−
Cy 3-sambubioside	+	+	−
Cy 3-sophoroside	+	−	+
Cy 3-(2^G-xylosylrutinoside)	+	+	−
Cy 3-(2^G-glucosylrutinoside)	+	−	+

From Harborne and Hall (1964).

Anthocyanins in red currant cv. Red Dutch at two maturity stages

(b) Anthocyanins	Pink		Dark red	
	(mg/100 g fresh wt)$^{-1}$	(%)	(mg/100 g fresh wt)$^{-1}$	(%)
Cy 3-glucoside	0.7	8.5	1.1	7.0
Cy 3-rutinoside	0.5	6.1	1.6	10.2
Cy 3-sambubioside	3.2	39.0	3.4	21.7
Cy 3-sophoroside	1.1	13.4	1.3	8.3
Cy 3-(2^G-xylosylrutinoside)	1.7	20.7	5.0	31.8
Cy 3-(2^G-glucosylrutinoside)	1.0	12.3	3.3	21.0
Total anthocyanins (mg (100 g fresh wt)$^{-1}$)	8.2		15.7	

Anthocyanin distribution in five red currant cultivars

(c) Anthocyanins	% of Total Anthocyanins				
	Red Lake	Jonkheer van Tets	Earliest of Fourlands	Red Dutch	Rondom
Cy 3-glucoside	2	3	5	10	10
Cy 3-rutinoside	16	17	16	9	8
Cy 3-sambubioside	9	11	10	29	31
Cy 3-sophoroside	—	—	4	9	8
Cy 3-(2^G-xylosylrutinoside)	73	69	37	28	29
Cy 3-(2^{G}glucosylrutinoside)	—	—	28	15	14
Total anthocyanins (mg(100 g fresh wt)$^{-1}$)	18.0	16.9	18.6	14.0	11.9

From Øydvin (1974).
Major anthocyanin patterns: (IV), (VII).

Carotenoids

Chlorophyll and carotenoid distribution in developing red currant Rondom during ripening

Content ($\mu g(g\,fresh\,wt)^{-1}$)	Unripe (Green)	Half Ripe (Green–Red)	Fully Ripe (Red)
Chlorophyll a	34.0	32.5	—
Chlorophyll b	11.2	9.0	—
Total carotenoids	11.2	12.5	3.6
Carotenoid Pattern	($\mu g(100\,g\,fresh\,wt)^{-1}$)		
β-Carotene	238.6	267.5	63.4
Mixture (unidentified)	52.6	40.0	13.3
Lutein	462.6	580.0	189.0
Chrysanthemaxanthin	7.8	15.0	6.5
Zeaxanthin	25.8	45.0	11.9
Mutatoxanthin	9.0	13.8	3.6
Auroxanthin	37.0	—	12.6
Isolutein	14.6	5.0	—
Antheraxanthin	7.8	16.3	1.8
trans-Luteoxanthin	42.6	33.4	4.3
trans-Violaxanthin	19.0	18.8	—
cis-Violaxanthin	14.6	8.8	—
cis-Luteoxanthin	5.6	—	5.0
Neoxanthin	131.0	132.5	19.5
Pigment 445	5.6	12.5	—
trans-Neochrome	13.4	47.6	21.2
cis-Neochrome	17.9	13.8	7.9
Mixture (unidentified)	14.5	—	—

From Gross (1982/83).
Major carotenoid pattern: (2) (modified).
Vitamin A value: 100 μg RE kg^{-1}.

Table A. 13 *Palmaceae DATE* (*Phoenix dactylifera* L.)

Carotenoids

Carotenoid distribution in two soft date cultivars Hayany and Barhee

	Hayany			Barhee
	Source 1, Harvest-ripe, Red	*Source 1, Ripened off the Tree, Turning Brown*	*Source 2, Ripe, Brown*	*Ripe, Brown*
Total carotenoids:				
(μg(g fresh wt)$^{-1}$)	12.7	12.6	11.4	13.2
(μg(g dry wt)$^{-1}$)	36.3	21.2	19.0	21.3
Carotenoid Pattern	% *of Total Carotenoids*			
Phytofluene	4.1	3.6	4.3	6.4
β-Carotene	9.1	11.6	11.4	10.8
Mutatochrome	1.2	traces	—	—
Mixture[a]	1.3	1.9	1.8	4.2
Lutein	42.6	41.3	43.5	39.0
Zeaxanthin	1.5	2.4	1.2	4.4
Mutatoxanthin	1.1	—	3.4	0.8
Isolutein	0.5	1.0	0.4	1.0
Unknown 440	1.0	—	0.5	1.0
Antheraxanthin	0.8	1.2	0.5	1.0
Luteoxanthin	2.3	1.0	1.9	2.1
trans-Violaxanthin	1.5	1.6	1.1	2.1
cis-Violaxanthin	3.0	3.4	3.2	3.0
Neoxanthin	30.0	29.0	21.0	14.6
Neochrome	—	1.0	3.0	3.0
Mixture (unidentified)	—	—	2.8	5.6

[a]Mixture of several pigments containing α-cryptoxanthin and its 5,6-epoxide.

Carotenoids

Carotenoid distribution in the semi-dry date Deglet Noor during ripening

	Pink (Unripe)	Yellow-brown	Brown (Ripened off the Tree)
Total carotenoids:			
(μg(g fresh wt)$^{-1}$)	11.6	9.2	9.2
(μg(g dry wt)$^{-1}$)	23.2	13.1	12.0

Carotenoid Pattern	% of Total Carotenoids		
Phytofluene	2.5	1.9	3.0
β-Carotene	6.9	6.5	5.9
Mixture (unidentified)	3.0	1.8	1.2
Lutein	45.7	50.1	52.7
Zeaxanthin	2.8	1.1	1.1
Mutatoxanthin	—	0.6	0.7
Isolutein	0.7	0.4	0.3
Antheraxanthin	0.8	0.4	0.7
trans-Luteoxanthin	2.3	1.6	1.2
trans-Violaxanthin	2.4	1.8	0.3
cis-Violaxanthin	3.1	2.4	1.0
cis-Luteoxanthin	0.5	0.7	1.7
Unknown 447	1.5	0.9	0.9
Neochrome	2.5	1.9	4.0
Neoxanthin	21.9	25.0	22.4
Mixture (unidentified)	3.4	2.9	2.9

From Gross *et al.*
Major carotenoid pattern: (2).
Vitamin A value: 100–240 μg RE kg^{-1}.

Table A. 14 Moraceae FIG (*Ficus carica* L.)

Anthocyanins

	Cultivar			
		Mission	Kadota	Calimyrna
Anthocyanins	Skin % of Total	Drupelets	Drupelets	Drupelets
Cy 3-rutinoside	75	+	+	+
Cy 3,5-diglucoside	11	−	−	−
Cy 3-glucoside	11	−	−	−
Pg 3-rutinoside	3	−	−	−

From Puech *et al.* (1975).
Major anthocyanin patterns: (V), (VI).

Carotenoids

Carotenoid distribution in black Mission figs

Carotenoid Pattern	% of Total Carotenoids
Phytoene	2.4
Phytofluene	0.7
α-Carotene	0.2
β-Carotene	9.9
Monols	2.1
Lutein	28.8
Zeaxanthin	1.1
Monoepoxide diols	2.0
Violaxanthin	40.0
Luteoxanthin	2.5
Valenciaxanthin	0.4
Sinensiaxanthin	2.0
Neoxanthin	7.1
Total carotenoids ($\mu g\, g^{-1}$)	8.5

From Curl (1964).
Major carotenoid patterns: (2), (5).
Vitamin A value: 160 μg RE kg^{-1}.

Table A. 15 Saxifragaceae GOOSEBERRY (*Ribes grossularia* L.)

Anthocyanins

Cy 3-glucoside
Cy 3-rutinoside

From Harborne (1967).
Major anthocyanin patterns: (I), (IV).

Carotenoids

Chlorophyll and carotenoid distribution in different cultivars of gooseberry (Ribes grossularia L.)

	Yellow Triumph Green	Blue Triumph Blue–violet	Achilles Green	Yellow Lauffener Yellow
Content ($\mu g(g\ fresh\ wt)^{-1}$)				
Chlorophyll *a*	8.3	13.0	15.7	8.7
Chlorophyll *b*	2.8	4.5	5.6	3.1
Total carotenoids	2.9	3.8	4.6	5.0
Carotenoid Pattern	\% of Total Carotenoids			
β-Carotene	29.2	29.4	25.1	40.2
Mixture[a]	1.0	1.8	—	9.6
Lutein	41.5	40.8	48.1	27.0
Zeaxanthin	1.6	3.4	1.6	1.5
Mutatoxanthin	—	1.3	0.4	1.3
Auroxanthin	—	2.7	—	—
Isolutein	2.5	0.6	1.0	0.6
Antheraxanthin	1.7	0.7	4.1	1.3
Luteoxanthin	2.8	5.1	2.0	3.2
trans-Violaxanthin	5.3	2.0	4.1	3.7
cis-Violaxanthin	3.0	0.6	2.0	1.6
Neochrome	1.0	2.2	1.0	1.8
trans-Neoxanthin	10.2	9.4	5.0	3.1
cis-Neoxanthin	—	—	5.7	5.1

[a] Mixture of 6 pigments, with mutatochrome and cryptoxanthin predominating.

Carotenoids

Chlorophyll and carotenoid distribution in developing gooseberry fruits (Ribes grossularia L.) Yellow Lauffener and leaves

	Fruit		Leaves
	Unripe (Green)	Ripe (Yellow)	(μg(g fresh wt)$^{-1}$)
(μg (g fresh wt)$^{-1}$)			
Chlorophyll a	9.1	8.7	1332
Chlorophyll b	3.2	3.1	400
Total carotenoids	2.1	5.0	254
Carotenoid Pattern[a]	% of Total Carotenoids		
β-Carotene	24.9	49.8	27.1
Lutein	40.9	29.8	37.9
Antheraxanthin	9.0	5.1	28.0
Violaxanthin	9.0	4.8	20.3
Neochrome	6.0	2.3	—
Neoxanthin	10.2	8.2	11.8

[a] Mixture from silica gel plates; the predominant pigment is mentioned.
From Gross (1982/83).
Major carotenoid patterns: (2), and (2), (4).
Vitamin A value: 150, 200 and 330 μg RE kg^{-1}, respectively.

Table A. 16 *Vitaceae GRAPE* (*Vitis* spp.)

Anthocyanins

Cy,	Pn,	Dp,	Mv 3-glucoside
Dp,	Pt,	Mv	3-(*p*-coumaroylglucoside)
Pn,	Dp,	Pt,	Mv 3,5-diglucoside
Dp,	Pt,	Mv	3-(*p*-coumaroylglucoside)-5-glucoside
Pn,	Dp,	Pt,	Mv 3-(caffeoylglucoside)

From Harborne (1967).

Anthocyanins		*Cabernet Sauvignon (Skin) (% of Total)*
Dp	3-glucoside	10.0
Cy	3-glucoside	1.3
Pt	3-glucoside	6.0
Pn	3-glucoside	5.3
Mv	3-glucoside	42.6
Dp	3-glucoside-acetate	2.6
Cy	3-glucoside-acetate	0.2
Pt	3-glucoside-acetate	2.1
Pn	3-glucoside-acetate	1.0
Mv	3-glucoside-acetate	20.5
Dp	3-glucoside-*p*-coumarate	0.5
Cy	3-glucoside-*p*-coumarate	0.1
Mv	3-glucoside-caffeoate	0.2
Pt	3-glucoside-*p*-coumarate	0.5
Pn	3-glucoside-*p*-coumarate	0.6
Mv	3-glucoside-*p*-coumarate	6.5

From Wulf and Nagel (1978).

APPENDIX

Anthocyanin distribution in the juice of American grapes

		Cultivar		
Anthocyanins		Concord	De Chaunac	Ives
Dp	3,5-diglucoside	+	+ +	+
Cy	3,5-diglucoside	+	+ +	+ +
Pt	3,5-diglucoside	+	+ +	+
Pn	3,5-diglucoside	+	+ +	+
Mv	3,5-diglucoside	+	+ +	+ +
Dp	3-glucoside	+ +	+	+
Cy	3-glucoside	+ +	+	+
Pt	3-glucoside	+	+	+
Pn	3-glucoside	+	+	+
Mv	3-glucoside	+	+	+ +
Dp	3(6-*O*-*p*-coumarylglucoside)-5-glucoside	+	tr.	+ +
Cy	3(6-*O*-*p*-coumarylglucoside)-5-glucoside	+	tr.	+
Pt	3(6-*O*-*p*-coumarylglucoside)-5-glucoside	+	tr.	+
Pn	3(6-*O*-*p*-coumarylglucoside)-5-glucoside	+	tr.	+
Mv	3(6-*O*-*p*-coumarylglucoside)-5-glucoside	+	tr.	+ +
Dp	3(6-*O*-*p*-coumarylglucoside)	+ +	−	+ +
Cy	3(6-*O*-*p*-coumarylglucoside)	+ +	−	+ +
Pt	3(6-*O*-*p*-coumarylglucoside)	+	−	+
Pn	3(6-*O*-*p*-coumarylglucoside)	+	−	+ +
Mv	3(6-*O*-*p*-coumarylglucoside)	+	−	+

+ + = major pigments; tr. = trace amounts.
Adapted from Williams *et al.* (1978).
Major anthocyanin patterns: (II), (VI), (VIII).

Carotenoids

Chlorophyll and carotenoid pigments in different cultivars of grapes (Vitis vinifera L.)

	Dabouki	Riesling	Thompson[a]
Content ($\mu g(g\ fresh\ wt)^{-1}$)			
Chlorophyll *a*	2.0	11.2	n.d.
Chlorophyll *b*	0.5	4.0	n.d.
Total carotenoids	1.0	3.5	1.8
Carotenoid Pattern	*% of Total Carotenoids*		
Phytofluene	—	—	0.2
α-Carotene	—	—	0.2
Unidentified	—	—	0.4
β-Carotene	33.7	29.2	32.3
Lutein	31.9	35.7	24.3
Zeaxanthin	4.1	7.7	1.2
Isolutein	1.9	1.3	6.6
Antheraxanthin	2.4	6.8	
Luteoxanthin	6.1	2.0	1.8
trans-Violaxanthin	4.4	5.0	17.4
cis-Violaxanthin	2.9	1.6	
Neoxanthin	12.6	10.7	13.5

n.d. = not determined.
[a] From Curl (1964); other data from Gross (1983).
Major carotenoid pattern: (2).
Vitamin A value: 60, 100 and 200 μg RE kg^{-1}, respectively.

Table A. 17 Rutaceae GRAPEFRUIT WHITE (*Citrus paradisi* Macfad.)

Carotenoids

Carotenoid distribution in flavedo and pulp of Marsh seedless grapefruit

Carotenoid Pattern	% of Total Carotenoids	
	Flavedo	Pulp
Phytoene	45.6	68.3
Phytofluene	16.2	12.7
β-Carotene	0.6	1.1
ζ-Carotene	1.0	1.4
β-Zeacarotene	0.5	—
Neurosporene	0.1	0.5
Mutatochrome	5.7	0.5
Cryptoxanthin 5,6-epoxide	1.5	—
Cryptoxanthin	2.0	0.9
Cryptoflavin	6.9	0.4
Lutein	1.9	—
Zeaxanthin	1.8	—
Mutatoxanthin	0.4	5.4
Antheraxanthin	1.2	0.6
Luteoxanthin	2.2	2.6
Violaxanthin	12.4	2.8
Neochrome	—	1.8
Total carotenoids (μg(g fresh wt)$^{-1}$)	1.4	0.24

From Romojaro *et al.* (1979).
Major carotenoid pattern: (5) (uncommon), containing large amounts of colourless carotenes.

Table A. 18 *Rutaceae GRAPEFRUIT PINK and RED (Citrus paradisi* Macfad.)

Carotenoids

Carotenoid distribution in peel and pulp of Ruby Red grapefruit

	% of Total Carotenoids	
Carotenoid Pattern	Pulp	Peel
Phytoene	16.0	47.0
Phytofluene	4.4	14.4
α-Carotene	—	0.1
β-Carotene	27.0	7.2
ζ-Carotene	3.5	7.2
γ-Carotene	0.8	0.4
Lycopene	40.0	11.0
Unknown 337	0.2	—
Unknown 335	0.6	0.5
α-Cryptoxanthin	0.2	0.1
Cryptoxanthin	1.1	1.4
Cryptoflavin	0.4	1.3
Rubixanthin	0.5	0.3
Cryptochrome	—	0.2
Lutein	0.3	0.9
Antheraxanthin	0.7	—
Violaxanthin	0.9	1.0
Luteoxanthins	0.4	1.8
Mutatoxanthin	0.4	0.2
Valenciachromes	0.2	0.3
Trollichrome-like	—	0.2
Auroxanthins	0.3	1.6
Total carotenoids (μg(g fresh wt)$^{-1}$)	8.2	10.4

From Curl and Bailey (1957).
Major carotenoid patterns: in pulp, (3), (4); in peel, (5), containing large amounts of colourless carotenes.
Vitamin A value: 380 μg RE kg^{-1}.

Carotenoids

Carotenoid distribution in different fruit parts of the Ruby Red grapefruit

	% of Total Carotenoids			
	Endocarp		Exocarp	
Carotenoid Pattern	Pulp	Septa	Flavedo	Albedo
Phytofluene	17.4	24.6	24.4	31.6
β-Carotene	10.4	13.8	17.1	21.9
γ-Carotene	0.4	—	—	—
Lycopene	70.4	60.4	17.7	43.3
Rubixanthin	0.2	—	—	—
Cryptoxanthin	—	—	2.8	—
Cryptoflavin	—	—	2.7	—
Citraurin	—	—	2.9	—
Lutein	0.1	—	5.8	—
Zeaxanthin	0.4	—	2.8	—
Mutatoxanthin	0.2	0.4	—	0.8
Antheraxanthin	0.5	0.4	5.5	—
trans-Violaxanthin	—	0.4	6.4	0.8
cis-Violaxanthin	—	—	9.6	—
Neoxanthin	—	—	2.4	—
Mixture (unidentified, trace amounts carotenoids)	—	—	—	1.6
Total carotenoids (μg(g fresh wt)$^{-1}$)	25.0	84.2	27.2	43.8

From Gross (unpublished results, 1983).
Major carotenoid patterns: in pulp, septa and albedo, (3), (4); in flavedo, (3), (4), (5).
Vitamin A value of the pulp: 400 μg RE kg^{-1}.

Table A. 19 Myrtle *GUAVA PINK* (*Psidium guajava* L.)
Carotenoids
Carotenoid distribution in pink-fleshed Brasilian guava

	% of Total Carotenoids		
	Cultivar		
	IAC-4	Undefined	
Carotenoid Pattern	Round-shaped	Round-shaped	Pear-shaped
β-Carotene	6.0	17.0	9.7
ζ-Carotene	tr.[a]	tr.	tr.
γ-Carotene	tr.	0.6	tr.
Zeinoxanthin (α-cryptoxanthin)	1.6	2.7	2.7
Lycopene	86.0	76.3	83.0
β-Carotene 5,6,5′,6′-diepoxide	tr.	0.1	0.5
Zeinoxanthin 5,8-epoxide	—	0.3	—
5,8-epoxy-5,8-dihydro-β,β-carotene-3,4,3′-triol[b]	6.4	3.0	4.1
Total carotenoids (μg(g fresh wt)$^{-1}$)	62.1	70.0	56.6

[a] tr. = traces.
[b] Tentative identification.
From Padula and Rodriguez-Amaya (1984).
Major carotenoid pattern: (3) (modified).
Vitamin A value: 620, 1980, and 915 μg RE kg^{-1} respectively.

Table A. 20 *Actinidiaceae* KIWIFRUIT or *CHINESE GOOSEBERRY*
(*Actinidia chinensis* Planch.)

Carotenoids

Quantitative changes of the pericarp pigments of Kiwifruit or Chinese gooseberry (*Actinidia chinensis*) cv. Bruno during ripening

	Oct. 10	Nov. 2	Nov. 10	Ripe (on the Tree)	Ripe (Ethephon)
Content ($\mu g (g\ fresh\ wt)^{-1}$)					
Chlorophyll *a*	19.3	17.3	16.9	17.1	14.4
Chlorophyll *b*	9.3	9.5	10.3	8.3	5.7
Total carotenoids	8.3	8.3	6.3	6.5	6.1

Carotenoid Pattern	% of Total Carotenoids				
β-Carotene	15.2	15.0	12.9	8.0	8.0
Unknown mixture (4 pigments)	0.9	1.8	1.8	3.5	1.6
Cryptoxanthin	—	—	—	3.2	3.1
Cryptoflavin	—	—	—	—	0.8
Lutein	48.7	45.7	46.6	45.9	44.4
Lutein-like	0.6	0.8	0.9	0.8	0.7
trans-Chrysanthemaxanthin	3.1	2.3	1.6	1.0	1.0
Zeaxanthin	1.2	1.1	1.5	3.0	3.2
cis-Chrysanthemaxanthin	1.7	1.7	1.5	1.1	0.7
Mutatoxanthin	1.1	1.2	1.5	1.5	1.8
Auroxanthin	—	2.8	1.3	3.5	2.5
Isolutein	1.1	0.7	1.0	0.5	1.2
Antheraxanthin	0.2	0.3	1.1	0.3	0.6
trans-Luteoxanthin	0.6	0.8	1.0	1.2	1.5
trans-Violaxanthin	1.2	0.7	1.3	1.1	3.2
cis-Violaxanthin	0.2	0.5	1.0	0.6	0.9
cis-Luteoxanthin	—	0.8	0.7	0.7	—
Unknown 440	0.3	0.5	0.4	0.1	0.3
Unknown 448	0.6	0.7	0.6	0.9	0.6
trans-Neochrome	2.7	3.3	2.3	6.1	2.0
cis-Neochrome	4.4	4.6	4.4	—	3.3
trans-Neoxanthin	12.7	11.8	13.4	13.4	14.4
cis-Neoxanthin	0.4	0.2	0.6	0.5	0.4
cis-Neochrome	1.3	0.6	0.9	1.0	1.2
Unknown mixture (9 pigments)	1.8	1.9	1.7	2.6	2.6

From Gross (1982).
Major carotenoid pattern: (2) (modified) containing neochrome as fifth main pigment.
Vitamin A value: 50 μg RE kg^{-1}.

Table A. 21 *Rutaceae KUMQUAT (Fortunella margarita* (Lour.) Swingle)
Carotenoids

Carotenoid distribution in pulp and peel of the ripe kumquat Fortunella margarita Nagami

	Pulp	*Peel*
Total Carotenoids (μg(g fresh wt)$^{-1}$)	8.4	172.0
Carotenoid Pattern	% of Total carotenoids	
Phytofluene	1.7	1.2
α-Carotene	—	0.1
β-Carotene	0.9	0.7
ζ-Carotene	2.8	0.3
Mutatochrome	—	0.3
Cryptoxanthin 5,6-epoxide	3.3	—
α-Cryptoxanthin	—	0.7
β-Cryptoxanthin	13.2	5.5
Cryptoflavin	5.1	0.8
Cryptoxanthin 5,6,5′,6′-diepoxide	—	0.7
trans-Cryptoxanthin 5′,6′-epoxide	—	1.6
cis-Cryptoxanthin 5′,6′-epoxide	—	1.4
trans-Cryptoflavin	—	3.2
cis-Cryptoflavin	0.5	0.5
Citraurin	4.6	3.8
Lutein	3.9	5.5
Zeaxanthin	3.6	1.1
Mutatoxanthin	6.2	1.0
trans-Antheraxanthin	4.3	2.1
cis-Antheraxanthin	4.5	2.2
trans-Luteoxanthin	9.6	7.4
trans-Violaxanthin	4.5	18.2
cis-Violaxanthin	19.9	31.7
cis-Luteoxanthin	1.9	2.2
trans-Neoxanthin	1.6	4.2
cis-Neoxanthin	3.2	3.7
Neochrome	4.7	—

Carotenoids

Quantitative pigment changes in the ripening kumquat Fortunella margarita Nagami

	Ripening Stages			
	1	2	3	4
Content ($\mu g(g\ fresh\ wt)^{-1}$)				
Chlorophyll *a*	20.8	9.7	5.0	—
Chlorophyll *b*	5.7	5.7	1.6	—
Total carotenoids	18.7	17.3	14.6	38.6

Carotenoid Pattern	% of Total Carotenoids			
Phytofluene	—	—	2.7	1.9
α-Carotene	1.8	4.2	1.9	0.2
β-Carotene	14.7	12.8	12.7	0.5
ζ-Carotene	—	—	—	0.3
Mutatochrome	—	—	—	0.2
Cryptoxanthin 5,6-epoxide	—	—	—	1.0
α-Cryptoxanthin	0.5	1.1	0.9	—
β-Cryptoxanthin	1.7	2.3	3.2	8.4
Cryptoflavin	0.9	1.8	2.4	1.8
Cryptoxanthin 5,6,5′,6′-diepoxide	—	—	—	0.4
trans-Cryptoxanthin 5′,6′-epoxide	—	—	—	1.6
cis-Cryptoxanthin 5′,6′-epoxide	—	—	—	0.6
trans-Cryptoflavin	—	—	—	1.7
cis-Cryptoflavin	—	—	—	0.6
Citraurin	0.9	1.2	3.5	6.1
Lutein	39.2	22.4	13.5	5.1
Zeaxanthin	4.4	2.3	2.3	1.5
Mutatoxanthin	—	0.8	4.6	1.7
trans-Antheraxanthin	—	3.1	2.1	1.5
cis-Antheraxanthin	—	1.2	1.8	2.2
trans-Luteoxanthin	—	4.7	5.9	8.2
trans-Violaxanthin	13.2	12.7	13.0	16.6
cis-Violaxanthin	9.0	12.8	12.5	28.8
cis-Luteoxanthin	—	—	—	3.1
trans-Neoxanthin	13.7	10.4	13.6	5.3
cis-Neoxanthin	—	4.3	2.7	—
Neochrome	—	1.9	0.7	0.7

From Huyskens *et al.* (1985).
Major carotenoid patterns: pulp and peel, (5), (8).
Vitamin A value (whole fruit): 480 μg RE kg^{-1}.

Table A. 22 Rutaceae *LEMON* (*Citrus limon* (L.) Burm.)

Carotenoids

Comparison of carotenoids in pulp and peel of lemon Eureka at two ripening stages

	% of Total Carotenoids			
	Mature-green		Yellow	
Carotenoid Pattern	Pulp	Peel	Pulp	Peel
Phytofluene	19.0	17.0	22.0	18.0
Unknown	—	—	1.6	3.2
α-Carotene	6.4	8.6	—	—
β-Carotene	3.8	13.0	8.9	6.8
ζ-Carotene	—	—	4.0	17.0
Phytofluene-like	—	—	0.5	2.3
β-Carotene 5,6-epoxide	—	—	14.0	—
Mutatochrome	1.3	2.8	2.1	1.5
α-Cryptoxanthin	2.6	3.2	1.8	3.1
Cryptoxanthin 5,6-epoxide	1.3	3.2	—	—
Cryptoxanthin	15.0	12.0	30.0	9.7
Cryptoflavins	5.7	1.4	2.6	14.8
Cryptoxanthin 5,6,5'6'-diepoxide	—	—	1.6	0.2
Cryptoxanthin 5,6,5',8'-diepoxide	—	—	—	1.2
Unknown 422	—	—	—	9.3
Unknown 432	2.6	4.3	—	—
Lutein	6.7	10.0	—	—
Zeaxanthin	—	2.0	—	—
Violaxanthin	11.5	3.7	1.1	9.9
Luteoxanthin	6.4	1.2	—	—
Auroxanthin	10.0	9.2	8.5	1.2
Trollein	8.0	4.0	—	—
trans-Neoxanthin	0.2	2.3	—	—
cis-Neoxanthin	0.6	2.1	—	—
Total carotenoids ($\mu g(g\ fresh\ wt)^{-1}$)	1.1	2.1	0.6	1.4

From Yokoyama and White (1967).
Major carotenoid patterns: pulp, (4), (5) (uncommon containing large amounts of phytofluene).
Vitamin A value: —.

Table A. 23 Sapindaceae *LITCHI* (*Litchi chinensis* Sonn.)
Anthocyanins

Changes in anthocyanin in peel in litchi during ripening

Anthocyanins	Ripening Stages[a]				
	1	2	3	4	5
Cy 3-glucoside	+	+	tr.	tr.	—
Cy 3-galactoside	+	+	++	++	++
Pg 3-rhamnoside	—	+	+	++	++
Pg 3,5-diglucoside	—	—	—	+	++

+, ++ = relative quantities; tr. = trace.
[a] Ripening stages: Stage 1 (red tinge) and stage 2 (one-quarter red) = begin; stage 3 = partially ripe (three-quarter red); stage 4 = fully ripe (fully red); stage 5 = overripe (deep red).
From Prasad and Jha (1978).
Major anthocyanin pattern: (III).

Table A. 24 *Rosaceae LOQUAT* (*Eriobotrya japonica* Thunb. Lindl.)
Carotenoids

Quantitative carotenoid distribution in the pulp and in peel of the loquat fruit (*Eriobotyra japonica*) Golden Nugget

	% of Total Carotenoids	
Carotenoid Pattern	Pulp	Peel
Phytofluene	1.0	0.3
cis-neo-β-Carotene	5.0	5.3
β-Carotene	33.0	50.0
γ-Carotene	6.1	5.2
Mutatochrome	1.4	1.4
Carbonyl 450	5.5	3.0
γ-Carotene-like	1.6	0.2
Cryptoxanthin 5,6,5',6'-diepoxide	2.2	0.4
Cryptoxcanthin 5,6-epoxide	2.8	0.6
Cryptoxanthin	22.0	4.7
Cryptoxanthin 5,6,5'8'-diepoxide	0.9	0.5
Cryptoflavin	1.6	0.6
Lutein	3.2	13.0
cis-Lutein	0.7	1.3
Isolutein	1.7	2.2
Violaxanthin	2.9	4.3
Crysanthemaxanthin	0.9	0.6
Luteoxanthin	1.4	2.5
Neochrome	1.0	0.5
Neoxanthin	4.9	3.2
Neoxanthin-like	0.2	0.4
Total carotenoid (μg(g fresh wt)$^{-1}$)	22.5	110.0

From Gross *et al.* (1973); Kobayashi *et al.* (1979).
[a] Major carotenoid pattern: pulp, (4); peel, (4) (modified).
Vitamin A value (Israeli cv.): pulp, 2100 μg RE kg^{-1};
  peel, 11 200 μg RE kg^{-1}.
  Japanese cvs: whole fruit, 1800 μg RE kg^{-1}.

Table A. 25 Rutaceae *MANDARIN* (*Citrus reticulata* Blanco)

Carotenoids

Quantitative changes of the flavedo pigments in Dancy tangerine and Clementine during ripening

	Dancy Tangerine			Clementine		
	Green	Colour Break	Ripe	Green	Colour Break	Ripe
Fruit diameter (cm)	4.9	5.3	6.5	4.5	4.7	5.1
Content (μg(g fresh wt)$^{-1}$)						
Chlorophyll *a*	249.2	86.0	—	73.7	39.9	—
Chlorophyll *b*	80.0	27.0	—	24.4	18.6	—
Total carotenoids	121.5	93.5	295.0	30.8	24.9	74.6

Carotenoid Pattern	% of Total Carotenoids					
Phytofluene	—	3.9	2.7	—	7.7	3.8
α-Carotene	11.7	4.2	0.2	10.5	3.7	0.1
β-Carotene	8.7	4.0	0.4	11.6	7.2	0.4
ζ-Carotene	—	tr.	0.3	—	—	0.8
Mutatochrome	—	—	—	—	—	1.2
γ-Carotene	—	—	tr.	—	—	0.2
β-Apo-8′-carotenal	0.1	0.2	0.7	—	—	1.8
α-Cryptoxanthin	1.6	—	—	1.2	1.5	—
β-Cryptoxanthin	1.8	4.5	13.6	5.3	9.7	6.2
Cryptoxanthin 5,6-epoxide	—	—	0.7	—	—	0.2
Cryptoflavin	—	0.8	1.3	—	—	—
Cryptoxanthin 5′,6′-epoxide	—	0.4	2.1	—	—	—
β-Citraurinene	—	—	—	4.1	6.8	15.7
β-Citraurin	—	3.0	8.2	2.3	8.3	32.5
Lutein	25.7	15.5	2.7	34.7	20.6	2.0
Zeaxanthin	1.1	0.8	1.5	1.1	1.3	1.2
Mutatoxanthin	0.2	—	—	0.4	0.7	1.1
trans-Antheraxanthin	1.7	2.3	2.3	1.8	1.9	2.1
cis-Antheraxanthin	—	—	6.2	—	—	—
Luteoxanthin	2.1	3.6	2.3	4.0	7.7	10.6
trans-Violaxanthin	20.7	19.6	12.5	4.3	3.8	5.5
cis-Violaxanthin	12.0	29.1	39.3	5.3	8.4	9.3
trans-Neoxanthin	11.1	5.4	1.8	12.8	10.4	5.3
cis-Neoxanthin	1.5	2.7	1.2	—	—	—
Neochrome	—	—	—	0.6	0.3	—

Carotenoids

Pigment distribution in the juice and flavedo of a mandarin hybrid (Citrus reticulata) cv. Michal during ripening

	Juice			Peel		
	Green	Colour Break	Ripe	Green	Colour Break	Ripe
Fruit diameter (cm)	4.70	4.85	5.10	4.70	4.85	5.10
Content (μg(g fresh wt)$^{-1}$)						
Chlorophyll *a*	—	—	—	240.0	52.6	—
Chlorophyll *b*	—	—	—	86.0	15.8	—
Total carotenoids	7.5	9.5	13.7	143.4	51.0	174.1
Carotenoid Pattern	% of Total Carotenoids					
Phytofluene	2.4	2.9	5.9	—	5.2	3.1
α-Carotene	0.8	0.5	0.5	9.7	2.4	0.2
β-Carotene	1.3	1.3	1.9	6.9	2.5	0.3
ζ-Carotene	4.1	4.2	3.3	—	—	0.4
δ-Carotene	—	—	—	—	—	0.1
Mutatochrome	—	—	—	—	—	0.5
Lycopene	—	—	0.6	—	—	—
β-Apo-8′-carotenal	—	—	—	0.7	0.7	1.3
α-Cryptoxanthin	1.2	1.9	0.9	1.9	—	—
β-Cryptoxanthin	35.0	35.6	41.0	—	3.1	6.4
Cryptoxanthin 5,6-epoxide	—	—	—	—	—	0.4
Cryptoxanthin 5′,6′-epoxide	—	—	—	—	—	0.3
β-Citraurinene	—	—	—	—	9.5	9.9
β-Citraurin	—	—	—	—	12.3	26.1
Lutein	12.7	10.3	6.5	23.5	12.8	2.6
Zeaxanthin	7.2	8.2	9.1	3.9	1.6	1.0
Mutatoxanthin	1.9	1.8	5.7	1.5	0.2	0.2
trans-Antheraxanthin	10.4	6.4	5.6	3.6	4.3	1.8
cis-Antheraxanthin	6.3	9.5	6.8	—	—	2.5
Luteoxanthin	—	3.5	4.8	5.8	5.6	9.1
trans-Violaxanthin	1.7	2.2	0.5	14.0	11.7	9.8
cis-Violaxanthin	7.3	5.8	3.1	11.0	18.2	19.8
trans-Neoxanthin	2.4	2.8	—	11.7	6.6	4.2
cis-Neoxanthin	0.9	0.5	—	3.6	3.3	—
Neochrome	—	—	0.2	—	—	—
trans-Shamoutixanthin	3.0	1.9	2.6	—	—	—
Shamoutichrome	—	—	0.3	—	—	—
Unknown	1.4	0.7	0.7	2.2	—	—

Table A.25 continued

Carotenoids

Quantitative changes of the flavedo pigments in Dancy tangerine during ripening

	June 16	June 30	July 14	July 28	Aug. 11	Aug. 25	Sept. 8	Sept. 22	Oct. 6	Oct. 20	Nov. 3*	Nov. 17	Dec. 1	Dec. 15	Dec. 29	Jan. 12
Fruit diameter (cm)	1.2	1.75	2.35	2.8	3.3	3.45	3.9	3.9	4.45	4.9	5.3	5.4	6.0	6.0	6.5	6.8
Content (μg(g fresh wt)$^{-1}$)																
Chlorophyll a	280.0	174.0	203.0	162.0	157.0	155.0	197.0	262.0	220.0	249.2	86.0	62.0	51.0	17.3	—	—
Chlorophyll b	120.0	73.0	97.0	62.0	63.0	54.0	53.0	93.0	88.0	80.0	27.0	20.0	12.0	4.5	—	—
Total carotenoids	93.0	55.0	60.0	49.0	50.0	51.0	56.0	59.4	93.0	121.5	93.5	152.4	195.0	260.0	295.0	315.0
Carotenoid Pattern							% of Total Carotenoids									
Phytofluene	—	—	—	—	—	—	—	—	—	—	3.9	3.2	1.4	2.1	2.7	2.5
α-Carotene	6.1	7.6	8.5	10.5	9.1	11.1	12.6	18.6	13.5	12.7	4.2	2.1	0.8	0.5	0.2	0.3
β-Carotene	14.0	12.8	10.2	14.8	14.9	13.7	13.5	15.7	9.7	9.7	4.0	2.3	1.6	0.8	0.4	0.3
ζ-Carotene	—	—	—	—	—	—	—	—	—	—	tr.	tr.	0.3	tr.	0.3	0.6
γ-Carotene	—	—	—	—	—	—	—	—	—	—	—	tr.	tr.	tr.	tr.	tr.
β-Apo-8'-carotenal	—	—	—	—	—	—	—	—	—	0.1	0.2	0.2	0.3	1.3	0.7	1.1

	1	2	3	4	5	6	7	8	9	10	11	12	13	14	15	16
OH-α-Carotene	1.6	3.3	2.9	5.2	6.1	3.5	1.6	1.9	1.5	1.6	—	4.5	—	6.6	—	7.8	—	11.6	—	13.6	—	13.8
Cryptoxanthin	2.5	1.5	—	—	—	—	—	0.9	2.2	1.8	—	—	—	—	—	—
Cryptoxanthin 5,6-epoxide	—	—	—	—	—	—	—	—	—	—	—	—	—	—	—	—
Cryptoflavin	—	—	—	—	—	—	—	—	—	—	0.8	0.8	1.4	—	0.5	0.9	0.7	0.8
Cryptoxanthin 5',6'-epoxide	—	—	—	—	—	—	—	—	—	—	—	—	—	0.7	1.0	1.3	1.2
Citraurin	—	—	—	—	—	—	—	—	—	—	—	—	—	—	—	—
Lutein	31.8	36.7	38.0	30.2	30.8	32.1	30.4	28.0	28.9	26.1	—	—	—	—	—	—	1.2	2.1	2.4
Lutein 5,6-epoxide	3.0	3.0	5.4	5.5	6.0	3.3	2.1	—	—	—	0.4	3.0	0.7	2.1	0.8	4.2	6.2	8.2	7.8
Zeaxanthin	5.8	1.2	—	2.9	—	1.5	1.9	3.2	3.7	1.1	3.0	13.5	6.6	6.1	3.7	2.7	2.3
Mutatoxanthin	—	—	—	—	—	—	—	—	0.3	0.2	2.0	0.8	—	1.9	—	1.0	0.5	—	1.5	—	1.3
trans-Antheraxanthin	1.9	2.5	—	2.9	—	—	6.1	3.0	5.5	1.7	0.8	2.3	1.9	5.1	5.8	2.3	0.6
cis-Antheraxanthin	—	—	—	—	—	—	—	—	—	—	2.3	—	5.2	—	—	6.2	3.2
Luteoxanthin	1.2	—	0.5	—	—	2.4	—	1.9	0.5	2.1	3.6	3.4	2.6	3.3	2.3	4.3
trans-Violaxanthin	9.4	15.0	13.4	11.1	12.3	11.8	20.7	11.2	23.8	20.7	19.6	19.1	18.3	15.1	12.5	4.0
cis-Violaxanthin	1.6	1.3	1.3	4.5	6.0	5.7	—	7.9	—	12.0	29.1	39.7	44.4	41.6	39.3	11.6
trans-Neoxanthin	12.7	10.4	17.9	12.0	14.5	14.1	10.0	6.4	8.1	11.1	5.4	3.3	2.6	2.0	1.7	39.8
cis-Neoxanthin	4.5	2.8	—	—	—	—	1.1	0.6	1.8	1.5	2.7	2.2	1.6	1.4	1.2	1.4
Shamoutixanthin-like	3.9	1.9	1.9	0.4	0.6	0.8	—	0.7	0.5	—	—	—	—	—	—	0.8

* Colourbreak

From Farin *et al.* (1983); Gross (1981).
Major carotenoid patterns: Juice, (4), (5); flavedo, (5), (8).
Vitamin A value: 500 μg RE (l juice)$^{-1}$.

APPENDIX

Table A. 26 *Anarcardiaceae MANGO* (*Mangifera indica* L.) **Anthocyanins**

Cultivar Haden
Pn 3-galactoside

From Proctor and Creasy (1969)
Major anthocyanin pattern: (I).

Carotenoids

Carotenoid distribution in Mango pulp of two different cultivars

	Alphonso[a]	Badami[b]		
	Fully Ripe	Unripe	Partially Ripe	Fully Ripe
Total carotenoids (μg(g fresh wt)$^{-1}$)	125.0	0.4	33.6	89.2
Carotenoid Pattern	% of Total Carotenoids			
Phytoene	3.7	1.27	11.84	6.32
Phytofluene	6.89	1.75	39.26	11.70
cis-β-Carotene	0.36	—	—	1.33
β-Carotene	59.50	57.47	33.55	50.64
ζ-Carotene	—	1.67	—	—
γ-Carotene	0.01	0.18	0.15	0.21
β-Carotene 5,6-epoxide	0.85	1.83	0.04	0.75
Mutatochrome	1.52	1.35	1.30	1.10
Cryptoxanthin	0.66	1.27	1.41	0.40
Cryptoflavin	—	—	—	0.06
Violaxanthin	1.25	—	—	—
cis-Violaxanthin	9.02	1.99	0.93	7.08
Unidentified—450	—	—	—	0.75
Lutein	—	4.97	—	—
Antheraxanthin	1.01	—	0.95	3.00
cis-Antheraxanthin	0.5	—	—	0.31
Zeaxanthin	0.01	—	0.33	0.29
Chrysanthemaxanthin	—	6.12	0.94	—
Unidentified—400	—	2.55	—	—
Luteoxanthin	11.25	3.12	1.38	1.93
Mutatoxanthin	0.76	9.44	2.85	3.73
Auroxanthin	2.71	5.02	5.07	10.40

[a] Jungalwala and Cama (1963).
[b] John *et al.* (1970).
Major carotenoid patterns: (4), (5).
Vitamin A value: 12 500 μg RE kg^{-1}.

Table A. 27 *Cucurbitaceae MELON (Cucumis melo L.) orange-fleshed*

Carotenoids

Carotenoid distribution in the pulp of an orange-fleshed muskmelon

Carotenoid Pattern	% of Total Carotenoids
Phytoene	1.5
Phytofluene	2.4
α-Carotene	1.2
β-Carotene	84.7
ζ-Carotene	6.8
Monols	.0·8
Lutein	1.0
Zeaxanthin	0.08
Violaxanthin	0.9
Luteoxanthins	0.15
Neoxanthin	0.23
Sinensiaxanthin-like	0.2
Total carotenoids (μg(g fresh wt)$^{-1}$)	20.2

From Curl (1966).
Major carotenoid pattern: (4) (modified).
Vitamin A value: 2800 μg RE kg^{-1}.

Table A. 28 *Cucurbitaceae MELON* (*Cucumis melo* L.) green-fleshed

Carotenoids

Carotenoid changes in mesocarp and exocarp of the green-fleshed muskmelon Galia during ripening

	Mesocarp			Exocarp		
	Ripening stage[a]			Ripening stage[b]		
	1	2	3	1	2	3
Carotenoid Pattern	(μg (100g fresh wt.)$^{-1}$)			(μg (10g fresh wt.)$^{-1}$)		
β-Carotene	71.0	48.8	37.8	136.0	86.0	24.0
Mixture (unknown)	2.1	2.5	—	—	6.0	2.0
Cryptoxanthin	—	—	4.0	—	—	7.0
Lutein	142.6	76.0	54.7	399.0	237.0	140.0
trans-Chrysanthemaxanthin	—	2.3	3.4	—	—	13.0
Zeaxanthin	4.2	2.9	5.3	17.0	24.0	15.0
cis-Chrysanthemaxanthin	—	—	4.0	18.0	—	11.0
trans-Isolutein	10.2	8.0	10.1	5.0	21.0	18.0
cis-Isolutein	—	—	4.2	—	—	12.0
Antheraxanthin	9.8	8.2	4.0	15.0	6.0	11.0
trans-Luteoxanthin	17.9	13.9	14.8	24.0	6.0	22.0
trans-Violaxanthin	36.1	16.2	26.8	17.0	19.0	6.0
cis-Violaxanthin	19.3	10.7	15.0	14.0	11.0	17.0
cis-Luteoxanthin	—	—	—	5.0	4.0	7.0
trans-Neoxanthin	16.1	5.5	1.9	105.0	40.0	11.0
cis-Neoxanthin	17.2	14.0	2.9	—	—	—
Neochrome	3.5	—	1.1	15.0	—	4.0
Total carotenoids	350.0	290.0	190.0	770.0	460.0	320.0

[a] Mesocarp: 1 = unripe (green); 2 = half-ripe (pale-green); 3 = fully ripe (whitish-green).
[b] Exocarp: 1 = unripe (deep green); 2 = colour break (green–yellow); 3 = fully ripe (yellow).
From Flügel and Gross (1982).
Major carotenoid pattern: mesocarp and exocarp, (2).
Vitamin A value: —.

Table A. 29 *Oleaceae OLIVE (Olea europaea L.)*
Anthocyanins

Anthocyanins	Cultivar Manzanillo	% Total
Cy 3-glucoside	+	15
Cy 3-rutinoside	+	
Cy 3-caffeylrutinoside	+	60
Cy 3-glucosylrutinoside	+	
Cy 3-(2^G glucosylrutinoside)	+	25
Cy 3-(2^G-glucosylrutinoside) caffeic acid ester	+	

From Maestro Durán and Vásquez Roncero. (1976)
Major anthocyanin patterns: (VII), (VIII).

Table A. 30 *Rutaceae ORANGE (Citrus sinensis (L.) Osbeck)*
Anthocyanins

Anthocyanins	Cultivar Moro Blood Orange
Pn 5-glucoside	+
Cy 3-glucoside	+
Cy 3,5-diglucoside	+
Pt 3-glucoside	+
Dp 3-glucoside and two unidentified pigments	+

From Licastro and Bellomo (1974).

Anthocyanins	Cultivar Moro
Cy 3-glucoside	+++
Dp, Pt, Pn, Pg mono- and diglucoside	+
Cy 3-(4″-acetyl)glucoside	+

From Maccarone *et al.* (1983).
Major anthocyanin patterns: (II), (VIII).

Carotenoids

Carotene distribution in pulp and peel of different orange cultivars

	Valencia			Washington Navel				Shamout
	Pulp[a]	Pulp[c]	Peel[a]	Pulp[b]	Pulp[c]	Peel[b]	Peel[d]	Pulp[c]
Carotenoid Pattern	% of Total Carotenoids							
Phytoene	4.0	4.2	3.1	4.3	5.6	12.1	—	5.7
Phytofluene	13.0	2.0	6.1	4.4	2.6	7.3	6.9	2.4
α-Carotene	0.5	1.0	0.1	0.1	0.7	0.03	0.2	0.4
β-Carotene	1.1	2.0	0.3	0.5	0.8	0.2	0.5	1.3
ζ-Carotene	5.4	3.0	3.5	8.5	1.8	8.4	0.8	1.4
Mutatochrome	—	tr.	—	—	—	—	0.3	—
α-Cryptoxanthin	1.5	3.5	0.3	0.5	1.8	0.2	—	1.5
Cryptoxanthin 5,6-epoxide	—	—	0.4	0.9	—	0.9	1.0	—
Cryptoxanthin	5.3	10.6	1.2	10.0	12.5	3.1	4.0	15.0
Cryptoflavin	0.5	0.3	2.0	0.4	0.5	4.7	8.0	0.7
Cryptoxanthin-diepoxides	—	—	—	—	0.2	—	3.4	0.1
Lutein	2.9	9.0	1.2	1.3	11.2	0.6	4.1	5.2
Isolutein	—	2.0	—	—	4.7	—	2.4	4.0
Chrysanthemaxanthin	—	—	—	—	—	0.2	—	—
Citraurin	—	1.9	—	—	1.3	5.0	5.8	2.6
Zeaxanthin	4.5	10.2	0.8	1.5	4.8	0.8	0.8	6.2
Antheraxanthins	5.8	9.3	6.3	11.6	6.2	3.1	—	6.8
Mutatoxanthins	6.2	14.0	1.7	0.5	13.0	0.6	1.5	15.0
Violaxanthins	7.4	1.8	44.0	45.6	1.7	26.4	17.0	3.0
Luteoxanthins	17.0	11.5	16.0	2.5	11.0	21.5	34.2	11.0
Auroxanthin	12.0	1.2	2.3	—	4.3	0.3	10.5	0.6
Shamoutixanthin	2.9	6.5	0.5	1.2	9.1	0.8	—	9.6
Shamoutichrome	4.0	2.8	0.9	0.2	3.5	1.8	—	1.5
Neoxanthins	—	2.5	—	0.5	2.2	—	1.7	4.0
Neochrome	—	—	—	—	—	—	2.6	—
Sinensiaxanthin	4.8	—	5.7	4.7	—	5.8	—	—
Total carotenoids (μg g^{-1} or ml)	15	12	120	23	6	67	50	8

[a] Curl and Bailey (1965); [b] Curl and Bailey (1961); [c] Gross et al. (1972); [d] Malachi et al. (1974).
Major carotenoid patterns: pulp, (4), (5), (6); peel, (5), (6).
Vitamin A value: in juice, 90, 140 and 200 μg RE l^{-1}, respectively.

Table A. 31 *Caricaceae PAPAYA (Carica papaya* L.)

Carotenoids

Carotenoid distribution in the yellow-fleshed Indian papaya

Carotenoid Pattern	% of Total Carotenoids
Phytoene	0.1
Phytofluene	0.1
β-Carotene	29.6
cis-β-carotene	0.3
Pigment-X	0.1
ζ-Carotene	2.9
γ-Carotene	0.1
β-Carotene 5,6-epoxide	0.2
Mutatochrome	0.1
Aurochrome	0.1
Cryptoxanthin	48.2
Cryptoflavin	12.9
Violaxanthin	3.4
cis-Violaxanthin	0.1
Unidentified—400	0.4
Unidentified—420	1.8
Antheraxanthin	0.1
Chrysanthemaxanthin	0.8
Neoxanthin	0.2
Total carotenoids (μg(g fresh wt)$^{-1}$)	13.8

From Subbarayan and Cama (1964); Yamamoto (1964). Major carotenoid patterns: yellow-fleshed, (4); red-fleshed, (3), (4).
Vitamin A values: yellow-fleshed, 1400–2000 μg RE kg^{-1}; red-fleshed 1160 μg RE kg^{-1}.

Table A. 32 *Passifloraceae PASSION FRUIT* (*Passiflora edulis* Sims.)

Anthocyanins

Anthocyanins	Rind	Pulp
Pg 3-glucosylglucoside[a]	+	−
Dp 3-glucoside[a]	−	+
Cy 3-glucoside[b]		

[a] From Harborne (1967).
[b] From Ishikura and Sugahara (1979).
Major anthocyanin pattern: (I).

Carotenoids

Main carotenoids in juice of passion fruit

Carotenoid Pattern	% of Total Carotenoids
Phytofluene	29
ζ-Carotene	37
β-Carotene	34
Total carotenoids (μg ml^{-1})	6.6

Additional minor pigments: β-apo-12′-carotenal, β-apo-8′-carotenal, cryptoxanthin, auroxanthin and mutatoxanthin.
From Leuenberger and Thommon (1972).
Major carotenoid pattern: uncommon?
Vitamin A value: 370 μg RE l^{-1}

Table A. 33 *Rosaceae PEACH (Prunus persica* L.)
Anthocyanins

	Cultivars		
Anthocyanins	*Elberta*	*Star King*	*Golden East*
Cy 3-glucoside	+	+	+

From Hsia *et al.* (1965).

Anthocyanins	*% of Total*
Cy 3-glucoside	90
Cy 3-rutinoside	10

From Ishikura (1975).
Major anthocyanin pattern: (I).

Carotenoids

Quantitative changes of carotenoid pattern in the mesocrp of Redhaven peaches during ripening

Carotenoid Pattern	22 June	29 June	10 July	18 July	25 July	1 Aug.	9 Aug.	17 Aug.	21 Aug.
				% of Total Carotenoids					
Phytofluene	—	—	—	—	—	1.0	5.7	8.7	10.3
β-Carotene	16.5	13.4	11.2	10.2	9.4	7.8	7.4	5.1	4.5
ζ-Carotene 4,9	3.4	—	—	—	—	—	—	—	—
Cryptoxanthin	—	—	0.9	2.0	2.7	4.5	8.6	10.3	14.5
Cryptoxanthin 5,6 monoepoxide	—	—	0.6	3.8	3.2	0.9	2.5	2.5	3.2
Cryptoxanthin 5,6-monoepoxide	—	—	1.3	—	—	0.7	—	1.8	0.7
Cryptoxanthin 5,6,5',8'-diepoxide	—	—	1.1	—	—	—	—	—	0.6
Lutein	17.5	17.0	13.9	10.5	7.6	3.8	2.2	2.4	2.0
Isolutein	1.4	—	4.8	3.9	—	13.2	8.3	9.1	7.6
trans-Luteoxanthin	3.9	3.6	—	—	9.6	—	—	—	—
Violaxanthin	31.5	37.8	41.1	48.3	49.2	54.6	51.6	38.4	34.6
Zeaxanthin	0.9	0.6	0.9	0.4	0.5	1.9	2.2	4.6	6.9
cis-Luteoxanthin	2.8	—	5.4	1.3	2.0	2.9	4.8	4.0	4.7
trans-Neoxanthin	22.0	23.7	18.0	18.0	14.3	8.6	5.7	4.9	4.5
cis-Neoxanthin	—	—	0.8	0.8	0.7	1.1	1.0	0.8	0.8
Neochrome	1.2	1.4	—	0.8	—	—	—	—	0.7
Unknown mixture	5.1	2.3	—	—	0.7	—	—	2.5	1.0
Total carotenoids (μg(g fresh wt)$^{-1}$)	6.2		4.8			9.0			4.7

From Gross (1979a).
Major carotenoid patterns: (4), (5) (modified).
Vitamin A value: 100 μg RE kg^{-1}.

Table A. 34 *Rosaceae PEAR* (*Pyrus communis* L.)
Anthocyanins

Anthocyanins	Pulp
Cy 3-galactoside	+

From Harborne (1967).

	Cultivars (in Peel)	
Anthocyanins	Painted Lady	Blakeney Red
Cy 3-galactoside	+	+

From Timberlake and Bridle (1971).

	Cultivars			
Anthocyanins	Devoe	Max-Red Bartlett	Seckel	Stark crimson
Cy 3-galactoside	+	+	+	+
Cy 3-arabinoside	+	+	+	+

From Francis (1970b).
Major anthocyanins pattern: (I).

Carotenoids

Chlorophyll and carotenoid distribution in the peel of two pear cultivars (Pyrus communis)

	Super Trévoux		Spadona
Cultivar	Unripe (Green)	Ripe (Yellow)	Ripe (Green)
Content (μg(g fresh wt)$^{-1}$)			
Chlorophyll a	29.3	3.2	31.7
Chlorophyll b	7.5	1.3	13.4
Total carotenoids	16.0	10.2	1.3
Carotenoid Pattern	*% of Total Carotenoids*		
Phytofluene	—	—	3.1
β-Carotene	23.6 (377.6)^a	6.0 (61.2)^a	19.0
β-Carotene 5,6-epoxide	—	1.3 (13.3)	—
Mutatochrome	—	—	3.4
Mixture (unidentified)	3.6 (57.6)	—	1.7
Cryptoxanthin	—	0.4	—
Cryptoxanthin 5,6-epoxide	—	0.2	—
Lutein	51.0 (816.0)	69.2 (705.8)	51.4
Zeaxanthin	1.5 (24.0)	2.8 (28.6)	3.0
Mutatoxanthin	1.3 (20.8)	—	—
Isolutein	—	1.9 (19.4)	1.2
Antheraxanthin	—	0.8	1.2
Luteoxanthin	3.7 (59.2)	1.0 (10.2)	3.8
trans-Violaxanthin	2.2 (35.2)	3.9 (67.3)	3.0
cis-Violaxanthin	—	2.7	3.6
Shamoutixanthin	1.5 (24.0)	1.6 (16.3)	—
trans-Neochrome	0.7	0.6	—
trans-Neoxanthin	9.7 (155.2)	3.8 (71.4)	4.3
cis-Neoxanthin	—	3.2	—
cis-Neochrome	1.2 (19.2)	0.5	—

^{a} μg (100 g fresh wt)$^{-1}$; isomers are totalized, trace amounts not included.
From Gross (1984).
Major carotenoid pattern: peel, (2).
Vitamin A value: 50–100 μg RE kg^{-1}.

Table A. 35 Solanaceae PEPPER (*Capsicum annuum* L.)

Carotenoids

Carotenoids in Capsicum annuum fruits at different stages of maturity

Carotenoid Pattern	% of Total Carotenoids					
	Green	Yellow	Pale Orange	Deep Red	Treated	CPTA
Phytofluene	—	6.3	3.6	4.8	3.9	19.0
α-Carotene	—	—	0.9	1.0	0.5	0.7
β-Carotene	18.3	9.0	7.7	6.8	6.9	2.1
β-Zeacarotene	—	—	tr.	3.8	2.1	0.7
ζ-Carotene	—	—	1.5	—	3.6	6.9
β-Carotene-5,6-epoxide	—	—	2.0	1.1	0.9	—
Lycopene	—	—	—	—	—	45.3
Cryptoxanthin	—	—	—	1.4	1.2	0.2
γ-Carotene	—	—	—	—	—	1.9
Neurosporene	—	—	—	—	—	0.7
Lutein-5,6-epoxide	1.8	2.3	—	—	—	—
Violaxanthin	22.0	41.3	44.7	32.4	22.9	6.9
cis-Lutein	—	5.9	3.2	4.6	0.8	—
trans-Lutein	43.0	4.8	5.4	4.6	1.3	8.2
Neoxanthin	14.9	8.0	5.2	4.9	2.4	—
cis-Zeaxanthin	—	—	—	4.2	6.2	—
trans-Zeaxanthin	—	4.6	1.3	9.3	6.2	4.1
Flavoxanthin	—	7.5	8.1	—	—	—
Capsanthin	—	8.8	10.5	14.2	30.7	—
Capsanthin-5,6-epoxide	—	—	1.7	2.5	6.3	—
Capsorubin	—	1.5	4.2	4.4	4.1	3.3
Total carotenoids (μg(g dry wt)$^{-1}$)	227.4	493.3	800.4	1353.2	2278.0	953.2

Fropm Valadon and Mummery (1977).
Major carotenoid patterns: (4), (6).
Vitamin A value: 6000 μg RE kg^{-1}.

Table A. 36 *Ebenaceae PERSIMMON* (*Diospyros kaki* Thunb.)

Carotenoids

Carotenoid distribution in three Japanese persimmon cultivars

	Cultivars		
	Fuyu	Honan Red	Tamopan
Total carotenoids (μg(g fresh wt)$^{-1}$)	65.0	80.0	61.0
Carotenoid Pattern	% of Total Carotenoids		
Phytofluene	4.2	2.6	0.5
α-Carotene	0.6	1.5	0.8
β-Carotene	4.6	6.4	7.5
ζ-Carotene	7.6	3.8	tr.
γ-Carotene	0.5	1.2	tr.
Lycopene	29.0	19.0	0.2
HO-α-carotene	2.3	0.1	2.0
Cryptoxanthin	31.0	38.0	28.0
Lutein	2.2	2.0	4.5
Zeaxanthin	6.8	11.0	12.0
Antheraxanthin	5.0	9.2	21.0
Mutatoxanthin	2.0	0.5	1.2
Violaxanthin	1.6	1.8	14.3
Luteoxanthin	—	—	1.7
Neoxanthin	1.0	1.0	2.0

From Brossard and Mackinney (1963).
Major carotenoid patterns: (3) and (3), (4), (5).
Vitamin A value: whole fruit, 2200–3500 μg RE kg^{-1}.

Carotenoids

Pigment changes in the peel of Persimmon (Diospyros kaki Thunb.) cv. Triumph during ripening on the tree

	Ripening Stages[a]						
	I	II	III	IV	V	VI	VII
Total content (μg(g fresh wt)$^{-1}$)							
Chlorophyll *a*	127.7	82.5	50.2	43.8	17.2	—	—
Chlorophyll *b*	31.6	21.5	13.2	12.1	5.0	—	—
Total carotenoids	96.4	70.4	45.6	58.3	66.2	92.0	310.0
Carotenoid Pattern	% of Total Carotenoids						
Phytofluene	—	—	—	—	—	—	4.2
α-Carotene	2.6	3.2	2.9	2.6	2.0	1.8	1.0
β-Carotene	17.9	16.7	13.8	11.3	9.3	9.7	3.4
ζ-Carotene	—	—	—	—	—	—	0.2
Mutatochrome	—	—	0.7	—	—	0.8	—
γ-Carotene	—	—	—	—	0.5	0.5	3.2
Lycopene	—	—	—	—	—	—	19.8
α-Cryptoxanthin	1.1	—	—	—	—	—	—
β-Cryptoxanthin	1.7	6.8	12.7	24.6	32.8	41.7	21.8
Cryptoxanthin 5,6-epoxide	—	—	—	1.1	1.4	2.3	2.2
Cryptoflavin	—	—	—	—	—	—	1.0
Lutein	30.8	23.3	21.3	12.1	8.8	6.5	9.1
Zeaxanthin	3.6	8.8	10.0	7.9	7.5	7.3	11.2
Mutatoxanthin	—	1.1	1.0	1.5	—	0.7	—
Isolutein	2.4	2.2	1.8	—	0.4	—	1.0
trans-Antheraxanthin	5.1	4.6	6.1	7.9	8.9	6.8	8.1
cis-Antheraxanthin	—	1.6	2.1	2.5	3.9	4.1	2.2
Luteoxanthin	2.0	6.0	3.7	4.2	—	2.4	—
trans-Violaxanthin	6.3	5.7	4.6	5.7	6.5	2.4	5.1
cis-Violaxanthin	9.1	6.2	7.5	7.9	8.8	4.2	4.4
Neoxanthin	15.4	13.8	11.8	10.7	9.2	8.8	5.2

[a] Ripening stages: at two-week intervals from unripe green I to harvest ripe VII.
From Ebert and Gross (1985).

Carotenoids

Carotenoid changes in pulp and peel of persimmon (Diospyros kaki Thunb.) cv. Triumph during post-harvest ripening

	Pulp			Peel		
	Ripening Stage[a]			Ripening Stage[a]		
	1	2	3	1	2	3
Total carotenoids (μg(g fresh wt)$^{-1}$)	18.3	59.0	68.0	128.0	366.0	491.0
Carotenoid Pattern	% of Total Carotenoids					
Phytofluene	—	—	1.1	—	—	0.4
α-Carotene	0.5	0.1	0.2	1.6	1.2	1.0
β-Carotene	8.6	2.9	3.9	9.4	7.6	6.7
ζ-Carotene	—	6.0	3.8	—	—	—
Mutatochrome	—	—	—	—	0.7	—
γ-Carotene	—	—	—	—	0.4	—
Lycopene	0.6	0.2	4.5	1.1	0.5	8.2
Cryptoxanthin	38.2	48.4	47.6	29.2	50.0	48.2
Cryptoxanthin 5,6-epoxide	2.7	4.0	3.9	0.9	1.2	1.9
Cryptoflavin	1.2	1.1	2.5	0.7	2.1	2.9
Lutein	0.4	0.4	0.2	12.4	5.5	4.1
Zeaxanthin	17.0	10.8	10.4	9.3	9.7	5.9
Mutatoxanthin	2.7	2.4	2.3	0.8	4.7	1.8
Isolutein	0.6	—	—	0.5	—	0.3
trans-Antheraxanthin	8.6	10.1	7.8	5.4	2.0	4.8
cis-Antheraxanthin	7.6	5.1	3.9	6.2	2.2	2.3
Luteoxanthin	1.3	0.5	1.8	1.7	4.8	1.9
trans-Violaxanthin	1.8	1.7	1.5	6.9	0.7	3.8
cis-Violaxanthin	2.9	2.5	2.4	6.7	1.5	2.0
Neoxanthin	3.6	3.8	2.2	7.2	5.2	3.0
Neochrome	1.7	—	—	—	—	0.8

[a] Ripening stage: 1 = harvest ripe; 2 = intermediate; 3 = fully ripe.
From Ebert (1984); Ebert and Gross (1985).
Major carotenoid patterns: (4) and (3), (4).
Vitamin A value: pulp, 3350 g RE kg^{-1}; peel, 9000–27 000 μg RE kg^{-1}.

Table A. 37 *Bromeliaceae PINEAPPLE (Ananas comosus* (L.) Merr.)

Carotenoids

Carotenoid distribution in fresh pineapple

Carotenoid Pattern	% of Total Carotenoids
Phytofluene	0.5
α-Carotene	0.5
β-Carotene	6.7
ζ-Carotene	0.9
Neurosporene	0.7
HO-α-Carotene	0.9
Cryptoxanthin	1.6
Lutein	2.5
cis-Lutein	1.0
Isolutein	0.3
Antheraxanthin	1.0
cis-Antheraxanthin	0.7
Flavoxanthin	0.3
Mutatoxanthin	0.6
cis-Violaxanthin	25.0
di-*cis*-Violaxanthin	27.0
Luteoxanthins	6.2
cis-Luteoxanthins	7.2
Auroxanthin	3.4
Neoxanthin	9.6
Shamoutixanthin-like	2.1
Neochromes	1.5
Sustance 407	0.8
Total carotenoids ($\mu g(g\ fresh\ wt)^{-1}$)	1.0

From Morgan (1966).
Major carotenoid patterns: (1), (5).
Vitamin A value: —.

Table A. 38 *Rosaceae PLUM (Prunus domestica* L.)

Anthocyanins

Cy 3-glucoside	+
Pn 3-glucoside	+
Cy 3-rutinoside	+
Pn 3-rutinoside	+

From Harborne (1967).

Carotenoids

Carotenoid distribution in three plum cultivars

Carotenoid Pattern	Saghiv (*Prunus domestica*)	Golden King (*Prunus salicina*)	Mirabelle (*Prunus insititia*)
	% of Total Carotenoids		
Phytofluene	7.7	6.0	5.3
α-Carotene	1.4	4.2	—
β-Carotene	25.0	33.6	16.4
Mutatochrome	—	3.2	1.8
Cryptoxanthin	2.7	6.8	2.9
trans-Cryptoflavin	—	—	1.1
Cryptoxanthin 5,6-epoxide	0.9	—	1.2
cis-Cryptoflavin	—	—	1.4
Cryptoxanthin 5′,6′-epoxide	0.6	—	0.6
Cryptoxanthin 5,6,5′,6′-epoxide	—	—	0.7
Cryptoxanthin 5,6,5′,8′-diepoxide	—	—	0.8
Violaxanthal	1.5	—	—
Lutein	10.2	21.9	23.4
Zeaxanthin	0.6	4.4	1.5
Chrysanthemaxanthin	—	1.2	—
Mutatoxanthin	0.5	3.0	1.4
Auroxanthin	—	0.7	—
Isolutein	—	—	1.0
trans-Antheraxanthin	1.2	—	3.2
cis-Antheraxanthin	1.8	—	—
Luteoxanthin	—	5.5	5.4
trans-Violaxanthin	4.8	—	6.3
cis-Violaxanthin	27.2	—	22.0
Persicachrome	0.8	—	—
Persicaxanthin	8.0	—	—
trans-Neochrome	—	3.7	—
cis-Neochrome	—	1.0	—
trans-Neoxanthin	2.5	3.9	3.6
cis-Neoxanthin	2.6	0.9	—
Total carotenoids (μg(g fresh wt)$^{-1}$	7.5	24.4	12.0

From Gross (1984).
Major carotenoid patterns: (2), (5), (8).
Vitamin A value: 340, 1730, and 410 μg RE kg^{-1}, respectively.

Table A. 39 *Punicaceae POMERGRANATE* (*Punica granatum* L.)

Anthocyanins

Anthocyanins	Seedcoats	Peels
Cy 3-glucoside[a]	+	+ +
Dp 3-glucoside	+	−
Cy 3,5-diglucoside	+	+ +
Dp 3,5-diglucoside	+	−
Pg 3-glucoside	+	>
Pg 3,5-diglucoside	+	>

[a] In order of decreasing amount in the seedcoats.
Du *et al.* (1975).
Major anthocyanin patterns: (II), (VI).

Carotenoids

Carotenoid Pattern	% of Total Carotenoids
Hydrocarbons	15.0
Monols	5.0
Diols and polyols	80.0
Total carotenoids (μg(g fresh wt)$^{-1}$)	0.2

From Curl (1964).
Major carotenoid patterns: (1), (2).
Vitamin A value: −.

Table A. 40 Rutaceae *PUMMELO WHITE* (*Citrus grandis* (L.) Osbeck.)

Carotenoids

Quantitative changes of endocarp pigments in the ripening pummelo Citrus grandis Goliath

	Ripening Stage[a]			
	1	2	3	4
Content (μg(g fresh wt)$^{-1}$)				
Chlorophyll *a*	0.9	0.5	0.3	0.2
Chlorophyll *b*	0.3	0.2	0.1	0.1
Total carotenoids	0.4	0.3	0.3	0.4

Carotenoid Pattern	% of Total Carotenoids			
Phytofluene	—	—	13.8	19.0
β-Carotene	28.8	26.5	22.9	12.5
Mutatochrome	—	1.8	3.4	6.1
Cryptoxanthin	—	1.7	0.3	4.1
Cryptoflavin	—	4.2	3.0	3.2
Unknown mixture	—	—	2.4	—
Lutein	41.7	34.1	30.4	23.0
Zeaxanthin	7.8	4.8	6.0	4.8
Mutatoxanthin	4.2	11.4	4.6	4.1
Auroxanthin	—	—	6.8	—
Isolutein	1.2	1.0	—	—
trans-Antheraxanthin	3.6	1.4	—	3.1
Valenciachrome	—	—	—	4.1
trans-Luteoxanthin	4.8	2.3	1.5	6.4
trans-Violaxanthin	4.8	—	0.9	1.9
cis-Violaxanthin	—	—	—	2.8
cis-Luteoxanthin	—	—	—	—
Neochrome	3.1	4.4	—	2.6
Shamoutixanthin	—	3.1	—	2.3
Shamoutichrome	—	3.3	—	—
Unknown mixture	—	—	4.0	—

[a] Ripening stage: 1 = unripe (green); 2 = colour break (green–yellow); 3 = almost ripe (pale yellow); 4 = fully ripe (lemon yellow).

Carotenoids

Quantitative changes of flavedo pigments in the ripening pummelo Citrus grandis Goliath

	Ripening Stage[a]			
	1	2	3	4
Content (μg(g fresh wt)$^{-1}$)				
Chlorophyll a	75.0	19.5	8.4	—
Chlorophyll b	16.5	6.0	3.1	—
Total carotenoids	26.0	9.1	9.5	5.0
Carotenoid Pattern	% of Total Carotenoids			
Phytofluene	—	8.4	12.4	67.2
β-Carotene	22.8	17.3	21.6	tr.
ζ-Carotene	—	tr.	—	3.5
Neurosporene	—	—	—	0.5
Mutatochrome	1.2	—	—	—
Unknown mixture	—	1.3	—	4.0
Cryptoxanthin	0.3	0.9	—	1.5
Cryptoflavin	—	—	—	3.0
Lutein	40.6	49.4	47.9	5.5
Zeaxanthin	2.8	0.7	2.3	1.1
Chrysanthemaxanthin	0.8	1.1	—	—
Isolutein	0.7	0.7	1.8	—
trans-Antheraxanthin	3.2	2.5	2.9	2.2
cis-Antheraxanthin	—	0.6	1.6	0.7
trans-Luteoxanthin	3.7	4.0	—	2.5
trans-Violaxanthin	10.8	3.7	2.0	2.6
cis-Violaxanthin	9.5	3.9	4.6	3.6
cis-Luteoxanthin	—	1.5	0.6	—
Neoxanthin	3.6	2.0	—	0.7
Neochrome	—	2.0	2.3	1.4

[a] Ripening stage: 1 = unripe (green) stage; 2 = color break (green–yellow); 3 = almost ripe (pale yellow); 4 = fully ripe (lemon-yellow).
From Gross *et al.* (1983).
Major carotenoid patterns: pulp, (2) (modified); peel, (2), (5) (uncommon) containing large amounts of colourless carotenes.
Vitamin A value: —.

Table A. 41 Rutaceae *PUMMELO RED* (*Citrus grandis*)
Carotenoids

Carotenoid distribution in different fruit parts of the red pummelo Chandler

	Endocarp		Exocarp	
	Pulp	Septa	Flavedo	Albedo
Carotenoid Pattern	% of Total Carotenoids			
α-Carotene	—	—	3.2	—
β-Carotene	5.1	6.2	12.0	1.3
Lycopene	90.4	89.7	0.9	95.1
Mixture (unidentified)	1.0	4.1	—	1.3
Cryptoxanthin	—	—	2.2	—
Lutein	2.1	—	31.8	2.0
Zeaxanthin	0.3	—	8.6	—
Isolutein	—	—	1.4	—
trans-Antheraxanthin	—	—	3.0	—
cis-Antheraxanthin	—	—	6.4	—
Luteoxanthin	—	—	5.4	—
trans-Violaxanthin	1.1	—	14.2	0.5
cis-Violaxanthin	—	—	8.0	—
Neoxanthin	—	—	2.2	—
Total carotenoids (μg(g fresh wt)$^{-1}$)	13.2	53.0	19.3	24.7

From Gross (unpublished data. 1983).
Major carotenoid patterns: pulp, septa, and albedo, (3); flavedo, (2), (5).
Vitamin A value: 100 μg RE kg^{-1} pulp.

Table A. 42 Rosaceae QUINCE (*Cydonia oblonga* Mill.)

Carotenoids

Carotenoid distribution in the peel of quince (Cydonia oblonga) Portugal

Carotenoid Pattern	% of Total Carotenoids
Phytofluene	43.5
β-Carotene	9.2
ζ-Carotene	2.1
Cryptoxanthin	4.3
Cryptoflavin	3.6
Cryptoxanthin 5,6,5',6'-diepoxide	1.6
Lutein	7.1
Zeaxanthin	3.1
Isolutein	0.3
Antheraxanthin	0.6
Luteoxanthin	2.1
trans-Violaxanthin	8.3
cis-Violaxanthin	7.0
Neoxanthins	7.2
Total carotenoids (μg(g fresh wt)$^{-1}$)	6.8

From Gross (unpublished results, 1980).
Major carotenoid pattern: (2) (modified) uncommon containing large amounts of colourless carotenes.
Vitamin A value: 150 μg RE kg^{-1}.

Table A. 43 Rosaceae RASPBERRY (*Rubus idaeus* L.)
Anthocyanins

(a)

Anthocyanins	Cultivars		
	September	Tweed	Preussen
Cy, Pg 3-glucoside	+	+	+
Cy, Pg 3-rutinoside	+	+	+
Pg, Cy 3-sophoroside	+	+	+
Pg, Cy 3-glucosylrutinoside	+	+	+

From Harborne and Hall (1964).

(b)

Anthocyanins	Cultivars	
	Durham	Heritage
Cy 3-sophoroside	+ + + +	+
Cy 3-glucoside	+ + +	+
Cy 3-rutinoside	+ +	−
Cy 3-glucosylrutinoside	+	−

From Francis (1972).

Anthocyanin changes during ripening of red raspberry cv. Meeker

(c)

Anthocyanins	Incipient	Pink	Prime ripe	Over-ripe
	% of Total			
Cy 3-glucoside	15	19	18	23
Cy 3-rutinoside	8	7	7	7
Cy 3-sophoroside	62	56	52	50
Cy 3-glucosylrutinoside	15	18	23	20
Cy 3,5-diglucoside	—	—	tr.	tr.
Pg 3-glucoside	—	—	—	tr.
Pg 3-sophoroside	—	tr.	tr.	tr.
Pg 3-glucosylrutinoside	—	—	—	tr.

tr. = trace amounts.
From Barritt and Torre (1975).

Anthocyanins

Anthocyanin distribution in red raspberries

Anthocyanins	Cultivars							
	New Hampshire	August Red	NHTx P15	Willamette	Lloyd George	Meeker	Glen Clova	Puyallup
	mg (100 g^{-1})							
Cy 3-glucoside	2.9	3.7	9.2	16.8	7.7	10.1	7.6	3.8
Cy 3-rutinoside	3.4	7.1	3.6	—	2.2	2.7	2.5	4.1
Cy 3-sophoroside	8.9	4.7	30.4	35.6	30.4	18.6	16.6	11.3
Cy 3-glucosyl-rutinoside	9.5	6.7	6.2	—	10.6	8.3	6.9	9.3
Cy 3,5-diglucoside	—	—	—	2.4	—	—	—	—
Pg 3-sophoroside	tr.	0.4	2.5	4.3	—	—	—	—
Pg 3-glucosyl-rutinoside	0.8	0.8	—	—	—	—	—	—
Total anthocyanins	25.5	23.4	51.9	59.1	40.9	39.7	33.6	28.5

tr. = trace amounts.
From Torre and Barritt (1977).
Major anthocyanin patterns: (V), (VII).

Table A. 44 *Rosaceae STRAWBERRY*
(*Fragaria* × *ananassa* Duch.)

Anthocyanins

Anthocyanins	
Pg 3-glucoside	+
Cy 3-glucoside	+

From Harborne (1967).

	% *of Total*
Pg 3-glucoside	70
Cy 3-glucoside	30

From Ishikura (1975).

Pg 3-glucoside
Cy 3-glucoside
Pg 3-glucoside isomer

From Wrolstad *et al.* (1970a).
Major anthocyanin pattern: (II).

Carotenoids

Quantitative changes of chlorophylls and carotenoids in developing strawberry fruits (Fragaria × ananassa) cv. Tenira

	Unripe	Partly Ripe	Fully Ripe
Content (μg(g fresh wt)$^{-1}$)			
Chlorophyll *a*	5.0	2.2	0.36
Chlorophyll *b*	1.8	0.7	0.2
Total carotenoids	1.9	0.8	0.4
Carotenoid Pattern	(μg(100 g)$^{-1}$)		
β-Carotene	43.7	17.5	3.4
Lutein	83.0	37.0	20.6
Zeaxanthin	8.6	3.0	—
Chrysanthemaxanthin	3.6	1.7	—
Isolutein	1.6	1.1	0.7
Antheraxanthin	3.8	2.2	2.4
Luteoxanthin	7.8	4.9	1.2
trans-Violaxanthin	8.7	2.6	3.4
cis-Violaxanthin	3.2	3.0	5.2
Neochrome	7.4	—	—
Neoxanthin	18.6	7.0	3.1
Total carotenoids	190.0	80.0	40.0

From Gross (1982c).
Major carotenoid pattern: (2) (modified).
Vitamin A value: —.

Table A. 45 *Solanaceae TOMATO (Lycopersicon esculentum* L.)
Carotenoids

Carotenoid distribution in three red tomato cultivars

Carotenoid Pattern	Early Redchief[a]	Summer Sunrise[b]	New Yorker[c]
	% of Total Carotenoids		
Phytoene	19.7	9.2	25.0
Phytofluene	8.4	7.8	5.0
β-Carotene	2.3	6.9	5.7
ζ-Carotene	2.0	1.7	1.1
γ-Carotene	0.5	0.5	0.8
Neurosporene	0.4	0.1	0.2
Lycopene	66.7	73.7	62.2
Total carotenoids (μg(g fresh wt)$^{-1}$)	950.0	1700.0	922.0

Recalculated from: [a] Koskitalo and Ormrod (1972);
[b] Raymundo *et al.* (1976);
[c] Lee and Robinson (1980).
Major carotenoid pattern: (3).
Vitamin A value: 200 μg RE kg^{-1}.

Carotenoids

Changes of main carotenes of ripening Homestead tomatoes

Carotenoid Pattern	Ripening Stages[a]					
	1	2	3	4	5	6
	% of Total Carotenoids					
Phytoene	—	—	20.0	10.7	10.8	9.0
Phytofluene	—	—	2.5	2.1	2.6	2.4
β-Carotene	100	75.0	55.0	17.0	9.8	2.6
ζ-Carotene	—	—	2.5	1.4	1.6	0.6
γ-Carotene	—	8.3	10.0	4.3	2.3	0.7
Lycopene	—	16.7	10.0	64.3	72.9	84.7
Total carotenes (μg(g fresh wt)$^{-1}$)	1.2	2.4	4.0	14.0	30.6	97.7

[a] Ripening stages: 1 = mature green; 2 = breaker; 3 = turning; 4 = pink; 5 = light red; 6 = red.
From Meredith and Purcell (1966).

Table A. 46 Cucurbitaceae *WATERMELON* (*Citrullus lanatus* (Thunb.)

Carotenoids

Carotenes in pulp of watermelons

Carotenoid Pattern	% of Total Carotenoids	
	Candy Red (Red-fleshed)	Orange Flesh Tendersweet (Orange-fleshed)
Phytoene	2.1	15.5
Phytofluene	1.4	5.9
α-Carotene	0.1	—
β-Carotene	4.0	4.1
cis-β-Carotene	0.1	—
ζ-Carotene	1.6	16.8
Mixture (unknown)	tr.	—
Proneurosporene	tr.	6.5
Mutatochrome	tr.	—
Pro-γ-carotene	tr.	—
Prolycopene	tr.	46.8
cis-γ-Carotene	0.4	—
Neurosporene	0.2	—
Neo-lycopene	7.6	—
Lycopene	73.7	4.4
Total carotenoids (μg(g fresh wt)$^{-1}$)	25.0	34.0

From Morgan (1963); Tomes and Johnson (1965).
Major carotenoid patterns: red, (3); orange, (7).
Vitamin A value: 80–370 μg RE kg^{-1}.

References

Aasen, A. J., Liaaen-Jensen, S. and Borch, G. (1971). The chirality of zeaxanthin from different sources. *Acta Chem. Scand.* **26**, 404–405.

Abdel-Gawad, H. and Romani, R. J. (1974). Hormone-induced reversal of colour change and related respiratory effects of ripening apricot fruits. *Physiol. Plant.* **32**, 161–165.

Abdel-Kader, A. S., Morris, L. L. and Maxie, E. C. (1966). Effect of growth regulating substances on the ripening and shelf-life of tomatoes. *Hortscience* **1**, 90–91.

Abou Aziz, A. B., El-Nabawy, S. M. and Zaki, H. A. (1975). Effect of different temperatures on the storage of papaya fruits and respirational activity during storage. *Sci. Hortic.* **3**, 173–177.

Aczél, A. (1973). Change of the carotenoid contents of apricot during canning process and storage. *Lebensm.-Wiss. u. Technol.* **6**, 36–37.

Adamovics, J. and Steinitz, F. R. (1976). High performance liquid chromatography of some anthocyanidins and flavonoids. *J. Chromatogr.* **129**, 464–465.

Adsule, P. G. and Roy, S. K. (1974). Studies on the physico-chemical characters of some important commercial varieties of mango of North India in relation to canning and freezing of slices. *J. Food Sci. Technol.* **11**, 269–273.

Aharoni, Y. (1968). "The effect of storage in modified atmospheres on the physiological process in the 'Shamouti' orange fruit (*Citrus sinensis* (L.) Osbeck." Ph.D. Thesis, The Hebrew University of Jerusalem (1968).

Aharoni, Y. and Houck, L. G. (1980). Improvement of internal color of oranges stored in oxygen-enriched atmospheres. *Sci. Hort.* **13**, 331–338.

Aljuburi, H., Huff, A. and Hsich, M. (1979). Enzymes of chlorophyll catabolism in orange flavedo. *Plant Physiol.* **63**, S–73.

Altman, L. J., Ash, L., Kowerski, R. C., Epstein, W. W., Larson, B. R., Rilling, H. C., Muscio, F. and Gregonis, D. E. (1972). Prephytoene pyrophosphate, a new intermediate in the biosynthesis of carotenoids. *J. Am. Chem. Soc.* **94**, 3257–3259.

Anderson, D. W., Guelfroy, D. E., Webb, A. D. and Kepner, R. E. (1970). Identification of acetic acid as acylating agent of anthocyanin pigments in grapes. *Phytochemistry*, **9**, 1579–1583.

Anderson, J. M., Waldron, J. C. and Thorne, S. W. (1978). Chlorophyll–protein complexes of spinach and barley thylakoids. *FEBS Lett.* **92**, 227–233.

Andrewes, A. G., Borch, G. and Liaaen-Jensen, S. (1974). On the absolute configuration of lutein. *Acta Chem. Scand.* **B28**, 139–140.

Aoki, S., Araki, C., Kaneo, K. and Katayama, O. (1970). L-phenylalanine ammonia-lyase activities in Japanese chestnut, strawberry, apple fruit and bracken. *J. Food Sci. Technol.* **17**, 507–511.

Arnon, D. I. (1949). Copper enzymes in isolated chloroplasts. Polyphenoloxidase in *Beta vulgaris*. *Plant Physiol.* **24**, 1–15.

REFERENCES

Bacon, M. T. (1965). Separation of chlorophyll *a* and *b* and related compounds by thin-layer chromatography on cellulose. *J. Chromatogr.* **17**, 322–326.

Ballinger, W. E., Maness, E. P. and Kushman, L. J. (1970). Anthocyanins in ripe fruit of the highbush blueberry. *J. Am. Soc. hort. Sci.* **95**, 283–285.

Ballinger, W. E., McClure, W. F., Nesbitt, W. B. and Maners, E. P. (1978). Light-sorting muscadine grapes (*Vitis rotundifolia*) Mich x for ripeness. *J. Am. Soc. hort. Sci.* **103**, 629–634.

Ballinger, W. E., Galletta, G. J. and Maness, E. P. (1979). Anthocyanins of fruits of *Vaccinium*, subgenera *Cyanococcus* and *Polycodium*. *J. Am. Soc. hort. Sci.* **104**, 504–557.

Banet, E., Romojaro, F. and Llorente, S. (1981). Evolution of photosynthetic pigments in flavedo and pulp of Marsh grapefruit. *An. Edaf. Agrobiol.* **XL**, 259–267.

Barmore, C. R. (1975). Effect of ethylene on chlorophyllase activity and chlorophyll content in calamondin rind tissue. *Hortscience*, **10**, 595–596.

Barrett, H. C. and Rhodes, A. M. (1976). A numerical taxonomic study of affinity relationship in cultivated *Citrus* and its close relatives. *Syst. Bot.* **1**, 105–136.

Barritt, B. H. and Torre, L. C. (1973). Cellulose thin-layer chromatographic separation of *Rubus* fruit anthocyanins. *J. Chromatogr.* **75**, 151–153.

Barritt, B. H. and Torre, L. C. (1975). Fruit anthocyanin pigments of red raspberry cultivars. *J. Am. Soc. hort. Sci.* **100**, 98–100.

Bartlett, L., Klyne, W., Mose, W. P., Scopes, P. M., Galasko, G., Mallams, A. K., Weedon, B. C. L., Szabolcs, J. and Toth, G. (1969). Optical rotatory dispersion of carotenoids. *J. Chem. Soc. C* **3**, 2527–2544.

Bate-Smith, E. C. (1948). Anthocyanins and flavones. *Biochem. J.* **43**, XLIX–L.

Bauernfeind, J. C. (Ed.) (1981). "Carotenoids as Colorants and Vitamin A Precursors." Academic Press, New York.

Bean, R. C. and Todd, G. W. (1960). Photosynthesis and respiration in developing fruits. I. $C^{14}O_2$ uptake by young oranges in the light and in the dark. *Plant Physiol.* **35**, 425–429.

Bean, R. C., Porter, G. G. and Barr, B. K. (1963) Photosynthesis and respiration in developing fruits. III. Variations in photosynthetic capacities during color change in citrus. *Plant Physiol.* **38**, 285–290.

Ben-Arie, R., Gross, J. and Sonego, L. (1982/83). Changes in ripening-parameters and pigments of the Chinese gooseberry (kiwi) during ripening and storage. *Sci. Hortic.* **18**, 65–70.

Ben-Arieh, H. and Guelfat-Reich, S. (1975). Early ripening of nectarines induced by succinic acid 2,2-dimethyl hydrazide (SADH) and 2-chloroethylphosphonic acid (CEPA). *Coll. Int. CNRS Paris*, No. 238, 109–113.

Ben-Aziz, A., Britton, G. and Goodwin, T. W. (1973) Carotene epoxides of *Lycopersicon esculentum*. *Phytochemistry* **12**, 2759–2764.

Ben-Shaul, Y. and Naftali, Y. (1969). The development and ultrastructure of lycopene bodies in chromoplasts of *Lycopersicum esculentum*. *Protoplasma* **67**, 333–344.

Beyer, P., Kreuz, K. and Kleinig, H. (1980). β-Carotene synthesis in isolated chromoplasts from *Narcissus pseudonarcissus*. *Planta* **150**, 435–438.

Beyer, P., Kreuz, K. and Kleinig, H. (1982). The site of carotenogenic enzymes in chromoplasts from *Narcissus pseudonarcissus* L. *Planta* **154**, 66–69.

Bickel, H. and Schultz, G. (1976). Biosynthesis of plastoquinone and β-carotene in isolated chloroplasts. *Phytochemistry* **15**, 1253–1255.

Birkofer, L., Kaiser, C. and Donike M. (1966). Trennung von Anthocyangemichen durch Komplexbildung an Aluminiumoxid. *J. Chromatogr.* **22**, 303–307.

REFERENCES

Blanke, M. M., Notton, B. A. and Hucklesby, D. P. (1986). Physical and kinetic properties of phosphoenolpyruvate carboxylase in developing apple fruit. *Phytochemistry* **25**, 601–606.

Boardman, N. K. and Thorne, S. W. (1971). Sensitive fluorescence method for the determination of chlorophyll a/chlorophyll b ratio. *Biochim. Biophys. Acta.* **253**, 228–231.

Bodea, C., Andrewes, A. C., Borch, G. and Liaaen-Jensen, S. (1978). Structure of the carotenoid physoxanthin. *Phytochemistry* **17**, 2037–2038.

Bogorad, L. (1976). Chlorophyll biosynthesis. In "Chemistry and Biochemistry of Plant Pigments" (T. W. Goodwin, Ed.) Vol. 1, 64–148. Academic Press, New York and London.

Braithwaite, G. D. and Goodwin, T. W. (1960). The incorporation of 1-^{14}C acetate, 2-^{14}C acetate and $^{14}CO_2$ into lycopene by tomato slices. *Biochem. J.* **26**, 1–5.

Bramlage, W. J., Devlin, R. M. and Smagula, J. M. (1972). Effect of preharvest application of ethephon on 'Early Black' cranberries. *J. Am. Soc. hort. Sci.* **97**, 625–628.

Braumann, Th. and Grimme, L. H. (1981). Reversed-phase high performance liquid chromatography of chlorophylls and carotenoids. *Biochim. Biophys. Acta.* **637**, 8–17.

Braumann, Th., Weber, G. and Grimme, L. H. (1982). Carotenoid and chlorophyll composition of light-harvesting and reaction centre proteins of the thylakoid membrane. *Photobiochem. Photobiophys.* **4**, 1–8.

Brisker, H. E., Goldschmidt, E. E. and Goren, R. (1976). Ethylene-induced formation of ABA in citrus peel. *Plant Physiol.* **58**, 377–379.

Britton, G. (1979). Carotenoid biosynthesis—a target for herbicide activity. *Z. Naturforsch.* **34C**, 979–985.

Britton, G. (1982). Carotenoid biosynthesis in higher plants. *Physiol. Vég.* **20**, 735–755.

Britton, G. and Goodwin, T. W. (1971) Biosynthesis of carotenoids. In "Methods in Enzymology", Vol. 18 C, 654–701.

Brossard, J. and Mackinney, G. (1963). The carotenoids of *Diospyros kaki* (Japanese persimmons). *J. Agric. Food Chem.* **11**, 501–503.

Bruinsma, J., Knegt, E. and Varga, A. (1975). The role of growth regulating substances in fruit ripening. *Coll. Int. CNRS Paris*, No. 238, 193–200.

Bubicz, M. (1965). Occurrence of carotenoids in fruits of *Berberis*. *Bull. Acad. Polon. Sci. Ser. Sci. Biol.* **13**, 251–255.

Buchecker, R. and Eugster, C. H. (1979). Search for the presence in egg yolk, in flowers of *Caltha palustris* and in autumn leaves of 3'-epilutein ($=$(3R, 3'5, 6'R)-β, ε-carotene-3,3'-diol) and 3'-O-didehydrolutein ($=$(3R, 6'R)-3-hydroxy-β, ε-carotene-3'-one). *Helv. Chim. Acta* **62**, 2817–2824.

Buchecker, R., Liaaen-Jensen, S. and Eugster, C. H. (1975). The identity of trollixanthin and trolliflor with neoxanthin. *Phytochemistry* **14**, 797–799.

Buchecker, R., Eugster, C. H. and Weber, A. (1978). Absolute configuration of α-doradexanthin and of fritschiellaxanthin, a new carotenoid from *Fritschiella tuberosa* Iyeng. *Helv. Chim. Acta* **61**, 1962–1968.

Budowski, P. and Gross, J. (1965). Conversion of carotenoids to 3-dehydroretinol (vitamin A_2) in the mouse. *Nature* **206**, 1254–1255.

Budowski, P., Ascarelli, I., Gross, J. and Nir, I. (1963). Provitamin A_2 from lutein. *Science* **142**, 969–971.

Burg, S. P. and Burg, E. A. (1962). Role of ethylene in fruit ripening. *Plant Physiol.* **37**, 179–189.

Bushway, R. J. and Wilson, A. M. (1982). Determination of α- and β-carotene in fruit and vegetables by high-performance liquid chromatography. *Can. Inst. Food. Sci. Technol. J.* **15**, 165–169.

Cabibel, M. and Ferry, P. (1979). Evaluation de la précision de la mesure de quelques caractéristiques physicochimiques de la tomate et choix de critères de maturité. *C.R. Acad. Agric.* **65**, 465–471.

Cabibel, M. and Ferry, P. (1980) Evolution de la teneur en caroténoïdes de la tomate en fonction des stades de maturation et des conditions culturales. *Ann. Technol. Agric.* **29**, 27–45.

Cabibel, M., Lapize, F. and Ferry, P. (1981). Determination of carotenes in tomatoes by high performance liquid chromatography. Application to some varieties of tomatoes. *Sci. Alim.* **I(4)**, 489–500.

Calvarano, I., Leuzzi, U. and Di Giacomo, A. (1972). Ricerche sulla composizione dei kumquats. *Essenze Deriv. Agr.* **43**, 131–142.

Camara, B. (1980). *In vitro* conversion of violaxanthin to capsorubin by a chromoplast enriched fraction of *Capsicum* fruits. *Biochem. Biophys. Res. Commun.* **93**, 113–117.

Camara, B. and Monéger, R. (1978). Free and esterified carotenoids in green and red fruits of *Capsicum annuum*. *Phytochemistry* **17**, 91–93.

Camara, B. and Brangeon, J. (1981). Carotenoid metabolism during chloroplast to chromoplast transformation in *Capsicum annuum* fruit. *Planta* **151**, 359–364.

Camara, B. and Monéger R. (1981). Carotenoid biosynthesis: *in vitro* conversion of antheraxanthin to capsanthin by a chromoplast enriched fraction of Capsicum fruits. *Biochem. Biophys. Res. Commun.* **99**, 1117–1122.

Camara, B. and Monéger, R. (1982). Biosynthetic capabilities and localization of enzymatic activities in carotenoid metabolism of *Capsicum annuum* isolated chromoplasts. *Physiol. Vég.* **20**, 757–773.

Camara, B., Payan, C., Escoffier, A. and Monéger, R. (1980) Les caròtenes du fruit du poivron (*Capsicum annuum*): mise en évidence du phytoène *trans* et du phytofluène *trans*; comparaison de la biosynthèse des différent carotènes. *C.R. Acad. Sci.* **291**, Serie D, 303–306.

Camara, B., Bardat, F. and Monéger, R. (1982a). Nature of the first C_{40} precursor in pepper (*Capsicum annuum* L.) fruit chromoplasts. *C.R. Acad. Sci.* **294**, Serie III, 649–652.

Camara, B., Bardat, F. and Monéger, R. (1982b) Sites of biosynthesis of carotenoids in *Capsicum* chromoplasts. *Eur. J. Biochem.* **127**, 255–258.

Camara, B., Bardat, F., Dogbo, O., Brangeon, J. and Monéger, R. (1983). Terpenoid metabolism in plastids. Isolation and biochemical characteristics of *Capsicum annuum* chromoplasts. *Plant Physiol.* **73**, 94–99.

Camire, A. L. and Clydestale, F. M. (1979). High pressure liquid chromatography of cranberry anthocyanins. *J. Food Sci.* **44**, 962–927.

Camm, E. L. and Towers, G. H. N. (1973). Phenylalanine ammonia-lyase. *Phytochemistry* **12**, 961–973.

Caprio, J. M. (1956). An analysis of the relation between regreening of Valencia oranges and mean monthly temperatures in southern California. *Proc. Am. Soc. hort. Sci.* **67**, 222–235.

Cecchi, H. and Rodriguez-Amaya, D. (1981). Carotenoids and vitamin A value in processed passion-fruit juice. *J. Ciencia Cult.* **33**, 72–76.

Chalmers, D. J. and Faragher, D. J. (1977). Regulation of anthocyanin synthesis in apple skin. II. Involvement of ethylene. *Aust. J. Plant Physiol.* **4**, 123–131.

Chalmers, D. J., Faragher, J. D. and Raff, J. W. (1973). Changes in anthocyanin synthesis as an index of maturity in red apple varieties. *J. hort. Sci.* **48**, 387–392.

Chen, L.-J. and Hrazdina, G. (1981). Structural aspects of anthocyanin–flavonoid complex formation and its role in plant color. *Phytochemistry* **20**, 297–303.
Chen, L.-J. and Hrazdina, G. (1982). Structural transformation reactions of anthocyanins. *Experientia* **38**, 1030–1032.
Cholnoky, L. and Szabolcs, J. (1960). Bemerkungen zur Struktur des Capsorubins. *Experientia* **16**, 483–484.
Cholnoky, L., Györgyfy, K, Nagy, E. and Panczel, M. (1955). Carotenoid pigments I. Pigments of red pepper. *Acta Chim. Acad. Sci. Hung.* **6**, 143–171.
Chuma Y., Nakagi, N. and Tagawa, A. (1982). Delayed light emission as a means of automatic sorting of tomatoes. *J. Fac. Agr. Kyushu Univ.* **26**, 221–234.
Clijsters, H. (1975). Relation possible entre l'activité photosynthetique de la pomme et sa maturation. *Coll. Int. CNRS Paris*, No. 238, 41–47.
Clough, J. M. and Pattenden, G. (1979). Naturally occurring poly-*cis*-carotenoids. Stereochemistry of poly-*cis*-lycopene and its congeners in 'Tangerine' tomato fruits. *J.C.S. Chem. Commun.* 616–619.
Coggins, C. W. and Lewis, L. H. (1962). Regreening of Valencia orange as influenced by potassium gibberellate. *Plant Physiol.* **37**, 625–627.
Coggins, C. W. and Hield, H. Z. (1962). Navel orange fruit response to gibberellate. *Proc. Am. Soc. hort. Sci.* **81**, 227–230.
Coggins, C. W., Hall, A. E. and Jones, W. W. (1981) The influence of temperature on regreening and carotenoid content of Valencia orange rind. *J. Am. Soc. hort. Sci.* **106**, 251–254.
Cohen, E. (1978). The effect of temperature and relative humidity during regreening on the colouring of Shamouti orange fruit. *J. hort. Sci.* **53**, 143–146.
Conradie, J. D. and Neethling, L. P. (1968). A rapid thin-layer chromatographic method for the anthocyanins in grape. *J. Chromatogr.* **34**, 419–420.
Coombe, B. G. and Hale, C. R. (1973) The hormone content of ripening grape berries and the effect of growth substance treatments. *Plant Physiol.* **51**, 629–634.
Costes, C. (1966). Biosynthèse du phytol des chlorophylles et du squelette tétraterpénique des caroténoïdes dans les feuilles vertes. *Phytochemistry* **5**, 311–324.
Costes, C., Burghoffer, C., Joyard, J., Block, M. and Douce, R. (1979). Occurrence and biosynthesis of violaxanthin in isolated spinach chloroplast envelope. *FEBS Lett.* **103**, 17–21.
Craker, L. E. (1971). Postharvest color promotion in cranberry with ethylene. *Hort. Sci.* **6**, 137–139.
Cran, D. G. and Possingham, J. V. (1973). The fine structure of avocado plastids. *Ann. Bot.* **37**, 993–997.
Creasy, L. L. (1976). Phenylalanine ammonia-lyase inactivating system in sunflower leaves. *Phytochemistry* **15**, 673–675.
Cruse, R. R., Lime, B. J. and Hensz, R. A. (1979). Pigmentation and color comparison of Ruby Red and Star Ruby grapefruit juice. *J. Agric. Food Chem.* **27**, 641–642.
Curl, A. L. (1953). Application of countercurrent distribution to Valencia orange juice carotenoids. *J. Agric. Food Chem* **1**, 456–460.
Curl, A. L. (1956). On the structure of hydroxy-α-carotene from orange juice. *Food Res.* **21**, 689–693.
Curl, A. L. (1959). The carotenoids of cling peaches. *Food Res.* **24**, 413–422.
Curl, A. L. (1960a). The carotenoids of apricots. *Food Res.* **25**, 190–196.
Curl, A. L. (1960b). The carotenoids of Japanese persimmons. *Food Res.* **25**, 670–674.
Curl, A. L. (1961). The xanthophylls of tomatoes. *J. Food. Sci.* **26**, 106–111.
Curl, A. L. (1962a). The carotenoids of red bell peppers. *J. Agric. Food Chem.* **10**, 504–509.

Curl, A. L. (1962b). The carotenoids of Meyer lemons. *J. Food Sci.* **27,** 171–176.
Curl, A. L. (1963). The carotenoids of Italian prunes. *J. Food Sci.* **28,** 623–626.
Curl, A. L. (1964a). The carotenoids of several low-carotenoid fruits. *J. Food Sci.* **29,** 241–245.
Curl, A. L. (1964b). The carotenoids of green bell peppers. *J. Agric. Food Chem.* **12,** 522–524.
Curl, A. L. (1965). The occurrence of β-citraurin and β-apo-8′-carotenal in the peels of California tangerines and oranges. *J. Food Sci.* **30,** 13–18.
Curl, A. L. (1966). The carotenoids of muskmelons. *J. Food Sci.* **31,** 759–761.
Curl, A. L. (1967). Apo-10′-violaxanthal, a new carotenoid from Valencia orange peels. *J. Food Sci.* **32,** 141–143.
Curl, A. L. and Bailey, G. F. (1956). Orange carotenoids. I. Comparison of Valencia orange peel and pulp. *J. Agric. Food Chem.* **4,** 156–159.
Curl, A. L. and Bailey, G. F. (1957a). The carotenoids of Ruby Red grapefruit. *J. Agric. Food Chem.* **5,** 63–68.
Curl, A. L. and Bailey, G. F. (1957b). The carotenoids of tangerines. *J. Agric. Food Chem.* **5,** 605–608.
Curl, A. L. and Bailey, G. F. (1961). The carotenoids of Navel oranges. *J. Food Sci.* **26,** 442–447.
Czygan, F.-C. (Ed.) (1980). "Pigments in Plants", 2nd edn. Fischer, Stuttgart.
Davies, B. H. (1976). Carotenoids. *In* "Chemistry and Biochemistry of Plant Pigments" (T. W. Goodwin, Ed.) Vol. 2, 38–165. Academic Press, London.
Davies, B. H., Mattews, S. and Kirk, J. T. O. (1970). The nature and biosynthesis of the carotenoids of different colour varieties of *Capsicum annuum*. *Phytochemistry* **9,** 797–805.
de la Torre Boronat, M. C. and Farré-Rovira, R. (1975). Los carotenoides del pimenton. *Anal. Bromatol.* **xxvii–2,** 179–196.
de Ritter, E. and Purcell, A. E. (1981). Carotenoid analytical methods. *In* "Carotenoids as Colorants and Vitamin A Precursors" (J. Ch. Bauernfeind, Ed.) 815–923. Academic Press, New York.
Dekazos, E. (1970). Quantitative determination of anthocyanin pigments during the maturation and ripening of red tart cherries. *J. Food Sci.* **35,** 242–244.
Dekazos, E. D. and Birth, G. S. (1970). A maturity index for blueberries using light transmittance. *J. Am. Soc. hort. Sci.* **95,** 610–614.
Denny, F. E. (1924). Hastening the colouring of lemons. *J. agric. Res.* **27,** 757–769.
Di Giacomo, A. and Rispoli, G. (1960). Variation of carotenoid content in orange juice during fruit ripening and preservation of the juice. *Conserve Deriv. Agrum.* **9,** 171–175.
Di Giacomo, A., Rispoli, G. and Aversa, M. C. (1968a). Detection of the addition of mandarin juice to orange juice: Analysis of the carotenoids present. *Ind. Conserve* **43,** 123–128.
Di Giacomo, A., Rispoli, G. and Tita, S. (1968b). Carotenoids of Italian tangerine juice. *Riv. Ital. Essenze, Profumi, Piante Off., Aromi, Saponi, Cosmet. Aerosol.* **50,** 64–67.
Diener, H.-A. and Naumann, W.-D. (1981). Influence of day and night temperatures on anthocyanin synthesis in apple skin. *Gartenbauwiss.* **46,** 125–132.
Dogbo, O., Camara, B. and Monéger, R. (1982). Chlorophyll synthetase in chloroplasts and chromoplasts from pepper (*Capsicum annuum* L) fruits. *C. R. Acad. Sci. Paris* **III,** 477–480.

Dörffling, K. (1970). Quantitative Veränderungen des Abscisinsäuregehaltes während der Fruchtentwicklung von *Solanum lycopersicum. Planta* **93**, 233–242.
Dostal, H. C. and Leopold, A. C. (1967). Gibberellin delays ripening of tomatoes. *Science* **158**, 1579–1580.
Drake, S. R., Proebsting, E. L. and Nelson, J. W. (1978). Influence of growth regulators on the quality of fresh and processed 'Bing' cherries. *J. Food Sci.* **43**, 1695–1697.
Drake, S. R., Proebsting, E. L. and Spayd, S. E. (1982). Maturity index for the color grade of canned dark sweet cherries. *J. Am. Soc. hort. Sci.* **107**, 180–183.
Du, C. T., Wang, P. L. and Francis, F. J. (1975). Anthocyanins of pomegranate *Punica granatum. J. Food Sci.* **40**, 417–418.
Durand, M. and Laval-Martin, D. (1974). Séparation en chromatographie sur couche minces des dérivés chlorophylliens tetrapyroliques et application à la recherche de ces derivés dans un fruit en cours de maturation. *J. Chromatogr.* **97**, 92–98.
Dyer, T. A. and Osborne, D. J. (1971). Leaf nucleic acids. II. Metabolism during senescence and effect of kinetin. *J. exp. Bot.* **22**, 552–560.
Eaks, I. L. and Dawson, A. J. (1979). The effect of vegetative ground cover and ethylene degreening on Valencia rind pigments. *J. Am. Soc. hort. Sci* **104**, 105–109.
Ebel, J. and Hahlbroch, K. (1982). Biosynthesis. *In* "The Flavonoids: Advances in Research" (J. B. Harborne and T. J. Marbry, Eds) 641–679. Chapman and Hall, London.
Ebert, G. (1984). "Farbstoffveränderungen während der Reife und Lagerung bei Persimmonen (*Diospyros kaki* Thunb)." Diplomarbeit der Reinischen Friedrich-Wilhelms Universitat.
Ebert, G. and Gross, J. (1985). Carotenoid changes in the peel of ripening persimmon (*Diospyros kaki*). cv. Triumph. *Phytochemistry*
Edgerton, L. J. and Blanpied, G. D. (1970). Interaction of succinic acid 2,2-dimethyl hydrazide, 2-chloroethylphosphonic acid and auxins on maturity, quality and abscision of apples. *J. Am. Soc. hort. Sci.* **95**, 664–666.
Edwards, R. A. and Reuter, F. H. (1967). Pigment changes during the maturation of tomato fruits. *Food Technol. Aust.* **19**, 352–357.
Eilati, S. K., Goldschmidt, E. E. and Monselise, S. P. (1969). Hormonal control of colour changes in orange peel. *Experientia* **25**, 209.
Eilati, S. K., Budowski, P. and Monselise, S. P. (1972). Xanthophyll esterification in the flavedo of citrus fruits. *Plant Cell Physiol.* **13**, 741–746.
Eilati, S. K., Budowski, P. and Monselise, S. P. (1975). Carotenoid changes in the 'Shamouti' orange peel during chloroplast–chromoplast transformation on and off the tree. *J. exp. Bot.* **26**, 624–632.
El-Zeftawi, B. M. (1976). Effects of ethephon and 2,4,5T on fruit size, rind pigments and alternate bearing of 'Imperial' mandarin. *Sci. Hort.* **5**, 315–320.
El-Zeftawi, B. M. (1977). Factors affecting pigment levels during ripening of Valencia orange. *J. hort. Sci.* **52**, 127–134.
El-Zeftawi, B. M. (1978a). Effects of ethephon, GA and light exclusion on rind, pigment, plastid ultrastructure and juice quality of Valencia oranges. *J. hort. Sci.* **53**, 215–223.
El-Zeftawi, B. M. (1978b). Chemical and temperature control on rind pigment of citrus fruit. *Proc. Int. Soc. Citriculture*, 33–36.
El-Zeftawi, B. M. (1980). Effects of gibberellic acid and cycocel on colouring and sizing of lemon. *Sci. Hort.* **12**, 127–181.
Englert, G., Brown, B. O., Moss, G. P., Weedon, B. C. L., Britton, G., Goodwin,

REFERENCES

T. W., Simpson, K. L. and Williams, R. J. H. (1979). Prolycopene a tetra-*cis* carotene with two hindered *cis* double. *J.C.S. Chem. Commun.* 545–547.

Erickson, L. C. (1961). Color development in Valencia oranges. *Proc. Am. Soc. hort. Sci.* **75**, 257–261.

Eskins, K. E., Scholfield, C. R. and Dutton, H. J. (1977). High-performance liquid chromatography of plant pigments. *J. Chromatogr.* **135**, 217–220.

Eugster, C. H. (1983). New carotenoid structures and stereochemistry. *In* "Carotenoid Chemistry and Biochemistry" (G. Britton and T. W. Goodwin, Eds.) 1–26. IUPAC, Pergamon Press, Oxford.

FAO/WHO Joint Expert Committee on Food Additives (1967). *WHO Tech. Rep. Ser.* **362**.

Faragher, J. D. and Chalmers, D. J. (1977). Regulation of anthocyanin synthesis in apple skin. III. Involvement of phenylalanine ammonia-lyase. *Aust. J. Plant Physiol.* **4**, 133–141.

Farin, D., Ikan, R. and Gross, J. (1983). The carotenoid pigments in the juice and flavedo of a mandarin hybrid (*Citrus reticulata*) cv. Michal during ripening. *Phytochemistry* **22**, 403–408.

Fiksdahl, A., Mortensen, S. T. and Liaaen-Jensen, S. (1978). High-pressure liquid chromatography of carotenoids. *J. Chromatogr.* **157**, 111–117.

Fischer, H. and Wenderoth, H. (1940). Optischaktives Hämotricarbonsäureimid aus Chlorophyll. *Ann. Chem.* **545**, 140–147.

Flora, L. F. (1978). Influence of leaf, cultivar and maturity on the anthocyanidin-3,5-diglucosides of muscadine grapes. *J. Food Sci.* **43**, 1819–1821.

Flügel. M. and Gross, J. (1982). Pigment and plastid changes in mesocarp and exocarp of ripening muskmelon *Cucumis melo* cv. Galia. *Angew. Botanik.* **56**, 393–406.

Fong, R. A., Kepner, R. E. and Webb, A. D. (1971). Acetic acid acylated anthocyanin pigments in the grape skin of a number of varieties of *Vitis vinifera. Am. J. Enol. Vitic.* **22**, 150–155.

Forkmann, G. (1980). The B-ring hydroxylation pattern of anthocyanin synthesis in pelargonidin and cyanidin producing lines of *Mathiola incana. Planta* **148**, 157–161.

Francis, F. J. (1970a) Color measurements in plant breeding. *Hortscience* **5**, 102–106.

Francis, F. J. (1970b). Anthocyanins in pears. *Hort. Sci.* **5**, 42.

Francis, F. J. (1972). Anthocyanins of 'Durham' and 'Heritage' raspberry fruits. *Hort. Sci.* **7**, 398.

Francis, F. J. (1980). Color quality evaluation of horticultural crops. *Hortscience* **15**, 58–59.

Francis, F. J., Harborne, J. B. and Barker, W. G. (1966). Anthocyanins in the lowbush blueberry, *Vaccinium angustifolium. J. Food Sci.* **31**, 583–587.

Frenkel, C. (1975). Role of oxidative metabolism in the regulation of fruit ripening. *Coll. Int. CNRS Paris* No. 238, 201–209.

Frenkel, C. and Dick, R. (1973). Auxin inhibition of ripening in Bartlett pears. *Plant Physiol.* **51**, 6–9.

Fritsch, H. and Grisebach, H. (1975) Biosynthesis of cyanidin in cell cultures of *Haplopappus gracilis. Phytochemistry* **14**, 2437–2442.

Fuchs, Y. and Cohen, A. (1969). Degreening of citrus fruits with ethrel. *J. Am. Soc. hort. Sci.* **94**, 617–618.

Fuleki, T. and Francis, F. J. (1967). The co-occurrence of monoglucosides and monogalactosides of cyanidin and peonidin in the American cranberry *Vaccinium macrocarpon. Phytochemistry* **6**, 1705–1708.

Fuleki, T. and Francis, F. J. (1968a). Quantitative methods for anthocyanins. 3. Purification of cranberry anthocyanins. *J. Food Sci.* **33**, 266–273.
Fuleki, T. and Francis, J. F. (1968b). Quantitative methods for anthocyanins. 4. Determination of individual anthocyanins in cranberry and cranberry products. *J. Food Sci.* **33**, 471–478.
Galler, M. and Mackinney, G. (1965). The carotenoids of certain fruits (apple, pear, cherry, strawberry). *J. Food Sci.* **30**, 393–395.
Geisler, G. and Radler, F. (1963). Entwicklungs-und Reifevorgänge an Trauben von *Vitis*. *Ber. Deutsch. Bot. Ges.* **76**, 112–119.
Geissman, T. A. and Jurd, L. (1955). The anthocyanin of *Spirodelia oligarrhiza*. *Arch. Biochem. Biophys.* **56**, 259–263.
Glover, J. and Redfearn, E. R. (1954). The mechanism of transformation of β-carotene into vitamin A *in vivo*. *Biochem. J.* **58**, XV.
Goedheer, J. C. (1979). Carotenoids in the photosynthetic apparatus. *Ber. Deutsch. Bot. Ges.* **92**, 427–436.
Goldschmidt, E. E. and Eilati, S. K. (1970). Gibberellin-treated Shamouti oranges: effects on coloration and translocation within peel of fruits attached to or detached from the tree. *Bot. Gaz.* **131**, 116–122.
Goldschmidt, E. E., Goren, R., Even-Chen, Z. and Bittner, S. (1973). Increase in free and bound abscisic acid during natural and ethylene-induced senescence of citrus fruit peel. *Plant Physiol.* **51**, 879–882.
Goldschmidt, E. E., Aharoni, Y., Eilati, S. K., Riov, J. W. and Monselise, S. P. (1978). Differential counteraction of ethylene effects by gibberellin A_3 and N_6-benzyladenine in senescing citrus peel. *Plant Physiol.* **59**, 193–195.
Gomkoto, G. (1977). Separation of anthocyanins with thin-layer chromatography. *Kertészetiegy. Közl.* **41**, 141–146.
Goodfellow D., Moss, G. P., Szabolcs, J., Toth, G. and Weedon, B. C. L. (1973). Configuration of carotenoid epoxides. *Tetrahedron Lett.* **40**, 3925–3928.
Goodwin, T. W. (1952) "The Comparative Biochemistry of the Carotenoids". Chapman and Hall, London.
Goodwin, T. W. (Ed.) (1976). "Chemistry and Biochemistry of Plant Pigments", Vols 1 and 2, 2nd edn. Academic Press, London.
Goodwin, T. W. (1982). "The Biochemistry of the Carotenoids", Vol. 1: Plants, 2nd edn. Chapman and Hall, London.
Goodwin, T. W. and Jamikorn, M. (1952). Biosynthesis of carotenoids in ripening tomatoes. *Nature* **170**, 104–105.
Gorski, P. M. and Creasy, L. L. (1977). Colour development in 'Golden Delicious' apples. *J. Am. Soc. hort. Sci.* **102**, 73–75.
Grierson, W., Ismail, F. H. and Oberbacher, M. F. (1972). Ethephon for postharvest degreening of oranges and grapefruit. *J. Am. Soc. hort. Sci.* **97**, 541–544.
Grisebach, H. (1980). Recent developments in flavonoid biosynthesis. *In* "Pigments in Plants" (F.-C. Czygan, Ed.) 2nd edn, 187–209. Fischer, Stuttgart.
Gross, J. 1967 "Provitamins A_2". Thesis. The Hebrew University of Jerusalem.
Gross, J. (1977). Carotenoid pigments in citrus. *In* "Citrus Science and Technology" (S. Nagy, P. E. Shaw and M. K. Veldhuis, Eds) Vol. 1, 302–354. AVI, Westport, CT.
Gross, J. (1979a). Carotenoid changes in the mesocarp of the Redhaven peach (*Prunus persica*) during ripening. *Z. Pflanzenphysiol.* **94**, 461–468.
Gross, J. (1979b). Changes of carotenoid pattern in peel of 'Golden Delicious' apples during storage. *Gartenbauwiss.* **44**, 27–29.

Gross, J. (1980). A rapid separation of citrus carotenoids by thin-layer chromatography. *Chromatographia* **13**, 572–573.
Gross, J. (1981). Pigment changes in the flavedo of Dancy tangerine (*Citrus reticulata*) during ripening. *Z. Pflanzenphysiol.* **103**, 451–457.
Gross, J. (1982a). Pigment changes in the pericarp of the Chinese gooseberry of Kiwi fruit (*Actinidia chinensis*) cv. Bruno during ripening. *Gartenbauwiss.* **47**, 162–167.
Gross, J. (1982b). Carotenoid changes in the juice of the ripening Dancy tangerine (*Citrus reticulata*). *Lebensm.-Wiss. u. Technol.* **15**, 36–38.
Gross, J. (1982c). Changes of chlorophylls and carotenoids in developing strawberry fruits (*Fragaria ananassa*) cv. Tenira. *Gartenbauwiss.* **47**, 142–144.
Gross, J. (1982/83). Chlorophyll and carotenoid pigments in *Ribes* fruits. *Sci. Hortic.* **18**, 131–136.
Gross, J. (1983). Pigment changes in the pericarp of the Chinese gooseberry (*Actinidia chinensis*) during ripening and storage. *Acta Hortic.* **138**, 187–194.
Gross, J. (1984). Carotenoid pigments in three plum cultivars. *Gartenbauwiss.* **49**, 18–21.
Gross, J. (1984). Chlorophyll and carotenoid pigments of grapes (*Vitis vinifera* L.) *Gartenbauwiss.* **49**, 180–182.
Gross, J. (1984). Chlorophyll and carotenoids in the peel of two pear cultivars. *Gartenbauwiss.* **49**, 128–131.
Gross, J. (1985). Carotenoid pigments in the developing cherry (*Prunus avium*) cv. 'Dönissen's Gelbe' *Gartenbauwiss.* **50**, 88–90.
Gross, J. and Budowski, P. (1966). Conversion of carotenoids into vitamin A_1 and A_2 in two species of freshwater fish. *Biochem. J.* **101**, 747–754.
Gross, J. and Costes, C. (1976). Nicotine effect on carotenogenesis in α-carotene containing banana leaves. *Physiol. Vég.* **14**, 427–435.
Gross, J. and Eckhardt, G. (1978). A natural apocarotenol from the peel of the ripe 'Golden Delicious' apple. *Phytochemistry* **17**, 1803–1804.
Gross, J. and Eckhardt, G. (1981). Structures of persicaxanthin, persicachrome and other apocarotenols of various fruits. *Phytochemistry* **20**, 2267–2269.
Gross, J. and Flügel, M. (1982). Pigment changes in peel of the ripening banana *Musa cavendishi*. *Gartenbauwiss.* **47**, 62–64.
Gross, J. and Lenz, F. (1979). Violaxanthin content as index of maturity in 'Golden Delicious' apples. *Gartenbauwiss.* **44**, 134–135.
Gross, J. and Ohad, I. (1983). *In vivo* fluorescence spectroscopy of chlorophyll in various unripe and ripe fruit. *Photochem. Photobiol.* **37**, 195–200.
Gross, J., Gabai, M. and Lifshitz, A. (1971). Carotenoids in juice of Shamouti orange. *J. Food Sci.* **36**, 466–473.
Gross, J., Gabai, M. and Lifshitz, A. (1972a). A comparative study of the carotenoid pigments in juice of Shamouti, Valencia and Washington oranges, three varieties of *Citrus sinensis*. *Phytochemistry* **11**, 303–308.
Gross, J., Gabai, M. and Lifshitz, A. (1972b). The carotenoids of the avocado pear. *J. Food Sci.* **37**, 589–591.
Gross, J., Gabai, M., Lifshitz, A. and Sklarz, B. (1973a). Carotenoids in pulp, peel and leaves of *Persea americana*. *Phytochemistry* **12**, 2259–2263.
Gross, J., Gabai, M., Lifshitz, A. and Sklarz, B. (1973b). Carotenoids of *Eriobotrya japonica*. *Phytochemistry* **12**, 1775–1782.
Gross, J., Gabai, M., Lifshitz, A. and Sklarz, B. (1974). Structures of some carotenoids from the pulp of *Persea americana*. *Phytochemistry* **13**, 1917–1921.

Gross, J., Carmon, M., Lifshitz, A. and Sklarz, B. (1975). Structural elucidation of some orange juice carotenoids. *Phytochemistry* **14**, 249–252.
Gross, J., Carmon, M., Lifshitz, A. and Costes, C. (1976). Carotenoids of banana pulp, peel and leaves. *Lebensm.-Wiss. u. Technol.* **9**, 211–214.
Gross, J., Zachariae, A., Lenz, F. and Eckhardt, G. (1978). Carotenoid changes in the peel of the 'Golden Delicious' apple during ripening and storage. *Z. Pflanzenphysiol.* **89**, 312–332.
Gross, J., Eckhardt, G., Sonnen, H. J. and Lenz, F. (1979). Effect of different nitrogen supply on flavedo carotenoids of calamondin fruit. *Gartenbauwiss.* **44**, 45–48.
Gross, J., Haber, O. and Ikan, R. (1983a). The carotenoid pigments of the date. *Sci. Hortic.* **20**, 251–257.
Gross, J., Ikan, R. and Eckhardt, G. (1983b). Carotenoids of the fruit of *Averrhoa carambola*. *Phytochemistry* **22**, 1479–1481.
Gross, J., Timberg, R. and Graef, M. (1983c). Pigment and ultrastructural changes in the developing pummelo (*Citrus grandis*). 'Goliath'. *Bot. Gaz.* **144**, 401–406.
Gross, J., Bazak, H., Blumenfeld, A. and Ben-Arieh, R. (1984) Changes of chlorophyll and carotenoid pigments in peel of 'Triumph' persimmon (*Diospyros kaki* L) induced by preharvest gibberellin (GA_3) treatment. *Sci. Hortic.* **24**, 305–314.
Grumbach, K. H. (1984). Does the chloroplast envelope contain carotenoids and quinones *in vitro*? *Physiol. Plant.* **60**, 180–186.
Grumbach, K. H. and Forn, B. (1980). Chloroplast autonomy in acetyl-coenzyme-A-formation and terpenoid biosynthesis. *Z. Naturforsch.* **35C**, 645–648.
Grumbach, K. H., Mungenast, P. and Ritz, J. (1982). Biosynthesis and degradation of chlorophylls in relation to the developmental stages of a plastid. *In* "Biochemistry and Metabolism of Plant Lipids" (J. F. G. M. Wintermans and P. J. Kuiper, Eds). Elsevier Biomedical Press, Amsterdam.
Guelfat-Reich, S. and Ben-Arieh, R. (1975). A comparison of the ripening effects of ethephon on Japanese plums and apricots. *Coll. Int. CNRS Paris*, No. 238, 105–108.
Hager, A. (1975). The reversible light-induced conversions of xanthophylls in the chloroplast. *Ber. Deutsch. Bot. Ges.* **88**, 27–44.
Hager, A. and Bertenrath, T. (1962) Verteilungschromatographische Trennung von Chlorophyllen und Carotinoiden grüner Pflanzen an Dünnschichten. *Planta* **58**, 564–568.
Hager, A. and Meier-Bertenrath, T. (1966). Extraction and quantitative determination of carotenoids and chlorophylls of leaves, algae and isolated chloroplasts with the aid of thin-layer chromatography. *Planta* **69**, 198–217.
Hale, C. R., Coombe, B. G. and Hawker, J. S. (1970). Effect of ethylene and 2-chloroethylphosphonic acid on the ripening of grapes. *Plant Physiol.* **45**, 620–623.
Haller, M. and Magness, J. R. (1944). "Picking maturity of apples". U.S. Department of Agriculture Circular No. 711, 23pp.
Hansen, P. (1970). ^{14}C studies on apples trees. VI. The influence of the fruit on the photosynthesis of the leaves and the relative photosynthetic yields of fruits and leaves. *Physiol. Plant.* **23**, 805–810.
Hansen, P. (1980). Yield components and fruit development in 'Golden Delicious' apples as affected by the timing of nitrogen supply. *Sci. Hortic.* **12**, 243–257.
Harborne, J. B. (1958). The chromatographic identification of anthocyanin pigments. *J. Chromatogr.* **1**, 473–488.
Harborne, J. B. (1967). "Comparative Biochemistry of the Flavonoids". Academic Press, London and New York.
Harborne, J. B. (1972). Evolution and function of flavonoids in plants. *In* "Recent

Advances in Phytochemistry" (V. C. Runeckles and J. E. Watkin, Eds) Vol. 4, 107–141. Appleton Century Crofts, New York.

Harborne, J. B. and Hall, E. (1964). Plant polyphenols. XIII. The systematic distribution and origin of anthocyanins containing branched trisaccharides. *Phytochemistry* **3**, 453–463.

Harborne, J. B. and Marbry, T. J. (Eds) (1982). "The Flavonoids: Advances in Research" Chapman and Hall, London.

Harborne, J. B., Marbry, T. J. and Marbry, H. (Eds) (1975). "The Flavonoids", Vols 1 and 2. Academic Press, New York.

Harris, W. M. and Spurr, A. R. (1969a). The red tomato. *Am. J. Bot.* **56**, 380–389.

Harris, W. M. and Spurr, A. R. (1969b). Chromoplasts of tomato fruits. Ultrastructure of low pigment and high-beta mutants. Carotene analyses. *Am. J. Bot.* **56**, 369–379.

Henze, J. (1983). "Ripening stages. Their determination in different apple cultivars". Unpublished research report.

Higby, W. K. (1963). Analysis of orange juice for total carotenoids and added β-carotene. *Food Technol.* **17**, 95–99.

Holden, M. (1976). Chlorophylls. *In* "Chemistry and Biochemistry of Plant Pigments" (T. W. Goodwin, Ed.) 2nd edn, 1–37. Academic Press, London.

Houck, L. G., Aharoni, Y. and Fouse, D. C. (1978). Color changes in orange fruits stored in high concentrations of oxygen and in ethylene. *Proc. Fla. St. hort. Soc.* **91**, 136–139.

Howes, C. D. (1974). Nicotine inhibition of carotenoid cyclization in *Cucurbita ficifolia* cotyledon. *Phytochemistry* **13**, 1469–1471.

Hrazdina, G. (1970). Column chromatographic isolation of the anthocyanidin-3,5-diglucosides from grapes. *J. Agric. Food Chem.* **18**, 243–245.

Hrazdina, G. (1982). Anthocyanins. *In* "The Flavonoids: Advances in Research" (J. B. Harborne and T. J. Marbry, Eds), 135–188. Chapman and Hall, London.

Hrazdina, G. and Franzese, A. J. (1974). Structure and properties of the acylated anthocyanins from *Vitis* species. *Phytochemistry* **13**, 225–229.

Hrazdina, G., Wagner, G. J. and Siegelman, H. W. (1978). Subcellular localization of enzymes of anthocyanin biosynthesis in protoplasts. *Phytochemistry* **17**, 53–56.

Hsia, C. L., Luh, B. S. and Chichester, C. O. (1965). Anthocyanins in freestone peaches. *J. Food Sci.* **30**, 5–12.

Huffman, W. A. H., Lime, B. J. and Scott, W. C. (1953). Processed juices from Texas red and pink grapefruit. A progress report. *Proc. Rio Grande Valley hort. Inst.* **7**, 102–105.

Huggard, R. L. and Wenzel, F. W. (1955). Color differences of citrus juices and concentrates using the Hunter color difference meter. *Food Technol.* **21**, 100–105.

Hulme, A. C. (Ed.) (1970/71). "The Biochemistry of Fruits and their Products," Vols 1 and 2. Academic Press, London.

Humphrey, A. M. (1980). Chorophyll. *Food Chem.* **5**, 57–67.

Hürter, C. (1982). Neue Methode zur objektiven Messung der Schalen—und Fruchtfleischfarbe. *Erwerbsobstbau* **24**, 72–74.

Huyskens, S. (1983). "Farbstoff und Plastidenveränderungen während der Reife bei Kumquat (*Fortunella margarita* (Lour) Swingle) cv. Nagami". Diplomarbeit, Friedrich Wilhems Universität, Bonn.

Huyskens, S., Timberg, R. and Gross, J. (1985). Pigment and plastid ultrastructural changes in kumquat (*Fortunella margarita*) 'Nagami' during ripening. *J. Plant Physiol.* **118**, 61–72.

Hyodo, H. (1971). Phenylalanine ammonia-lyase in strawberry fruits. *Plant Cell Physiol.* **12**, 989–991.
Ishikura, N. (1975). A survey of anthocyanins in fruits of some angiosperms. I. *Bot. Mag. Tokyo* **88**, 41–45.
Ishikura, N. and Sugahara, K. (1979) A survey of anthocyanins in fruits of some angiosperms. II. *Bot. Mag. Tokyo* **92**, 157–161.
Isler, O. (Ed.) (1971). "Carotenoids". Birkhäuser, Basel.
Ismail, M. A., Biggs, R. H. and Oberbacher, M. F. (1967). Effect of gibberellic acid on colour change in the rind of three orange cultivars (*Citrus sinensis* Blanco). *J. Am. Soc. hort. Sci.* **91**, 143–149.
Itoo, S. (1980). Persimmon. *In* "Tropical and Subtropical Fruits" (S. Nagy and P. Shaw, Eds) 442–468. AVI, Westport, CT.
IUPAC (1967). Main lectures presented at the First International Symposium on Carotenoids other than Vitamin A. *Pure appl. Chem.* **14**, 215–278.
IUPAC (1969). Main lectures presented at the Second International Symposium on Carotenoids other than Vitamin A. *Pure appl. Chem.* **20**, 365–553.
IUPAC (1971). Tentative Rules of the Nomenclature of Carotenoids. In "Carotenoids" (O. Isler, Ed.) 851–864.
IUPAC (1973). Plenary lectures presented at the Third International Symposium on Carotenoids other than Vitamin A. *Pure appl. Chem.* **35**, 1–130.
IUPAC (1974). "Nomenclature of Carotenoids". Butterworth, London.
IUPAC (1976). Main lectures presented at the Fourth International Symposium on Carotenoids. *Pure appl. Chem.* **47**, 97–243.
IUPAC (1979). Main lectures presented at the Fifth International Symposium on Carotenoids. *Pure appl. Chem.* **51**, 435–675.
IUPAC (1982). "Carotenoid Chemistry and Biochemistry", *Proc 6th Int. Symp. on Carotenoids* (G. Britton and T. W. Goodwin, Eds). Pergamon Press, Oxford.
Iwahori, S. and Lyons, J. M. (1970). Maturation and quality of tomatoes with preharvest treatments of 2-chlorotheylphosphonic acid. *J. Am. hort. Soc.* **95**, 88–91.
Jahn, O. L. (1973). Degreening citrus fruit with postharvest applications of (2-chloroethyl) phosphonic acid (ethephon). *J. Am. Soc. hort. Sci.* **98**, 230–233.
Jahn, O. L. (1974a). Degreening of Florida lemons. *Proc. Fla. St. hort. Soc.* **87**, 218–221.
Jahn, O. L. (1974b). Ethylene concentrations and temperatures for degreening Florida citrus fruit. *Proc. 1st Int. Citrus Congress*, Spain.
Jahn, O. L. (1975). Comparison of instrumental methods for measuring ripening changes of intact tomato fruit. *J. Am. Soc. hort. Sci.* **100**, 688–691.
Jahn, O. L. and Young, R. (1976). Changes in chlorophyll a, b and the a/b ratio during color development in citrus fruits *J. Am. Soc hort. Sci.* **101**, 416–418.
Jahn, O. L., Chase, W. G. and Cubbedge, R. H. (1969). Degreening of citrus fruits in response to varying levels of oxygen and ethylene. *J. Am. Soc. hort. Sci.* **94**, 123–125.
Jahn, O. L., Chase, W. G. and Cubbedge, R. H. (1973). Degreening response of 'Hamlin' oranges in relation to temperature, ethylene concentration and fruit maturity. *J. Am. Soc. hort. Sci.* **98**, 177–181.
Jeffrey, S. W., Douce, R. and Benson, A. A. (1974). Carotenoid transformations in the chloroplast envelope. *Proc. Natl Acad. Sci. U.S.A.* **71**, 807–810.
Jen, J. J. (1974). Influence of spectral quality of light on pigment system of ripening tomatoes. *J. Food Sci.* **39**, 907–910.
Jensen, A. and Liaaen-Jensen, S. (1959). Quantitative paper chromatography of carotenoids. *Acta Chem. Scand.* **13**, 1863–1868.

Johjima, T. and Ogura, H. (1983). Analysis of tomato carotenoids by thin-layer chromatography and a *cis* form γ-carotene newly identified in tangerine tomato. *J. Jap. Soc. hort Sci.* **52**, 202–209.

John, J., Subbarayan, C. and Cama, H. R. (1970). Carotenoids in 3 stages of ripening mango. *J. Food Sci.* **35**, 262–265.

Jones, W. W. and Embleton, T. W. (1959). The visual effect of nitrogen nutrition in fruit quality of Valencia orange. *Proc. Am. Soc. hort. Sci.* **73**, 234–236.

Jungalwala, F. B. and Cama, H. R. (1963). Carotenoids in mango (*Mangifera indica*) fruit. *Ind. J. Chem.* **1**, 36–40.

Kader, A. A. and Morris, L. L. (1978). Tomato fruit color measured with an Agstron ES-W reflectance spectrophotometer. *Hortscience* **13**, 577–578.

Kamsteeg, J., van Brederode, J. and van Nigtewecht, G. (1980). The pH-dependent substrate specificity of UDP-glucose: anthocyanidin 3-rhamnosylglucoside, 5-*O*-glucosyl transferase in petals of *Silene dioica*: the formation of anthocyanidin 3,5-glucosides. *Z. Pflanzenphysiol.* **96**, 87–93.

Kargl, T. E., Quackenbush, F. W. and Tomes, M. L. (1960). The carotene polyene system in a strain of tomatoes high in delta-carotene and its comparison with eight other tomato strains. *Proc. Am. Soc. hort. Sci.* **75**, 574–578.

Karrer, P. and Jucker, E. (1948). "Carotinoide". Birkhäuser, Basel.

Karrer, P., Morf, R., von Krauss, E. and Zubrys, A. (1932). Pflanzenfarbstoffe XXXIX. Vermischte Beobachtungen über Carotinoide aus Kakifrüchten. *Helv. Chim. Acta* **15**, 490–493.

Kasmire, R. F., Rappaport, L. and Hay, D. (1970). Effects of (2-chloroethyl) phosphonic acid in ripening of cantaloupes. *J. Am. Soc. hort. Sci.* **95**, 134–137.

Katayama, T., Nakayama, T. O. M., Lee, T. H. and Chichester, C. O. (1971). Carotenoid transformations in ripening apricots and peaches. *J. Food Sci.* **36**, 804–806.

Kefford, J. F. and Chandler, B. V. (1970). "The Chemical Constituents of Citrus Fruits". Academic Press, New York.

Khan, M. and Mackinney, G. (1953). Carotenoids of grapefruit. *Plant Physiol.* **28**, 550–552.

Kho, F. F. (1978). Conversion of hydroxylated and methylated dihydroflavonols into anthocyanins in a white flowering mutant of *Petunia hybrida*. *Phytochemistry* **17**, 245–248.

Kho, F. F., Bolsman-Louwen, A. C., Vuik, J. C. and Bennink, G. J. H. (1977). Anthocyanin synthesis in a white flowering mutant of *Petunia hybrida*. II. Accumulation of mutant dihydroflavonol intermediates. *Planta* **135**, 109–118.

Kirk, J. T. O. and Juniper, B. E. (1967). The ultrastructure of chromoplasts of different colour varieties of Capsicum. *In* "Biochemistry of Chloroplasts" (T. W. Goodwin, Ed.) Vol. 2, 691–701. Academic Press, London and New York.

Kirk, J. T. O. and Tilney-Bassett, R. A. E. (1978a) "The Plastids. Their Chemistry, Structure, Growth and Inheritance", 2nd edn. Elsevier, Amsterdam.

Kirk, J. T. O. and Tilney-Bassett, R. A. E. (1978b). Chloroplast structure. *In* "The Plastids. Their Structure Growth and Inheritance", 2nd edn, 127–189. Elsevier, Amsterdam.

Kitagawa, H., Sugiura, A. and Sugiyama, M. (1966). Effects of gibberellic spray on storage quality of Kaki. *Hortscience* **1**, 59–60.

Kitagawa, H., Adachi, S. and Tarutani, T. (1974). Studies on the coloring of Satsuma mandarin. I. The relationship of a method of ethylene treatment and degreening. *J. Jap. Soc. hort. Sci.* **40**, 190–194.

Kitagawa, H., Kawada, K, and Tarutani, T. (1977). Degreening of Satsuma mandarin in Japan. *Proc. Int. Soc. Citriculture.* **1**, 219–223.
Kleinig, H. and Nitsche, H. (1968). Carotinoidester – Muster in gelben Blütenblättern. *Phytochemistry* **7**, 1171–1175.
Kliewer, M. W. (1970). Effect of day temperature and light intensity on coloration of *Vitis vinifera* L. grapes. *J. Am. Soc. hort. Sci.* **95**, 693–697.
Kliewer, M. W. (1977). Influence of temperature, solar, radiation and nitrogen on coloration and composition of 'Emperor' grapes. *Am. J. Enol. Vitic.* **28**, 96–103.
Kliewer, M. W. and Torres, R. E. (1972). Effect of controlled day and night temperatures on grape coloration. *Am. J. Enol. Vitic.* **23**, 71–77.
Knee, M. (1972). Anthocyanin, carotenoid and chlorophyll changes in the peel of Cox's Orange Pippin apples during ripening on and off the tree. *J. exp. Bot.* **23**, 184–196.
Kobayashi, K., Iso, H., Nishiyama, K. and Akuta, S. (1978). The relationship between carotenoid pigments and colors of fresh and canned loquat fruits (*Eriobotrya japonica* Lindl). *Nippon Shokuhin Kogyo Gakkaishi* **25**, 426–430.
Koch, J. and Haase-Sajak, E. (1965) Natural coloring material of citrus fruits. Orange and mandarin orange carotenoids. *Z. Lebensm. Unters. Forsch.* **126**, 260–271.
Koskitalo, L. N. and Ormrod, D. P. (1972). Effects of sub-optimal ripening temperatures of the color quality and pigment on composition of tomato fruit. *J. Food Sci.* **37**, 56–59.
Kufner, R. B. (1980). The biological degradation of chlorophyll in senescent tissues. *In* "Pigment in Plants" (E.-C. Czygan, Ed.) 2nd edn, 308–313. Fischer, Stuttgart.
Kuhn, R. and Grundmann, C. (1932). Die Konstitution des Lycopins. *Ber. Deutsch. Chem. Ges.* **65**, 1880–1889.
Kursanov, A. L. (1934). Die Photosynthese grüner Früchte und ihre Abhängigkeit von der normalen Tätigkeit der Blätter. *Planta* **22**, 240–250.
Kushman, L. J. and Ballinger, W. E. (1975). Relation of quality indices of individual blueberries to photoelectric measurements of anthocyanin content. *J. Am. Soc. hort. Sci.* **100**, 561–564.
Kushwaha, S. C., Suzue, G., Subbarayan, C. and Porter, J. W. (1970). The conversion of phytoene ^{14}C to acyclic, monocyclic and dicyclic carotenes and the conversion of lycopene 15,15'-^{3}H to mono- and dicyclic carotenes by soluble enzyme systems obtained from plastids of tomato fruits. *J. Biol. Chem.* **245**, 4708–4717.
Kvåle, A. (1971). The effect of different nitrogen levels of the trees on pigment content, ground colour and soluble solids of the apple cultivars Prins. Gravenstein, James Grieve and Ingrid Marie. *Acta Agric. Scand.* **21**, 207–213.
Lakshmanan, M. R., Chansang, N. and Olson, J. A. (1972). Purification and properties of carotene 15,15'-dioxygenase of rabbit intestine. *J. Lipid Res.* **13**, 477–482.
Lang, K. (Ed., (1963). "Carotine und Carotinoide", *Wiss. Veröff. Deutsch. Ges. Ernähr.* Steinkopff, Darmstadt.
Laval-Martin, D. (1969). Variations de la teneur en chlorophylles au cours de la maturation et en fonction de la température chez la poire Passe-Crassane. *Physiol. Vég.* **7**, 251–259.
Laval-Martin, D. (1974). The maturation of the 'cherry' tomato fruit. Evidence by freeze-etched studies of the evolution of chloroplasts in two classes of chromoplasts. *Protoplasma* **82**, 33–59.
Laval-Martin, D., Quennemet, J. and Monéger, R. (1975). Pigment evolution in *Lycopersicon esculentum* fruits during growth and ripening. *Phytochemistry* **14**, 2357–2362.

Laval-Martin, D., Farineau, J. and Diamond, J. (1977). Light versus dark carbon metabolism in cherry tomato fruits. I. Occurrence of photosynthesis, study of the intermediates. *Plant Physiol.* **60**, 872–876.
Lederer, E. (1934) "Les Caroténoïdes des Plantes". Hermann, Paris.
Lee, C. Y. and Robinson, R. W. (1980). Influence of the crimson gene on vitamin A content of tomato. *Hortscience* **15**, 260–261.
LeLous, J., Majoie, B., Morinière, J.-L. and Wulfert, E. (1975). Etudes des flavonoïdes de *Ribes nigrum*. *Ann. Pharm. Fr.* **33**, 393–399.
Lenz, F. and Gross, J. (1979). Fruchtqualitätsmerkmale bei 'Golden Delicious' in Abhängigkeit vom Fruchtbehang. *Erwerbsobstbau* **21**, 173–174.
Lenz, F. and Noga, G. (1982). Photosynthese und Atmung bei Apfelfrüchten. *Erwerbsobstbau.* **24**, 198–200.
Lessertois, D. and Monéger, R. (1978). Evolution des pigments pendant la croissance et la maturation du fruit de *Prunus persica*. *Phytochemistry* **17**, 411–415.
Leuenberger, U. and Stewart, I. (1976). β-Citraurinene, a new C_{30}-citrus carotenoid. *Phytochemistry* **15**, 227–229.
Leuenberger, F. J. and Thommen, H. (1972). The presence of carotenoids within *Passiflora edulis*. *Z. Lebensm.-Unters.-Forsch.* **149**, 279–282.
Leuenberger, U., Stewart, I. and King, R. W. (1976). Isolation of β-citraurol, a C-30 carotenoid in citrus. *J. Org. Chem.* **41**, 891–892.
Liaaen-Jensen, S. (1971). Isolation, Reactions. *In* "Carotenoids" (O. Isler, Ed.) 61–188. Birkhäuser, Basel.
Liaaen-Jensen, S. (1979). Carotenoids—a chemosystematic approach. *Pure appl. Chem.* **51**, 661–675.
Liaaen-Jensen, S. (1980). Stereochemistry of naturally occurring carotenoids. *Fortschr. Chem. Org. Naturst.* **39**, 123–172.
Licastro, F. and Bellomo, A. (1974). Identification of anthocyanins in Moro oranges. *Chem. Abstr.* **80**, 58 557.
Lichtenthaler, H. K. (1971). The unequal synthesis of the lipophilic plastidquinones in sun and shade leaves of *Fagus silvatica* L. *Z. Naturforsch.* **26B**, 832–841.
Lichtenthaler, H. K. (1981). Adaption of leaves and chloroplasts to high quanta fluence rates. *In* "Photosynthesis VI. Photosynthesis and Productivity, Photosynthesis and Environment" (G. Akoyunoglou, Ed.), *Balaban Int. Sci. Serv.*, 273–287.
Lichtenthaler, H. K. and Grumbach, K. H. (1974). Kinetic of lipoquinone and pigment synthesis in green *Hordeum* seedlings during an artificial day–night rhythm with a prolonged dark phase. *Z. Naturforsch.* **29C**, 532–540.
Lichtenthaler, H. K., Burkhardt, G., Kuhn, G. and Prenzel, U. (1981). Light-induced accumulation and stability of chlorophylls and chlorophyll-proteins during chloroplast development in radish seedlings. *Z. Naturforsch.* **36C**, 421–430.
Lichtenthaler, H. K., Prenzel, U. and Kuhn, G. (1982). Carotenoid composition of chlorophyll-carotenoid-proteins from radish chloroplasts. *Z. Naturforsch.* **37C**, 10–12.
Lime, B. J., Stephens, T. S. and Griffiths, F. P. (1954). Processing characteristics of colored Texas grapefruit. I. Color and maturity studies of Ruby Red grapefruit. *Food Technol.* **8**, 566–569.
Ljubešić, N. (1968). The fine structure of chloroplasts during the yellowing and regreening of leaves. *Protoplasma* **66**, 369–379.
Ljubešić, N. (1984). Structural and functional changes of plastids during yellowing and regreening of lemon fruits. *Acta Bot. Croat.* **43**, 25–30.

REFERENCES

Looney, N. E. (1968). Light regimes within standard size apple trees as determined spectrophotometrically. *Proc. Am. Soc. hort. Sci.* **93**, 1–6.
Lynn, D. Y. C. and Luh, B. S. (1964). Anthocyanin pigments in Bing cherries. *J. Food Sci.* **29**, 735–743.
Lynn, D. Y. C., Co, L. and Schanderl, S. H. (1967). The occurrence of 418 and 444 nm chlorophyll-type compounds in some green plant tissue. *Phytochemistry* **6**, 145–148.
Maccarone, E., Maccarone, A., Perini, G. and Rapisarda, P. (1983). Anthocyanins in the Moro orange. *Ann. Chim. (Rome)* **73**, 533–539.
Mackinney, G. (1937). Carotenoids of the peach. *Plant Physiol.* **12**, 216–218.
Mackinney, G. (1941). Absorption of light by chlorophyll solutions. *J. Biol. Chem.* **140**, 315–322.
Mackinney, G. and Jenkins, J. A. (1952). Carotenoid differences in tomatoes. *Proc. Natl Acad. Sci. U.S.A.* **38**, 48–52.
Maestro Durán, R. and Vásquez Roncero, A. (1976). Anthocyanic pigments in ripe Manzanillo olives. *Grasas y Aceites* **27**, 237–243.
Magness, J. R. (1928). Observation on color development in apple. *Proc. Am. Soc. hort. Sci.* **25**, 289–292.
Malachi, T., Gross, J., Lifshitz, A. and Sklarz, B. (1974). Flavedo carotenoid pigments of ripe Washington-Navel orange. *Lebensmitt.-Wiss.u.-Technol.* **7**, 330–334.
Mancini, F. and Schianchi, G. F. (1963). Chemical and biological estimation of the carotene content in fresh and processed Italian apricots. In "Carotine und Carotinoide" (K. Lang, Ed.) 401–406. Dietrich Steinkopff, Darmstadt.
Markakis, P. (Ed.) (1982). "Anthocyanins as Food Colors". Academic Press, New York.
Märki-Fischer, E., Buchecker, R. and Eugster, C. H. (1984). Reinvestigation of the carotenoids from *Rosa foetida*, structures of 12 novel carotenoids; stereoisomeric lutoexanthins, auroxanthins, latoxanthins and latochromes. *Helv. Chim. Acta* **67**, 2143–2154.
Märki-Fischer, E. and Eugster, C. H. (1985). Karpoxanthin and 6-Epikarpoxathin. *Helv. Chim. Acta* **68**, 1704–1707.
Mathews, R. F., Locascio, S. J. and Ozaki, H. Y. (1975). Ascorbic acid and carotene contents of pepper. *Proc. Fla. St. hort. Soc.* **88**, 263–265.
Mathews-Roth, M. M., Wilson, T., Fujimori, E. and Krinsky, N. I. (1974). Carotenoid chromophore length and protection against photosensitization. *Photochem. Photobiol.* **19**, 217–222.
Matus, Z., Baranyai, M., Tóth, G. and Szabolcs, J. (1981). Identification of oxo, epoxy and some *cis*-carotenoids in high-performance liquid chromatography. *Chromatographia* **14**, 337–340.
McDermott, J. C. B., Brown, D. J., Britton, G. and Goodwin, T. W. (1974). Alternative pathways of zeaxanthin biosynthesis in *Flavobacterium* species. *Biochem. J.* **144**, 231–243.
McGlasson, W. B. (1970). The ethylene factor. In "The Biochemistry of Fruits and their Products" (A. C. Hulme Ed.) Vol. I, 475–519. Academic Press, London.
Meier, D. and Lichtenthaler, H. K. (1981). Ultrastructural development of chloroplasts in radish seedlings grown at high- and low-light conditions in the presence of the herbicide bentazon. *Protoplasma* **107**, 195–207.
Melin, C., Moulet, A., Dupin, J. F. and Hartman, C. (1977). Phenylalanine-ammoniaque lyase et composés phénoliques au cours de la maturation de la cerise. *Phytochemistry* **16**, 75–78.

Meredith, F. I. and Purcell, A. E. (1966). Changes in the concentration of carotenes of ripening Homestead tomatoes. *Proc. Am. Soc. hort. Sci.* **89,** 544–548.

Meredith, F. I. and Young, R. H. (1971). Changes in lycopene and carotene content of 'Redblush' grapefruit exposed to high temperature. *Hortscience* **6,** 233–234.

Milborrow, B. V. (1967). The identification of (+)-abscisic II in plants and measurements of its concentrations. *Planta* **76,** 93–113.

Milborrow, B. V. (1974a). The chemistry and physiology of abscisic acid. *Ann. Rev. Plant. Physiol.* **25,** 259–374.

Milborrow, B. V. (1974b). Biosynthesis of abscisic acid by a cell-free system. *Phytochemistry* **13,** 131–136.

Milborrow, B. V. (1982). Stereochemical aspects of carotenoid biosynthesis. *In* "Carotenoid Chemistry and Biochemistry" (G. Britton and T. W. Goodwin, Eds) 279–295. IUPAC, Pergamon Press, Oxford.

Miller, E. V. and Winston, J. R. (1939). Investigation of the development of color in citrus fruits. *Proc. Fla. St. hort. Soc.* **52,** 87–90.

Miller, E. V., Winston, J. R. and Shomer, H. A. (1940). Physiological studies of plastid pigments in rinds of maturing oranges. *J. Agric. Res.* **60,** 259–267.

Miniati, E. (1981). Anthocyanin pigment in the pistachio nut. *Fitoterapia* **LII,** 267–271.

Mohr, W. P. (1979). Pigment bodies in fruits of crimson and high pigment lines of tomatoes. *Ann. Bot.* **44,** 427–434.

Moll, W. A. W. and Stegwee, D. (1978). The activity of triton X-100 soluble chlorophyllase in liposomes. *Planta* **140,** 75–80.

Molnár, P. and Szabolcs, J. (1979). Alkaline permanganate oxidation of carotenoid epoxides and furanoxides. *Acta Chim. Acad. Sci. Hung.* **99,** 155–173.

Molnár, P. and Szabolcs, J. (1980). β-Citraurin epoxide, a new carotenoid from Valencia orange peel. *Phytochemistry* **19,** 633–637.

Monselise, S. P. (1973). "Citrus Fruits as Raw Materials in the Preparation of Juice and Other Products". Gunther Hempel, Brunswick.

Moreshet, S. and Green, G. C. (1980). Photosynthesis and diffusion conductance of the Valencia orange fruit under field conditions. *J. exp. Bot.* **31,** 15–27.

Morgan, R. C. (1966). Chemical studies on concentrated pineapple juice. 1. Carotenoid composition of fresh pineapples. *J. Food Sci.* **31,** 213–217.

Morgan, R. C. (1967). The carotenoids of Queensland fruits – Carotenes of the watermelon (*Citrullus vulgaris*). *J. Food Sci.* **32,** 275–278.

Morton, A. D. (1967). Thin-layer chromatography of anthocyanins from black currant juice. *J. Chromatogr.* **28,** 480–481.

Moskowitz, A. H. and Hrazdina, G. (1981). Vacuolar contents of fruit subepidermal cell from *Vitis*. *Plant Physiol.* **68,** 686–692.

Moss, G. P. and Weedon, B. C. L. (1976). Chemistry of the carotenoids. *In* "Chemistry and Biochemistry of Plant Pigments" (T. W. Goodwin, Ed.) 2nd edn, Vol. 1, 149–224. Academic Press, London.

Moutounet, M. (1976). Les caroténoïdes de la prune d'Ente et du pruneau d'Agen. *Ann. Technol. Agric.* **26,** 73–84.

Mullick, D. B. (1969). Thin-layer chromatography of anthocyanidins. 1. Technique and solvents for two-dimensional chromatography. *J. Chromatogr.* **39,** 291–301.

Mustárdy, L. A., Machowicz, E. and Faludi-Daniel, A. (1976). Light induced structural changes of thylakoids in normal and carotenoid deficient chloroplasts of maize. *Protoplasma* **88,** 65–73.

Nitsch, J. P. (1970). Hormonal factors in growth and development. *In* "The Biochemistry of Fruits and Their Products". (A. C. Hulme, Ed.) Vol. 1, 427–472. Academic Press, London.

Nitsche, H. (1972). Mimulaxanthin—a new allenic xanthophyll from the petals of *Mimulus guttatus*. *Phytochemistry* **11**, 401–404.

Noga, G. (1981). "Ursachen der Rauhschaligkeit und mangelhaften Ausfärbung bei Satsumamandarinen (*Citrus unshiu* Marc)". Dissertation, Friedrich Wilhelms Universität, Bonn.

Noga, G. and Lenz, F. (1983). Separation of citrus carotenoids by reversed-phase high-performance liquid chromatography. *Chromatographia 17*, 139–142.

Nybom, N. (1968). Cellulose thin layers for anthocyanin analysis with special reference to the anthocyanins of black raspberries. *J. Chromatogr.* **38**, 383.

Okombi, G., Billot, J. and Hartmann, C. (1980). Les caroténoïdes de la cerise (*Prunus avium*). cv. Bigarreau Napoléon. Evolution au cours de la croissance et de la maturation. *Fruits* **35**, 313–320.

Olsen, K. L. and Couey, H. M. (1975). Golden Delicious apples. Factors in production and maintenance of high quality. *Coll. Int. CNRS Paris*, No. 238, 63–66.

Øydvin, J. (1974). Inheritance of four cyanidin-3-glycosides in red currant. *Hort. Res.* **14**, 1–7.

Padula, M. and Rodriguez-Amaya, D. B. (1984). Carotenoid composition and general properties of the guava cultivar IAC-4. Changes on processing and storage of the juice. *J. Sci. Food Agric.*

Palmer, L. S. (1922). Carotenoids and related pigments. "The Chemical Catalogue", New York.

Papageorgiou, G. (1975). Chlorophyll fluorescence. *In* "Bioenergetics of Photosynthesis" (Govindjee, Ed.) 319–371. Academic Press, New York.

Papastephanou, C., Barnes, F. J., Briedis, A. V. and Porter, J. W. (1973). Enzymatic synthesis of carotenes by cell-free preparations of several genetic selections of tomatoes. *Arch. Biochem. Biophys.* **157**, 415–425.

Park, Y., Morris, M. M. and Mackinney, G. (1973). On chlorophyll breakdown in senescent leaves. *J. Agric. Food Chem.* **21**, 279–281.

Paynter, V. A. and Jen, J. J. (1976). Comparative effects of light and ethephon on the ripening of detached tomatoes. *J. Food Sci.* **41**, 1366–1369.

Pecheur, J. and Ribaillier, D. (1975). La régulation de la maturation des fruits par l'étéphon. *Coll. Inst. CNRS Paris*, No. 238, 121–131.

Pennisi, L., Scuderi, H. and Muratore, A. (1955). Studies on the maturity of Sicilian oranges. *Riv. Agric.* **1**, 402.

Peto, R., Doll, R., Buckley, J. D. and Sporn, M. B. (1981). Can dietary β-carotene materially reduce human cancer rates? *Nature* **290**, 201–208.

Petzold, E. N. and Quackenbush, F. W. (1960). Zeinoxanthin, a crystalline carotenol from corn gluten. *Archs. Biochem. Biophys.* **86**, 163–165.

Pfander, H. (1979). Synthesis of carotenoid glycosylesters and other carotenoids. *Pure appl. Chem.* **51**, 565–580.

Phan, C. T. (1970). Photosynthetic activity of fruit tissues. *Plant Cell Physiol.* **11**, 823–825.

Phan, C. T. (1975). Occurrence of active chloroplasts in the internal tissues of apples. Their possible role in fruit maturation. *Coll. Int. CNRS Paris*, No. 238, 49–55.

Philip, T. (1973a). Nature of xanthophyll esterification in grapefruit. *J. Agric. Food Chem.* **21**, 963–964.

Philip, T. (1973b). The nature of carotenoid esterification in tangerines. *J. Food Sci.* **38**, 1032–1034.

Philip, T. (1973c). Nature of xanthophyll esterification in citrus fruits. *J. Agric. Food Chem.* **21**, 964–966.

Pifferi, P. G. and Vaccari, A. (1980). Purification of anthocyanins of traces of metals by ion-exchange resin. *Lebensm.-Wiss. u. Technol.* **14**, 85–86.

Pirrie, A. J. and Mullins, M. G. (1976). Changes in anthocyanin and phenolics content of grape leaf and fruit tissues treated with sucrose, nitrate and abscisic acid. *Plant Physiol.* **58**, 468–472.

Pirrie, A. J. and Mullins, M. G. (1977). Interrelationships of sugars, anthocyanins, total phenols and dry weight in the skin of grape berries during ripening. *Am. J. Enol. Vitic.* **28**, 204–209.

Pirrie, A. J. and Mullins, M. G. (1980). Concentration of phenolics in the skin of grape berries during fruit development and ripening. *Am. J. Enol. Vitic.* **31**, 34–36.

Porter, J. W. and Lincoln, R. E. (1950). The mechanism of carotene biosynthesis. *Arch. Biochem. Biophys.* **27**, 390–403.

Possingham, J. V., Coote, M. and Hawker, J. S. (1980). The plastids and pigments of fresh and dried Chinese gooseberries (*Actinidia chinensis*). *Ann. Bot.* **45**, 529–533.

Prabha, T. N., Ravindranath, B. and Patwardhan, M. V. (1980). Anthocyanins of avocado (*Persea americana*) peel. *J. Food Sci. Technol.* **17**, 241–242.

Prasad, U. S. and Jha, O. P. (1978). Changes in pigmentation patterns during litchi ripening: flavonoid production. *Plant Biochem. J.* **5**, 44–49.

Pratt, H. K. (1971). Melons. *In* "The Biochemistry of Fruits and their Products" (A. C. Hulme, Ed.) Vol. 2, 207–232. Academic Press, London.

Prenzel, U. and Lichtenthaler, H. K. (1982). High performance liquid chromatography of prenylquinones, prenylvitamines and prenols. *J. Chromtogr.* **272**, 7–19.

Proctor, J. T. A. (1974). Color stimulation in attached apples with supplementary light. *Can. J. Plant. Sci.* **54**, 499–503.

Proctor, J. T. A. and Creasy, L. L. (1969). The anthocyanin of the mango fruit. *Phytochemistry* **8**, 2108.

Proctor, J. T. A. and Creasy, L. L. (1971). Effect of supplementary light on anthocyanin synthesis in 'McIntosh' apples. *J. Am. Soc. hort. Sci.* **96**, 523–526.

Pruthi, J. S. and Lal, G. (1958). Carotenoids in passion fruit juice. *Food Res.* **23**, 505–510.

Puech, A. A., Rebeiz, C. A., Catlin, P. B. and Crane, J. C. (1975). Characterization of anthocyanins in the fig (*Ficus carica* L.) fruits. *J. Food Sci.* **40**, 775–779.

Puech, A. A., Rebeiz, C. A. and Crane, J. C. (1976a). Pigment changes associated with application of ethephon ((2-chloroethyl) phosphonic acid) to fig (*Ficus carica* L.) fruits. *Plant Physiol.* **57**, 504–509.

Puech, A. A., Rebeiz, C. A. and Crane, J. C. (1976b). Characterization of major plastid pigments in skin of 'Mission' fig fruits. *J. Am. Soc. hort. Sci.* **101**, 392–394.

Purcell, A. E., Carra, J. H. and de Gruy, I. V. (1963). The development of chromoplasts and carotenoids in colored grapefruit. *J. Rio Grande Valley hort. Soc.* **17**, 123–127.

Purcell, A. E., Young, R. H., Schultz, E. F. and Meredith, F. I. (1968). The effect of artificial climate on the internal fruit color of Redblush grapefruit. *Proc. Am. Soc. hort. Sci.* **92**, 170–178.

Purvis, A. C. (1980). Sequence of chloroplast degreening in calamondin fruit as induced by ethylene and $AgNO_3$. *Plant Physiol.* **66**, 624–627.

Purvis, A. C. and Barmore, C. R. (1981). Involvement of ethylene in chlorophyll degradation on peel of citrus fruit. *Plant Physiol.* **68**, 854–856.

Qureshi, A. A., Andrewes, A. G., Qureshi, N. and Porter, J. W. (1974). The enzymatic conversion of cis-^{14}C phytofluene, trans-^{14}C phytofluene and trans-ζ-^{14}C carotene to more unsaturated acyclic, monocyclic and dicyclic carotenes by a cell free preparation of red tomato fruits. *Arch. Biochem. Biophys.* **162**, 93–107.

Randerath, K. (1963). "Thin Layer Chromatography". Academic Press, New York.

Rasmussen, G. K. (1975). Cellulose activity, endogenous abscisic acid and ethylene in four citrus cultivars during maturation. *Plant Physiol.* **56**, 765–767.

Raymundo, L. C., Griffiths, A. E. and Simpson, K. L. (1967). Dimethyl sulfoxide and biosynthesis of carotenoids in detached tomatoes. *Phytochemistry* **6**, 1527–1532.

Raymundo, L. C., Griffiths, A. E. and Simpson, K. L. (1970). Biosynthesis of carotenoids in the tomato fruit. *Phytochemistry* **9**, 1239–1245.

Raymundo, L. C., Chichester, C. O. and Simpson, K. L. (1976). Light dependent carotenoid synthesis in the tomato fruit. *J. Agric. Food Chem.* **24**, 59–64.

Rebeiz, C. A. and Lascelles, J. (1982). Biosynthesis of pigments in plants and bacteria. *In* "Photosynthesis", Vol. 1, 699–780. Academic Press, London and New York.

Rebeiz, C. A., Belanger, F. C., Freyssinet, G. and Saab D. B. (1980) Chloroplast biogenesis. XXIX. The occurrence of several novel chlorophyll a and b chromophores in higher plants. *Biochim. Biophys. Acta* **590**, 234–247.

Recommended Dietary Allowances (1974). 8th edn, 20–54. Food and Nutrition Board, National Research Council, National Academy of Sciences, Washington, D.C.

Reese, R. L. and Koo, R. C. J. (1975). Effects of N and K fertilization on internal and external fruit quality of three major Florida cultivars. *J. Am. Soc. hort. Sci.* **100**, 425–428.

Reid, M. S. (1975). The role of ethylene in the ripening of some unusual fruits. *Coll. Int. CNRS Paris*, No. 238, 177–182.

Reid, M. S., Lee, T. H., Pratt, H. K. and Chichester, C. O. (1970). Chlorophyll and carotenoid changes in developing muskmelons. *J. Am. Soc. hort. Sci.* **95**, 814–815.

Reitz, H. J. and Koo, R. C. J. (1960). Effect of nitrogen and potassium fertilization on yield, fruit quality and leaf analysis of Valencia orange. *Proc. Am. Soc. hort. Sci.* **75**, 244–252.

Reuther, W. (1973). Climate and citrus behavior. *In* "Citrus Industry", Vol. 3, 280–337.

Reuther, W. and Smith, P. F. (1952). Relation of nitrogen, potassium and magnesium to some fruit qualities of 'Valencia' oranges. *Proc. Am. Soc. hort. Sci.* **59**, 1–12.

Rhodes, M. J. C. and Wooltorton, L. S. C. (1967). The respiration climacteric in apple fruits. The action of hydrolytic enzymes in peel tissue during the climacteric period in fruit detached from the tree. *Phytochemistry* **6**, 1–12.

Ribéreau-Gayon, P. (1967). Etude de mecanismes de synthèse et transformation de l'acide malique, de l'acide tartrique et de l'acide citrique chez *Vitis vinifera*. *Phytochemistry* **7**, 1471–1482.

Ribérau-Gayon, P. (1982). The anthocyanins of grapes and wines. *In* "Anthocyanins as Food Colors" (P. Markakis, Ed.) 209–244. Academic Press, New York.

Rigbey, B., Dana, M. N. and Binning, L. K. (1977). Ethephon sprays and cranberry fruit color. *Hort. Sci.* **7**, 82–83.

Rohrback, R. P. and McClure, W. E. (1978). A production capacity conveyor for small fruit sorting: the M-belt. *Trans. Am. Soc. Agric. Eng.* **21**, 1092–1095.

Romojaro, F., Banet, E. and Llorrente, S. (1979). Carotenoides en flaveda y pulpa de pomelo Marsh. *Rev. Agroquim. Tecn. Aliment.* **19**, 385–392.

Rønneberg, H., Borch, G., Liaaen-Jensen, S., Matsutaka, H. and Matsuno T. (1978). Animal carotenoids. 16. Tunaxanthin. *Acta Chem. Scand.* **B32**, 621–622.

Rosso, S. W. (1967). The ultrastructure of chromoplast development in red tomatoes. *J. Ultrastruct. Res.* **25**, 307–357.

Rotstein, A., Gross, J. and Lifshitz, A. (1972). Changes in the pulp carotenoid pigments of the ripening Shamouti orange. *Lebensm.-Wiss. u. Technol.* **5**, 140–143.

Roy, S. K. (1973). A simple and rapid method for estimation of total carotenoid pigments in mango. *J. Food Sci. Technol.* **10**, 45.

Rüdiger, W., Benz, J. and Guthoff, C. (1980). Detection and partial characterization of activity of chlorophyll synthetase in etioplast membranes. *Eur. J. Biochem.* **109**, 193–200.

Rudnicki, R., Machnik, S. and Pieniazec, J. (1968). Accumulation of abscisic acid during ripening of pears (Clapp's Favourite) in various storage conditions. *Bull. Acad. Pol. Sci.* **16**, 509–512.

Sabater, B. and Rodriguez, T. (1978). Control of chlorophyll in detached leaves of barley and oat through effect of kinetin on chlorophyllase levels. *Physiol. Plant* **43**, 274–276.

Saleh, N. A., Poulton, J. E. and Grisebach, H. (1976). UDP-glucose: Cyanidin 3-O-glucosyltansferase from red cabbage seedlings. *Phytochemistry* **15**, 1865–1868.

Sandhu, S. S. and Dhillon, B. S. (1982). Relation between growth pattern, endogenous growth hormones and metabolites in the developing fruit of 'Sharbati' peach. *Ind. J. Agric. Sci.* **52**, 302–310.

Sanger, J. E. (1971). Quantitative investigations of leaf pigments from their inception in bud through autumn coloration to decomposition in falling leaves. *Ecology* **52**, 1075–1089.

Sapers, G. M., Jones, S. B. and Maher, G. T. (1983). Factors affecting the recovery of juice and anthocyanin from cranberries. *J. Am. Soc. hort. Sci.* **108**, 246–249.

Sapers, G. M., Phillips, J. G., Rudolf, H. M. and Di Vito, A. M. (1983). Cranberry quality: Selection procedure for breeding programs. *J. Am. Soc. hort. Sci.* **108**, 241–246.

Sapozhnikov, D. J., Krasovskaya, T. A. and Mayerskaya, A. N. (1957). Changes observed in the relation between the main carotenoids in the plastids of green leaves exposed to light. *Dokl. Akad. Nauk. U.S.S.R.* **113**, 465–467.

Saquet-Barel, H., Comte, F., Kiepferle, H. and Crouzet, J. (1982) Purification and concentration of anthocyanins by adsorption on porous polymers. *Lebensm.-Wiss. u. Technol.* **15**, 199–202.

Saucedo, C. V., Esparza, F. T. and Lakshminarayana, S. (1977). Effects of refrigerated temperatures on the incidence of chilling injury and ripening quality of mango fruit. *Proc. Fla. St. hort. Soc.* **90**, 205–210.

Schachter, S. (1977). Sulle caratteristiche analitiche dei succhi di mandarino prodotti in Calabria e Basilicata. *Essenze-Deriv. Agrumari.* **47**, 381–399.

Schaller, D. R. and von Elbe, J. H. (1971). The carotenoids in Montmorency cherries. *J. Food Sci.* **36**, 712–713.

Schanderl, S. H. and Lynn, D. Y. C. (1966). Changes in chlorophylls and spectrally related pigments during ripening of *Capsicum frutescens*. *J. Food Sci.* **31**, 141–145.

Schmidt-Stohn, G., Preisel, H. G. and Krebs, O. (1980). Investigations on *Codiaeum varriegatum* var. pictum. II. A method for the quantitative estimation of leaf pigments. *Gartenbauwiss.* **45**, 56–61.

Schneider, H. A. W. (1975). Chlorophylls: Aspects of biosynthesis and its regulation. *Ber. Deutsch. Bot. Ges.* **88**, 83–123.

Schneider, M. M., Hampp, R. and Ziegler, H. (1977). Envelope permeability to possible precursors of carotenoid biosynthesis during chloroplast–chromoplast transformation. *Plant Physiol.* **60**, 518–520.
Schoch, S., Lempert, U., and Rüdiger, W. (1977). On the last step of chlorophyll biosynthesis. Intermediates between chlorophyllide and phytol containing chlorophyll. *Z. Pflanzenphysiol.* **83**, 427–436.
Schulz, H. (1976). Bestimmung des Pflück-und Auslagerungszeitpunktes von Kernobstfrüchten mit Hilfe des Farbumschlages der Grundfarbe von grün nach gelb. *Gartenbau* **23**, 371–374.
Schwartz, S. J. and von Elbe, J. H. (1983). High performance liquid chromatography of plant pigments—A review. *J. Liq. Chromatogr.* **5**, 43–73.
Scott, P. C. and Leopold, A. C. (1968). Opposing effect of gibberellin and ethylene. *Plant Physiol.* **42**, 1021–1022.
Šesták, Z. (1975). Pigments of plastids and photosynthetic chromatophores. *J. Chromatogr. Libr.* **3**, 1039–1049.
Šesták, Z. (1977). Photosynthetic characteristics during ontogenesis of leaves. 1. Chlorophylls. *Photosynthetica* **11**, 367–448.
Shaw, P. E. (1980). Loquat. In "Tropical and Subtropical Fruits" (S. Nagy and P. E. Shaw, Eds) 479–491. AVI, Westport, CT.
Shawa, A. Y. (1979). Effect of ethephon on color abscission and keeping quality of 'McFarlin' cranberry. *Hort. Sci.* **14**, 168–169.
Shimizu, S. and Tanaki, H. (1963). Chlorophyllase of tobacco plants. II. Enzymic phytylation of chlorophyllide and pheophorbide *in vitro*. *Archs. Biochem. Biophys.* **102**, 152–158.
Shimokawa, K. (1976). Preferential degradation of chlorophyll *b* in ethylene treated fruits of 'Satsuma' mandarin. *Sci. Hortic.* **11**, 253–256.
Shimokawa, K. (1979). Preferential degradation of chlorophyll *b* in ethylene-treated fruits of 'Satsuma' mandarins. *Sci. Hortic.* **11**, 253–256.
Shimokawa, K. and Sakanoshita, A. (1977). Ethylene-induced chloroplast senescence of Satsuma mandarin (*Citrus unshiu* Marc). *Bull. Fac. Agric. Miyayawi Univ.* **24**, 27–33.
Shimokawa, K., Sakanoshita, A. and Horiba, K. (1978a). Ethylene induced changes of chloroplast structure in Satsuma mandarin (*Citrus unshiu* Marc). *Plant Cell Physiol.* **19**, 229–236.
Shimokawa, K., Shimada, S. and Yaeo, K. (1978b). Ethylene-enhanced chlorophyllase activity during degreening of *Citrus unshiu* Marc. *Sci. Hort.* **8**, 129–135.
Shomer, I. (1975). "Structure, development and adhesion of juice sacs in citrus fruits segments". Thesis, The Hebrew University, Jerusalem.
Shrikhande, A. J. and Francis, F. J. (1973). Anthocyanin pigments of sour cherries. *J. Food Sci.* **38**, 649–651.
Shulman, Y. and Lavee, S. (1973). The effect of cytokinins and auxins on anthocyanin accumulation in green Manazanillo olives. *J. exp. Bot.* **24**, 665–661.
Shulman, Y. and Lavee, S. (1976). Endogenous cytokinins in maturing Manzanillo olive fruits. *Plant Physiol.* **57**, 490–492.
Siefermann-Harms, D. (1981). The role of carotenoids in chloroplasts of higher plants. In "Biogenesis and Function of Plant Lipids" (P. Mazliak, *et al.*, Eds) 331–340.
Siefermann-Harms, D., Hertzberg, S., Borch, G. and Liaaen-Jensen, S. (1981). Lactucaxanthin, and ε,ε-carotene-3,3'-diol from *Lactuca sativa*. *Phytochemistry* **20**, 85–88.

Sigelman, H. W. and Hendricks, S. B. (1958). Photocontrol of anthocyanin synthesis in apple skin. *Plant Physiol.* **33,** 185–190.
Simpson, D. J. and Lee, T. H. (1976). The fine structure and formation of fibrils of *Capsicum annuum* L. chromoplasts. *Z. Pflanzenphysiol.* **77,** 127–138.
Simpson, K. L. and Chichester, C. O. (1981). Metabolism and nutritional significance of carotenoids. *Ann. Rev. Nutr.* **1,** 351–374.
Simpson, K. L., Lee, J. C., Rodriguez, D. A. and Chichester, C. O. (1976). Metabolism in senescent and stored tissues. *In* "Chemistry and Biochemistry of Plant Pigments" (T. W. Goodwin, Ed.) 780–842. Academic Press, London and New York.
Simpson, D. J., Baqar, M. R. and Lee, T. H. (1977). Fine structure and carotenoid composition of the fibrillar chromoplasts of *Asparagus officinalis* L. *Ann. Bot.* **41,** 1101–1108.
Sistrom, W. R., Griffiths, M. and Stanier, R. Y. (1956). The biology of a photosynthetic bacterium which lacks colored carotenoids. *J. Cell Comp. Physiol.* **48,** 473–515.
Sitte, P. (1977). Chromoplasten-bunte Objekte der modernen Zellbiologie. *Biologie in unserer Zeit.* **7,** 65–74.
Sitte, P., Falk, H. and Liedvogel, B. (1980). Chromoplasts. *In* "Pigments in Plants" (F.-Ch. Czygan, Ed.) 2nd edn, 117–148. Gustav Fischer, Stuttgart.
Smith, J. H. C. and Benitez, A. (1955) Chlorophylls. *In* "Modern Methods of Plant Analysis (K. Paech, and M. V. Tracey, Eds) Vol. IV, 142–196. Springer, Berlin.
Smith, R. H. and Luh, B. S. (1965) Anthocyanin pigments in the hybrid grape variety Rubired. *J. Food Sci.* **30,** 995–1005.
Sonnen, H. D., Lenz, F. and Gross, J. (1979). Influence of root temperature on carotenoid development in the peel of *Citrus unshiu* (Marc) and *Citrus madurensis* (Louv). *Gartenbauwiss.* **44,** 49–52.
Soost, R. K. and Burnett, R. H. (1961). Effect of gibberellin on yield and characteristic of 'Clementine' mandarin. *Proc. Am. Soc. hort. Sci.* **77,** 194–201.
Spayd, S. E. and Morris, J. R. (1980). Prediction of sensory color of strawberry jam. *Ark. Farm. Res.* **29,** 12.
Spurgeon, S. L. and Porter, J. W. (1983). Biosynthesis of carotenoids. *In* "Biosynthesis of Isoprenoid Compounds" (J. W. Porter and S. L. Spurgeon, Eds) 1–122. J. Wiley & Sons, Chichester.
Spurr, A. R. and Harris, W. M. (1968). Ultrastructure of chloroplasts and chromoplasts in *Capsicum annuum.* I. Thylakoid membrane changes during fruit ripening. *Am. J. Bot.* **55,** 1210–1224.
Stahl, E. (1967). "Thin Layer Chromatography. A Laboratory Handbook," 2nd. ed. Springer, Berlin.
Stearns, Ch. and Young, C. T. (1942). The relation of climate conditions to color development in citrus fruit. *Proc. Fla. St. hort. Soc.* **55,** 59–61.
Stewart, I. (1973). Carotenoids in citrus. *In* "Congreso Mundial de Citricultura" (O. Carpena, Ed.) 325–330. Int. Soc. Citricult.
Stewart, I. (1977a). High performance liquid chromatographic determination of provitamin A in orange juice. *J. Assoc. Off. Anal. Chem.* **60,** 132–136.
Stewart, I. (1977b). Provitamin A and carotenoid content of citrus juices. *J. Agric. Food Chem.* **25,** 1132–1137.
Stewart, I. (1980). Color as related to quality in citrus. *In* "Citrus Nutrition and Quality" (S. Nagy and J. A. Attaway, Eds), *ACS Symp. Ser.* 143, 129–149.
Stewart, I. and Leuenberger, U. B. (1976). Citrus color. *Alimenta* **15,** 33–36.

Stewart, I. and Wheaton, T. A. (1971a). Continuous separation of carotenoids by liquid chromatography. *J. Chromatogr.* **55**, 325–336.
Stewart, I. and Wheaton, T. A. (1971b). Effects of ethylene and temperature on carotenoid pigmentation of citrus peel. *Proc. Fla. State hort. Soc.* **84**, 9–11.
Stewart, I. and Wheaton, T. A. (1972). Carotenoids in citrus: their accumulation induced by ethylene. *J. Agr. Food Chem.* **20**, 448–449.
Stewart, I. and Wheaton, T. A. (1973). Conversion of β-citraurin to reticulataxanthin and β-apo-8′carotenal to citranaxanthin during the isolation of carotenoids from citrus. *Phytochemistry* **12**, 2947–2951.
Stewart, P. S., Bailey, P. A. and Beven, J. L. (1983). High-performance liquid chromatographic separation of carotenoids in tobacco and their characterization with the aid of a microcomputer. *J. Chromatogr.* **282**, 589–593.
Strack, D. and Mansell, R. L. (1975). A method for the separation of anthocyanins on polyamide column. *J. Chromatogr.* **109**, 325–331.
Strack, D., Akavia, N. and Reznik, H. (1980). High performance liquid chromatographic identification of anthocyanins. *Z. Naturforsch.* **35C**, 533–538.
Strain, H. H. (1966). Fat-soluble chloroplast pigments: their identification and distribution in various Australian plants. *In* "Biochemistry of Chloroplasts" (T. W. Goodwin, Ed.) Vol. I, 387–406. Academic Press, London and New York.
Strain, H. H. and Sherma, J. (1972). Investigations of the chloroplast pigments of higher plants, green algae and brown algae and their influence upon the invention, modifications and applications of Tswett's chromatographic method. *J. Chromatogr.* **73**, 371–397.
Straub, O. (1976). Key to carotenoids. *In* "Lists of Natural Carotenoids". Birkhäuser, Basel.
Strmiska, F. and Holčiková, K. (1970). Die Veränderungen des Carotinoidgehaltes bei der Lagerung und Verarbeitung von Obst und Gemüse. *Flüss. Obst.* **37**, 46–53.
Subbarayan, C. and Cama, H. R. (1964). Carotenoids in *Carica papaya* (papaya fruit). *Ind. J. Chem.* **2**, 451–454.
Subbarayan, C., Kushwaha, S. C., Suzue, G. and Porter, J. W. (1970). Enzymatic conversion of isopentenyl pyrophosphate-4-^{14}C and phytoene-^{14}C to acyclic carotenes by an ammonia sulfate precipitated spinach enzyme system. *Arch. Biochem. Biophys.* **137**, 547–557.
Sun, B. H. and Francis, F. J. (1968). Apple anthocyanins: identification of cyanidin 7-arabinoside. *J. Food Sci.* **32**, 647–649.
Swift, J. E. and Milborrow, B. V. (1981). Stereochemistry of allene biosynthesis and the formation of the acetylenic carotenoid diadinoxanthin and peridinin (C_{37}) from neoxanthin. *Biochem. J.* **199**, 69–74.
Szabolcs, J. and Rónai, A. (1969). The structure of α-cryptoxanthin and the identity of zeinoxanthin with α-cryptoxanthin. *Acta Chim. Acad. Sci. Hung.* **61**, 309–313.
Takeguchi, C. A. and Yamamoto, H. Y. (1968). Light induced oxygen-18 uptake by epoxy xanthophylls in New Zealand spinach leaves (*Tetragonia expansa*). *Biochim. Biophys. Acta* **153**, 459–465.
Tan, S. C. (1979). Relationships and interactions between phenylalanine ammonia-lyase, phenylalanine ammonia-lyase inactivating system and anthocyanin in apples. *J. Am. Soc. hort. Sci.* **104**, 581–586.
Taylor, R. F. and Ikawa, M. (1986). Gas chromatography, gas-chromatography-mass spectrometry and high-pressure liquid chromatography of carotenoids and retinoids. *In* "Methods in Enzymology" (D. B. Cormick and L. D. Wright, Eds), Vol. 67, 233–267.

Temkin-Gorodeiski, N., Zisman, U., Gur, G., Daos, A. and Grinberg, S. (1983). Degreening of bell peppers for export. *Hasadeh* **64**, 58–63.

Testoni, A. (1979). Effects of growth regulators on persimmon fruits. *Ann. Inst. Sperim. Valor. Tecnol. Prod. Agr.* **10**, 63–70.

Tetley, T. V. (1931). The morphology and cytology of the apple fruit with special reference to the Bramley's seedling variety. *J. Pomol. hort. Sci.* **9**, 278–297.

Thaler, H. and Schulte, K. E. (1940). Die Carotinoide der gelben Pfirsiche. *Biochem. Z.* **306**, 4–5.

Thiman, K. V., Edmondson, Y. H. and Radner, B. S. (1951). The biogenesis of anthocyanin III. The role of sugars in anthocyanin formation. *Arch. Biochem. Biophys.* **34**, 305–323.

Thommen, H. (1962). The occurrence of 8'-apo-β-carotenal in the skin and juice of fresh oranges. *Naturwiss.* **22**, 517–518.

Thomas, J. R. and Gausman, H. W. (1977). Leaf reflectance vs. leaf chlorophyll and carotenoid concentrations of eight crops. *Agron. J.* **69**, 799–802.

Thomas, J. R. and Oerther, G. F. (1972). Estimating nitrogen content of sweet pepper leaves by reflectance measurements. *Agron. J.* **64**, 11–13.

Thomas, P. and Janave, M. T. (1975). Effects of gamma irradiation and storage temperature on carotenoids and ascorbic acid content of mangoes on ripening. *J. Sci. Food Agric.* **26**, 1503–1512.

Thomas, R. L. and Jen, J. J. (1975a). Red light intensity and carotenoid biosynthesis in ripening tomatoes. *J. Food Sci.* **40**, 566–568.

Thomas, R. L. and Jen, J. J. (1975b). Phytochrome-mediated carotenoid biosynthesis in ripening tomatoes. *Plant Physiol.* **56**, 452–453.

Thomson, W. W. (1966) Ultrastructural development of chromoplasts in Valencia oranges. *Bot. Gaz.* **127**, 133–139.

Thomson, W. W. and Whatley, J. M. (1980). Development of non green plastids. *Ann. Rev. Pl. Physiol.* **31**, 375–394.

Thomson, W. W., Lewis, L. N. and Coggins, C. W. (1967). The reversion of chromoplasts to chloroplasts in Valencia oranges. *Cytologia* **32**, 117–124.

Thornber, J. P., Alberte, R. S., Hunter, F. A., Schiozawa, J. A. and Kan, H. S. (1976). The organization of chlorophyll in the plant photosynthetic unit. *In* "Chlorophyll-proteins, Reaction Centers and Photosynthetic Membranes" (J. H. Olson and G. Hind, Eds) Vol. 28, 132–148. Brookhaven National Laboratory, Upton, New York.

Thornber, J. P., Markwell, J. P. and Reinmann, S. (1979). Plant chlorophyll–protein complexes: recent advances. *Photochem. Photobiol.* **29**, 1205–1216.

Timberlake, C. F. (1981). Anthocyanins in fruits and vegetables. *In* "Recent Advances in Biochemistry of Fruits and Vegetables" (J. Friend and M. J. C. Rhodes Eds.) 233–247. Academic Press, New York.

Timberlake, C. F. and Bridle, P. (1971). The anthocyanins of apples and pears: the occurrence of acyl derivatives. *J. Sci. Food Agric.* **22**, 509–513.

Timberlake, C. F. and Bridle, P. (1975). The Anthocyanins. *In* "The Flavonoids" (J. B. Harborne, T. J. Marbry and H. Marbry, Eds), Vol. 1, 214–266. Academic Press, New York.

Timberlake, C. F. and Bridle, P. (1982). Distribution of anthocyanins in food plants. *In* "Anthocyanins as Food Colors" (P. Markakis, Ed.), 126–162. Academic Press, New York.

Todd, G. W., Bean, R. C. and Propst, B. (1961). Photosynthesis and respiration in developing fruits. II. Comparative rates at various stages of development. *Plant Physiol.* **36**, 69–73.

Tomes, M. L. (1963). Temperature inhibition of carotene synthesis in tomato. *Bot. Gaz.* **124**, 180–185.
Tomes, M. L. and Johnson, K. W. (1965). Carotene pigments of an orange-fleshed watermelon. *Proc. Am. Soc. hort. Sci.* **87**, 438–442.
Tomes, M. L., Quackenbush, F. W. and Kargl, T. E. (1956). Action of the gene B in biosynthesis of carotenes in tomato. *Bot. Gaz.* **117**, 248–253.
Tomes, M. L., Johnson, K. W. and Hess, M. (1963). The carotene pigment content of certain red fleshed watermelon. *Proc. Am. Soc. hort. Sci.* **82**, 460–464.
Torre, L. C. and Barritt, B. H. (1977). Quantitative evaluation of *Rubus* fruit anthocyanin pigments. *J. Food Sci.* **42**, 488–490.
Trombly, H. H. and Porter, J. W. (1953). Additional carotenes and a colorless polyene of Lycopersicon species and strains. *Archs Biochem. Biophys.* **43**, 443–457.
Trudel, M. J. and Uzbun, T. L. (1971). Influence of potassium on carotenoid content of tomato fruit. *J. Am. Soc. hort. Sci.* **96**, 763–765.
Tswett, M. (1906). Adsorptionsanalyse und chromatografische Methode: Anwendung auf die Chemie des Chlorophylls. *Ber. Deutsch. Bot. Ges.* **24**, 384–393.
Turrell, F. M., Weber, J. R. and Austin, S. W. (1961). Chlorophyll content and reflection spectra of citrus leaves. *Bot. Gaz.* **123**, 10–15.
Tóth, G. and Szabolcs, J. (1981). Occurrence of some mono-*cis*-isomers of asymmetric C_{40}-carotenoids. *Phytochemistry* **20**, 2411–2415.
Trudel, M. J. and Ozbun, T. L. (1970). Relationship between chlorophylls and carotenoids of ripening tomato fruits as influenced by potassium nutrition. *J. exp. Bot.* **21**, 881–886.
Tsumaki, T., Yamaguchi, M. and Hori F. (1954). The carotenoid contents of the persimmon in ripening. *Sci. Rep. Fac. Sci. Kyushu Univ.* **2**, 35–42.
Umeda, K., Tanaka, Y. and Sator, S. (1971a). Citrus carotenoids. VI. Redder carotenoid group contained in the peel of *Citrus unshiu*. *Chem. Abstr.* **75**, 156573.
Umeda, K., Tanaka, Y. and Ohira K. (1971b) Citrus carotenoids. VII. Carotenoid pattern of *Citrus unshiu* flesh and purity analysis of orange juice. *Chem. Abstr.* **74**, 17183.
Vaccari, A. and Pifferi, P. G. (1978). New solvents for paper and silica gel thin-layer chromatography of anthocyanins. *Chromatographia* **11**, 193–196.
Valadon, L. R. G. and Mummery, R. S. (1967). Carotenoids of some apples. *Ann. Bot.* **31**, 497–503.
Valadon, L. R. G. and Mummery, R. S. (1977). Carotenoids of lilies and of red pepper: biogenesis of capsanthin and capsorubin. *Z. Pflanzenphysiol.* **82**, 407–416.
Vavich, M. G. and Kemmerer, A. R. (1950). The carotenes of cantaloupes. *Food Res.* **15**, 494–497.
Vecchi, M., Englert, G., Maurer, R. and Meduna, V. (1981). Separation and characterization of the *cis*-isomers of β,β-carotene. *Helv. Chim. Acta* **64**, 2746–2758.
Vendrell, M. (1969). Reversion of senescence: effect of 2,4-dichlorphenoxy acid and indole acetic acid on respiration, ethylene production and ripening of banana fruits. *Aust J. biol. Sci.* **22**, 601–610.
Vendrell, H. and Naito, R. (1970). Acceleration and delay of ripening in banana fruit tissue by gibberellic acid. *Aust. J. biol. Sci.* **23**, 553–559.
Verbeek, L. and Lichtenthaler, H. K. (1973). The influence of nitrogen deficiency on the lipoquinone and isoprenoid synthesis of chloroplasts of *Hordeum vulgare* L. *Z. Pflanzenphysiol.* **20**, 245–258.
Vernon, L. P. (1960). Spectrophotometric determination of chlorophylls and phaeophytins in plant extracts. *Anal. Chem.* **32**, 1144–1150.

Vogele, A. C. (1937). Effect of environmental factors upon the color of the tomato and the watermelon. *Plant Physiol.* **12**, 929–955.
von Elbe, J. H. and Schaller, D. R. (1968). Hydrochloric acid in isolating anthocyanin pigments from Montmorency cherries. *J. Food Sci.* **33**, 439–440.
von Elbe, J. H., Bixby, D. G. and Moore, J. D. (1969). Electrophoretic comparison of anthocyanin pigments in eight varieties of sour cherries. *J. Food Sci.* **34**, 113–115.
von Hess, D. and Meyer, C. (1962). Dünnschichtchromatographie von Anthocyanin. *Z. Naturforsch.* **17B**, 853–854.
von Loeseke, H. W. (1929). Quantitative changes in the chloroplast pigments in the peel of bananas during ripening. *J. Am. Chem. Soc.* **51**, 2439–2441.
Vrhovec, B. and Wrischer, M. (1970). The effect of amitrole on the fine structure of developing chloroplasts. *Acta Bot. Croat.* **29**, 43–49.
Wade, N. L. and Brady, C. J. (1971). Effects of kinetin on respiration, ethylene production and ripening of bananas. *Aust. J. Biol. Sci.* **24**, 165–167.
Walles, B. (1972). An electron microscope study on photodestruction of plastid ribosomes in β-carotene-deficient mutants of *Helianthus annuus* L. *Protoplasma* **75**, 215–227.
Watada, A. E. and Abbott, J. A. (1975). Objective methods of estimating anthocyanidin content for determining color grade of grapes. *J. Food Sci.* **40**, 1278–1279.
Weaver, R. J. and McCune, S. B. (1960). Influence of light color development in *Vitis vinifera* grapes. *Am. J. Enol. Vitic.* **11**, 179–184.
Weaver, R. J. and Pool, R. M. (1971). Effect of (2-chloroethyl) phosphonic acid (ethephon) on maturation of *Vitis vinifera* L. *J. Am. Soc. hort. Sci.* **96**, 725–727.
Weedon, B. C. L. (1967). Carotenoids: recent advances. *Chem. Br.* **3**, 424–432.
Weedon, B. C. L. (1971). Stereochemistry. *In* "Carotenoids" (O. Isler, Ed.), 267–323. Birkhäuser, Basel.
Wellmann, E. and Baron, D. (1974). Phytochrome control of enzymes involved in flavonoid synthesis in cell suspension of parsley. *Planta* **119**, 161–164.
Wheaton, T. A. and Stewart, I. (1973). Optimum temperature and ethylene concentrations for postharvest development of carotenoid pigments in citrus. *J. Am. Soc. hort. Sci.* **98**, 337–340.
White, R. C., Jones, I. D. and Gibbs, E. (1963). Determination of chlorophylls, chlorophyllides, pheophytins and pheophorbides in plant material. *J. Food Sci.* **28**, 431–436.
Whitfield, D. M. and Rowan, K. S. (1974) Changes in the chlorophylls and carotenoids of leaves of *Nicotiana tabacum* during senescence. *Phytochemistry* **13**, 77–83.
Wild, A. (1979). Physiologie der Photosynthese höherer Pflanzen. Die Anpassung an die Lichtbedingungen. *Ber. Deutsch. Ges.* **92**, 341–364.
Wilkinson, M., Sweeney, J. G. and Iacobucci, G. A. (1977). High pressure liquid chromatography of anthocyanidins. *J. Chromatogr.* **132**, 349–351.
Williams, M., Hrazdina, G., Wilkinson, M. M., Sweeney, J. G. and Iacobucci, G. (1978). High-pressure liquid chromatographic separation of 3-glucosides, 3.5-diglucosides, 3-(6-O-p-coumaryl) glucosides and 3-(6-O-p-coumaryl) glycoside)-5-glucosides of anthocyanidins. *J. Chromatogr.* **155**, 389–398.
Williams, R. J. H., Britton, G. and Goodwin, T. W. (1967). The biosynthesis of cyclic carotenes. *Biochem. J.* **105**, 99–105.
Willstätter, R. and Everest, A. E. (1913). Anthocyane. I. Uber den Farbstoff der Kornblume. *Liebig's Ann.* **401**, 189–232.
Willstätter, R. and Mieg, W. (1907). Untersuchungen über Chlorophyll. IV. Uber die gelben Begleiter des Chlorophylls. *Ann. Chem.* **355**, 1–28.

Willstätter, R. and Stoll, A. (1913). "Untersuchungen über Chlorophyll", Springer, Berlin.
Winston, J. R. (1955). The coloring or degreening of mature citrus fruits with ethylene. *U.S. Dept of Agric. Circ.* **961**.
Winterstein, A., Studer, A. and Ruegg, R. (1960). New results of carotenoid research. *Chem. Ber.* **93**, 2951–2965.
Woodward, J. R. (1972). Physical and chemical changes in developing strawberry fruits. *J. Sci. Fd Agr.* **23**, 465–473.
Workman, M. (1963). Color and pigment changes in Golden Delicious and Grimes Golden apples. *J. Am. Soc. hort. Sci.* **83**, 149–161.
Wrolstad, R. E. and Putnam, T. P. (1969). Isolation of strawberry anthocyanin pigments by adsorption on insoluble polyvinylpyrrolidone. *J. Food Sci.* **34**, 154–155.
Wrolstad, R. E., Kidrum, K. I. and Amos, J. F. (1970a). Characterization of an additional pigment in extracts of strawberries *Fragaria*. *J. Chromatogr.* **50**, 311–318.
Wrolstad, R. E., Putnam, T. P. and Varseveld, G. W. (1970b). Color quality of frozen strawberries: effect of anthocyanin, pH total acidity and ascorbic acid variability. *J. Food Sci.* **35**, 448–452.
Wulf, L. W. and Nagel, C. W. (1978). High-pressure liquid chromatographic separation of anthocyanins of *Vitis vinifera*. *Am. J. Enol. Vitic.* **29**, 42–49.
Yamamoto, H. Y. (1964). Comparison of the carotenoids in yellow and red-fleshed *Carica papaya*. *Nature* **201**, 1049–1050.
Yamamoto, H. Y., Chichester, C. O. and Nakayama, T. O. M. (1962). Biosynthetic origin of oxygen in the leaf xanthophylls. *Arch. Biochem. Biophys.* **96**, 645–649.
Yamamoto, R., Osima, Y. and Goma, T. (1932). Carotene in mango fruit (*Mangifera indica* L.). *Chem. Abstr.* **27**, 349.
Yokoyama, H. and Vandercook, C. E. (1967). Citrus carotenoids. I. Comparison of carotenoids of mature-green and yellow lemons. *J. Food Sci.* **32**, 42–48.
Yokoyama, H. and White, M. J. (1966). Citrus carotenoids. VI. Carotenoid pigments in the flavedo of Sinton citrangequat. *Phytochemistry* **5**, 1159–1173.
Yokoyama, H. and White, M. J. (1967). Carotenoids in the flavedo of Marsh seedless grapefruit. *J. Agric. Food Chem.* **15**, 693–696.
Yokoyama, H. and White, M. J. (1968). Citrus carotenoids. VIII. The isolation of semi-β-carotenone and β-carotenone from citrus relatives. *Phytochemistry* **7**, 1031–1034.
Yokoyama, H., Coggins, C. W. and Henning, G. L. (1971). The effect of CPTA on the formation of carotenoids in citrus. *Phytochemistry* **10**, 1831–1834.
Yokoyama, H., Debenedict, C., Coggins, C. W. and Hennig, G. L. (1972). Induced color changes in grapefruit and orange. *Phytochemistry* **11**, 1721–1723.
Yokoyama, H., Guerrero, H. C. and Boettger, H. (1972). Recent studies on the structures of citrus carotenoids. *In* "The Chemistry of Plant Pigments" (C. O. Chichester, Ed.), 1–7. Academic Press, New York and London.
Young, L. B. and Erikson, L. C. (1961). Influence of temperature on colour change in Valencia oranges. *Proc. Am. Soc. hort. Sci.* **78**, 178–200.
Zakaria, M., Simpson, K., Brown, P. R. and Krstulovic, A. (1979). Use of reversed-phase high-performance liquid chromatographic analysis for the determination of provitamin A carotenes in tomatoes. *J. Chromatogr.* **176**, 109–117.
Zechmeister, L. (1934). "Carotenoids". Springer, Berlin.
Zechmeister, L. (1962). "*Cis-trans* Isomeric Carotenoids, Vitamins A and Arylpolyenes". Springer, Vienna.

Zechmeister, L. and Tuzson, P. (1930). Der Farbstoff der Wasser-Melone. *Ber. Deutsch. Chem. Ges.* **63,** 2881–2883.

Zechmeister, L. and Tuzson, P. (1933). Mandarin pigment. *Z. Physiol. Chem.* **221,** 278–280.

Zechmeister, L. and Tuzson, P. (1937). Polyene pigment of the orange. II. *Chem. Ber.* **70,** 1966–1969.

Zelles, L. (1967). "Untersuchungen über den Farbstoffgehalt der Schale von Äpfeln und Birnen während der Vegetationsperiode und unter verschiedenen Lagerbedingungen". Dissertation, Rheinische Friedrich-Wilhelms-Universität, Bonn.

Index

Abscisic acid, 55–56
Absolute configuration of carotenoids, 99–101
Absorption spectra/maxima, identification by; *see also* Spectrophotometry
  anthocyanins, 64, 65, 70
  carotenoids, 96–98, 99, 100, 102, 103
    in banana, 194–196
    in carambola, 199
    identification by, 110
  chlorophylls, 2, 4,
    determination, 5–6
    in vivo, 14
Actinidia chinensis, see Chinese gooseberry
Acyclic carotenes, 90, 91
Acylated anthocyanins, 63, 70, 77, 213, 214
Acyl moieties of anthocyanins, 63, 77
  identification, 71, 77
Allenic cartenoids, 90, 96, 116
Allylic apocarotenols, 132–135, 180
Allylic hydroxyl test, 103
δ-Aminolevulinic acid metabolism, 15, 16
Analytical methods
  anthocyanins, 67–71
  carotenoids, 104–110
  chlorophylls, 5–9
  non-destructive, 6–8, 67
Ananas comosus, see Pineapple
Antheraxanthin, *see also* Zeaxanthin, derivatives
  metabolism, 118, 132
Anthocyanidins
  biosynthesis, 71–73
  denomination of, 59, 60
  structure, 61
Anthocyanins
  analysis, 67–71
  biosynthesis, 71–74
  during ripening, 79–81
  in fruits, 74–83, 189–255 *passim*
  characterization, 76–77

  distribution, 189–255 *passim*
  factors affecting, 82–85
  localization, 74–75
  patterns (of distribution), 77, 188
  function, 75
  properties, 64–71
    chemical, 64–67
    spectroscopic, 64, 65, 70
    structure, 59–63
    pH-related changes, 64, 66
Apocarotenals, citrus, 129–131
C_{30}-Apocarotenoids, citrus, 127–129
Apocarotenols, allylic, 132–135, 180
Apples (*Malus domestica*)
  anthocyanin content
    colour development, optimal regimes, 82
    factors affecting, 82, 84
    of various cultivars, 189
  carotenoids, 136, 190–191
    changes during ripening, 167, 169–170, 177, 178, 179, 181, 184–185, 186, 190–191
    factors affecting, 184–185, 186
    maturity index, violaxanthin level, 179
    patterns, 136, 169–170, 190–191
    Golden Delicious, 23, 35, 43, 58, 136, 167, 169, 170, 175, 184, 185, 190, 191
    photosynthetic rates, 35, 43
Apricot (*Prunus armeniaca*)
  carotenoids, 136–137, 192
    during storage, 185
    patterns, 137, 192
Auroxanthin, violaxanthin isomerization to, 101, 102
  structure, 94
Auxins, 56
Averrhoa carambola, see Carambola
Avocado (*Persea americana*)
  anthocyanins, 76, 193
  carotenoids, 137–138, 193
    patterns in leaves, 123
    patterns in peel, 123

Avocado (*Persea americana*) (*cont.*)
  patterns in pulp, 138, 193
  chlorophyll level, variation with depth, 23, 30
  fluorescence emission spectra, 35, 40
  plastid ultrastructure, 30, 32, 33

Banana (*Musa* × *paradisiaca*)
  carotenoids, 127, 138, 194–196
  during ripening, 168
  esterification, 178
  patterns in leaves, 123
  patterns in peels, 123, 195, 196
  patterns in pulp, 138, 194
Biosynthesis,
  anthocyanins, 71–74
    during ripening, 79–81
  carotenoids, 111–116, 124, 126, 129, 166–186
    during ripening, 166–181 *passim*
  chlorophyll, 15–30
    during ripening, 22–30
Blackberry (*Rubus* spp.)
  anthocyanins, 197
  carotenoids, 138, 197
  patterns, 138, 197
Black currant (*Ribes nigrum*)
  anthocyanins, 205
  ethylene effect, 84
Blueberry (*Vaccinium* spp.)
  anthocyanins, 198
  carotenoids, 138, 198
  patterns, 138, 198

Cantaloupe, *see* Melon
Capsicum spp.,
  unique carotenoids; capsanthin, capsorubin, cryptocapsin
  metabolism, 131–132
  structure, 133
Carambola (Star fruit; *Averrhoa carambola*)
  carotenoids, 139, 163, 199
  patterns, 139, 199
    in ripe and unripe fruit, 199
Carbohydrate, *see* Glycoside patterns; Sugar; Sugar moieties
Carbon dioxide metabolism, 37–38
Carica papaya, *see* Papaya

Carotene(s), hydrocarbons
  acyclic, 90, 91
  cyclic
    cyclization mechanism, 115
    formation, 112, 114
    structure, 92–96
α-Carotene
  derivatives, 95
  structure, 95
    in fruits, 123, 189–258, *passim*
    in leaves, 123
β-Carotene
  derivatives, 92–94
  in fruits, 189–258 *passim*
  in chloroplasts, 122
  isomers, absorption spectra, 100
  metabolism
    conversion into vitamin A
      in animals, 119–121
      during fruit ripening, 169, 170, 171, 173
  structure, 92
γ-Carotene
  distribution, 192–257 *passim*
  structure, intermediate in β-carotene biosynthesis, 114, 115
δ-Carotene
  distribution, 227
  structure, intermediate in α-carotene biosynthesis, 114, 115
ζ-Carotene
  absorption spectrum, 97
  accumulation in ripe fruits, 169–171, 173–194
  chromophore, 97, 101
  colour, 98
  in fruits, 189–258 *passim*
  major pigments in carambola, 139, 199, 163
  structure, 91
Carotenemia, 121
Carotenodermia, 121
Carotenogenesis, *see* Biosynthesis
Carotenoids, 87–186
  absorption spectra, *see* Absorption spectra
  analysis, 104–111
  apocarotenoids, 127–135, 165
  biosynthesis, 111–116, 124, 126, 129, 131, 166–181

INDEX

classification, 90
commercial uses, 121
definition and nomenclature, 88–90
distribution, *see* Distribution
factors affecting, 182–186
in fruits, 126–186
function, 116–121
historical background, 87–88
localization
  in chloroplasts, 122, 123, 124
  in chromoplasts, 124, 125, 126
medical use, 121, 122
metabolism
  in animals, 118–121
  in chloroplasts, 122–124
  in chromoplasts, 124–126
patterns of distribution, 135, 136–161 *passim*, 162–163, 164, 169–175, 188, 190–255 *passim*
properties, 92–104
  chemical tests, 101–104
  physical, 92, 96, 98–101
provitamins A, 119, see page 300
recent publications reviewing, 88
semi-systematic names, 91–96, 194–196
structure, 88–96, 128, 130, 132, 133
unique, 127–135, 165
β-Carotenone, 131, 132
Chalcone metabolism, 72, 73
Chalcone synthase activity, 72
De Chaunac grape, 75
Chemical properties
  anthocyanins, 64–67
  carotenoids, 101–104
  chlorophylls, 4
Cherries (*Prunus* spp.)
  sour (*P. cerasus*), 78, 79, 80
    anthocyanins, 78
    changes during ripening, 80, 81, 201
    ethylene effect, 84
    patterns, 76, 78, 79, 200–201
  sweet (*P. avium*), 139, 202
    carotenoids, 139–140, 202
      during development, 202
      patterns, 139, 202
    chlorophylls, 202
Cherry tomato, 37, 160, 167
Chinese gooseberry, *see* Kiwi fruit (*Actinidia chinensis*)

2-Chloroethyl phosphonic acid, *see* Ethephon
Chlorophyll(s), 1–58
  analytical methods, 5–9
  biosynthesis, 15–20
  degradation, 20–22
  during ripening, 24, 27, 28
  energy transfer to, 116, 117
  localization, 9
  in fruits, 22–35, *see also various fruits*
    changes during ripening, 22–30
    content in various fruits, 23
    factors affecting, 42–58
  function, 13
  novel, 20
  properties, 2–9
    chemical, 4
    spectroscopic, 2–4, 14
  ripe fruits containing, 30, 31
  structure, 1–2,3
Chlorophyll *a*
  absorption spectrum, 4
  during ripening, 24–30 *passim*
  properties, 2, 4
  structure, 1–2, 3
Chlorophyll *a/b*-protein complex, 13–14
Chlorophyll *a/b* ratios, 12
  degreening rate related to, 44
  during ripening, 25–30 *passim*
  spectrofluorometric determinations, 6
Chlorophyllase activities, 16–17, 18
  during ripening, 27–29
  ethylene-induced increases, 49–53
Chlorophyll *b*
  absorption spectrum, 4
  during ripening, 24–30 *passim*
  properties, 2, 4
  structure, 2, 3
Chlorophyllide
  metabolism, 16, 17–18, 19
  structure, 19
Chlorophyll-protein complexes, 13–14
  identification, 14
Chlorophyll synthetase activity, 17, 18
Chloroplasts, 9–14, *see also* Plastids
  biosynthetic capabilities, 124
  carotenoids, 122–124, 162, 164, 172–174
  in ripe fruit, 144, 176
  in unripe fruit, 22

Carotenoids (*cont.*)
  chemical composition, 12
  chromoplasts to, transformation, 42, 44–49, 166, 167, 175
  function, 12–13
  regreening, transformation of chromoplasts into chloroplasts in leaves, citrus flavedo, 44, 45
  types, 123, 124
  ultrastructure, 10–11, 21, 32, *35*, *36*, 39, 48
Chromatography, 1, 8–9, *see also* Column chromatography; High performance liquid chromatography; Paper chromatography; Thin layer chromatography (adsorption and partition)
  anthocyanins, 68–70
  carotenids, 106–110
  chlorophylls, 8–9
  historical development, 1, 8–9
  two dimensional, 70
Chromophores of carotenoids, 96–98
Chromoplasts, 124–126, *see also* Plastids
  biosynthetic capabilities, 126
  carotenoid patterns, correlation with ultrastructural changes, 126, 166, 174, 175
  chloroplast to, transformation, 44–49 *passim*, 166, 175–177 *passim*
  to chloroplast transformation, 44–45
  gerontoplast, type of chromoplast in green fleshed muskmelon, 31
  types, 124–126, 177
Cinamic acids, 63
cis-isomers of carotenoids, 99, 180
Citrangequat-kumquat hybrids, carotenoids, 145
Citraurin, 109, 127–128, 135
  biosynthesis, 129, 171
  fruits containing, 145, 147, 148, 165, 226, 227, 229, 235
  epoxides, 128, 129
  ethylene effect, 182
  genetic studies with, 165
  reduction to citraurol, 102, 103
  synthesis, temperature-dependent, 184
Citraurinene, 128, 129
  ethylene effect, 182
  fruits containing, 148, 165, 227
  genetic significance, 165
Citraurol, 129, 135, 165
  semisynthesis of, 102, 103, 129
  structure, 128
Citrullus lanatus, see Watermelon
Citrus fruits, *see also* Grapefruit, Lemon, Mandarins, Oranges, Pummelo, Satsuma mandarins, Tangerine
  carotenoids, 164–166
    colourless carotenes, 163
    during ripening, 166–168, 171, 173, 174–175
    esterification, 177
    factors affecting biosynthesis, 182–186
    unique, 127–132
  degreening, 44
  ethylene action, 45–54 *passim*
  genetic studies, 165–166
  gibberellin action, 54–55
  plastid changes during ripening, 46–49, 50–53, 175–176
  progenitors, 158, 165–166
  regreening, 44–45, 47
  temperature requirements, 57, 184
Classification, *see also* Taxonomy
  of anthocyanin patterns, 77, 188
  of carotenoid patterns, 135, 162–163, 188
Colorants, carotenoids as, 121
Colorimetry, *see* Spectrophotometry
Colour
  carotenoids, absorption spectra related to, 98
  fruit
    anthocyanins, 59, 64, 75
    assessment, 67
    carotenoid content relationship, 162, 178–179
    changes during ripening, 22–24
    development, optimal regimes, 82, 83, 182–185
    external/internal, assessment, 6–8
    fruit load effects on, 186
    improvement, temperature dependent, 184
    nutrient effects on, 58, 83–84, 186
    oxygen effects on, 183

Colour break temperatures, 57
Column chromatography, 8–9, 68–69, 107
  anthocyanins, 68–69
  carotenoids, 107
  chlorophylls, 8–9
Commercial uses of carotenoids, 121
Cranberry (*Vaccinium macrocarpon*)
  anthocyanin distribution, 74, 78, 203
  anthocyanin localization (in Pilgrim), 74
  carotenoids, 140, 204
  patterns, 140, 204
Cryptoflavin, 139
  major pigment in carambola, 139, 199
  provitamin A activity, 139, 199
  structure, 93
α-Cryptoxanthin, 131
  absolute configuration, 131
  provitamin A activity—devoid of, 131
  structure, 95
β-Cryptoxanthin, structure, 93
  in fruits containing pattern (4), 135, 189–258 *passim*
  level, as an index of maturity, in persimmon, 179
  major pigment in *Citrus reticulata*, persimmon, 147, 148, 155, 226–229, 243
  synthesis during ripening, in peach mesocarp, 170
Crystal(s), chromoplasts containing, 125
Crystallization of carotenoids, 92
Cucumis melo, see Melon
Cultivars, differentiation by their pigment patterns, 79, 165
Currants, see Black currant; Red currant
Cyanidin
  absorption spectrum, 64, 65
  structure, 61
Cyanidin 3-(6-*p*-coumarylglucoside)-5-glucoside structure, 63
Cyclic carotenes, *see* Carotenes
Cyclization mechanism in carotenoids, 112, 114–115
  inhibitors, 115
Cycloheximide effects on chlorophyll degradation, 54
Cydonia oblonga, see Quince

Cytokinins, 56
  effects, 56, 85

Dancy tangerine, 23, 108–109, 129, 147, 148, 168, 171, 172, 173, 176, 226, 228–229
Date (*Phoenix dactylifera*) carotenoids, 140–141, 208–209
  during ripening, 208–209
  patterns, 140, 208–209
Definitions concerning carotenoid structure, 88–90
Degradation
  chloroplast carotenoids, during ripening, 169–174
  chlorophyll, 20–22
    during ripening, 22–30 *passim*
    ethylene-enhanced, 44, 54
    temperature effects, 57
Degreening in citrus fruits, 44
Delayed light emission method, 8
Delphinidin
  absorption spectrum, 65
  structure, 61
Densitometry, pigment quantitation by, 71
Determination (of content), 71
  anthocyanins, 68, 71, 77–78
  carotenoids, 105–106
  chlorophyll, 5–8
Developing fruit, photosynthesis in, 35–39, *see also* Ripening
Diospyros kaki, see Persimmon
Distribution of pigments, 187–258, *see also various fruit*
  anthocyanins, 76–83 *passim*; 189–255 *passim*
  patterns, 77, 188
  carotenoids, 122–166, 190–255 *passim*
  patterns, 135, 136–161 *passim*, 162–163, 164, 169–175, 188, 190–255 *passim*
  chlorophyll, 22–35, 202, 207, 211, 212, 215, 220, 222, 226–229, 241, 244, 249, 250, 256
Drugs, carotenoids as, 121, 122

End-groups of carotenes, 89, 90

Energy transfer, carotenoids to chlorophylls, 116–117
Epoxidation reaction in the violaxanthin cycle, 118
Epoxide test, 101, 102
Epoxy carotenoids
  metabolism, 5,6- and 5,8-epoxides, 116
  mixture, characteristic of pattern (5), 135, 163, 188
  structure, 92–96 *passim*
Eriobotrya joponica, see Loquat
Esters
  anthocyanins, see Acylated anthocyanins
  carotenoids
    analysis, omitting saponification, 104, 105
    hydrolysis (saponification), 105
    in banana, 194–196
    in mature fruits, 177
  chlorophylls *a* and *b*, 2, 3, 17
  chlorophyllide-phytyl ester
    biosynthesis, 16–18
    hydrolysis, 4, 20
Esterification
  carotenoids
    hydroxyl test, acetylation, 102–103
    in fruits, during ripening, 177–178, 181
  of chlorophyllide, 16–18
Ethephon (2-chloroethyl phosphonic acid), 43, 45
  treatment with, anthocyanin accumulation promoted by, 84–85, 182
Ethylene, 42–54
  mode of action, 45–49, 84–85, 182
  use, 43, 44
Etioplasts
  avocado, plastid ultrastructure, 30, 33, 137
  "etioplast" carotenoid pattern, 138, 176
Extraction
  anthocyanins, 67–68
  carotenoids, 104–105
  chlorophyll, 5

Fertilizers
  fruit, effects of
    on anthocyanin levels, 84
    on carotenoid levels, 185–186
    on chlorophyll, 58
Ficus carica, see Fig
Fig (*Ficus carica*)
  anthocyanins, 210
  ethephon effect, 84, 85
  carotenoids, 141, 210
  patterns, 141, 210
Flavan nucleus, 60
Flavonoids
  anthocyanins, class of, 59, *see also* Anthocyanins
  anthocyanins, complexing with other, 65–67
  biosynthesis, 71–74
  structure, 59
Fluorescence emission spectroscopy, *see* Spectrofluorometry
Fortunella margarita, see Kumquat
Fragaria × *ananassa*, see Strawberry
Function
  anthocyanins, 75
  carotenoids, 116–121
  chlorophyll, 12–14

Genetic studies, use of carotenoids in, 165–166
Geranylgeraniol, gradual reduction to phytol, 18
  -pyrophosphate, intermediate in carotenoid biosynthesis, 112, 113
Gerontoplasts, green-fleshed muskmelon, 31, 176
Giberellin
  effect, 54–55, 182
  practical use, 55, 85
Glycoside patterns of anthocyanins, 60, 63, 76, 77
  changes during ripening, 81
Golden Delicious apples, 35, 43, 58, 136, 167, 169–170, 179, 184–185, 190–191
Gooseberry (*Ribes grossularia*), *see also* Chinese gooseberry
  anthocyanins, 211
  carotenoids, 141, 211–212
    during development, 212
    patterns, 141, 211, 212

chlorophyll, 211, 212
Grana (thylakoids), 9–11, 32, 34, 36, 45–47, 49, 53, 167, 175, 176
Grape(s) (*Vitis vinifera*)
  anthocyanin distribution/content, 70, 75, 76, 77–79, 213–214
    factors affecting, 83, 84
    patterns, use in taxonomy, 78–79
  carotenoids, 141, 215
    patterns, 141, 215
  chlorophyll, 215
  De Chaunac, 75
Grapefruit (*Citrus paradisi*) white and coloured
  coloured (pink/red), 142–143, 169, 175, 217–218
    albedo, 218
    during ripening, 175
    peel (flavedo), 217
    pulp, 217, 218
    septa, 218
    temperature effects, 184
  white
    carotenoids, 142, 216
    during ripening, 169, 174, 175
    flavedo, 216
    flavedo, plastid ultrastructure during ripening, 45, 48–49
    patterns, 142, 169, 174, 175, 216
    pulp, 216
  progenitors, 158, 165–166
Guava (*Psidium guajava*) carotenoids, 143–144, 219
  patterns, 144, 219

Harvest, post, *see also* Storage
  carotenoids
    ripening after, 181
    in apple, 191
    in persimmon, 245
  temperature effects after, 57, 184–185
High performance/pressure liquid chromatography, 9
  anthocyanins, 70
  carotenoids, 108, 110
  chlorophyll, 9
  reversed-phase, 108, 109–110
Hormones, *see* Phytohormones
HPLC, *see* High performance/pressure liquid chromatography
Hues, 7
Hybrids, citrus, 147, 148, 158, 165–166, 227
  progenitors of, 158
Hydrolysis
  anthocyanins, 17
  carotenoids, saponification, 105
  chlorophyll, 4
Hydroxyl test, carotenoids, 102–103

Identification
  anthocyanins, sugar moieties, 70–71
  carotenoids, 110
Indole acetic acid, 56
Ion-exchange resins, anthocyanin separation, 68–69
β-Ionone carotenoid series, 92–94
ε-Ionone carotenoid series, 95
Isolation, *see* Extraction
Isomerization of carotenoids, 99, 101
Isomers of carotenoids, 99–101 *passim*
  cis-, 180
  identification, 110
Isopentenylphyrophosphate metabolism, 111–112, 113
Isoprene units, joining, 88, 89

Japanese plum, *see* Loquat

Ketocarotenoids, *Capsicum*, 131–132, 133
Kinetin, 56
Kiwifruit (*Actinidia chinensis*)
  carotenoids, 144, 220
    during ripening, 220
    patterns, 144, 220
  chlorophylls
    content during ripening, 31, 220
    uneven distribution in pericarp, 23, 31, 33
  chloroplast ultrastructure, 33, 39
  fluorescence emission spectra
    leaves, 14, 15
    fruit, 35, 41, 42
Kumquat (*Fortunella margarita*), 144–145

Kumquat (cont.)
  carotenoids, 145, 221
    during ripening, 171, 222
    patterns, 145, 171, 221
    and pulp and peel, 221
  chlorophyll content, 222
  chloroplasts of flavedo, 10
  citrangequat, hybrid of, 145

Lambert–Beer law, 5, 68
Lamellar body, during ripening, 176
  in citrus pulp chromoplasts, 176
Lemon (*Citrus limon*)
  carotenoids, 146, 223
    during ripening, 223
    patterns, 146, 223
    pulp and peel, 223
  progenitors of, 158, 165–166
  ripening pattern, 169
  yellowing and regreening, ultrastructural changes in, 44, 46–47
Light
  damage by, protective effects of carotenoids, 117
  effects of
    on anthocyanin content, 82–83
    on carotenoid content, 182–183
    on chlorphyll content, 56–57
  transmittance/reflectance, measure of fruit colour, 7–8
Litchi (*Litchi chinensis*)
  anthocyanin changes during ripening, 81, 224
  3,5-diglucosides, content, 76
Load, fruit of tree, effects on fruit colour, 186
Localization
  anthocyanins, 74–75
  carotenoids
    in chloroplasts, 122–124
    in chromoplasts, 124–126
  chlorophylls
    in chloroplasts of
      green leaves, 9
      unripe fruits, 122
      green fleshed pulp of ripe fruits, 30–35
Loquat (Japanese plum; *Eriobotrya japonica*)
  carotenoids, 146–147, 225

  patterns, 147, 225
  pulp and peel, 225
Lutein (dihydroxy-α-carotene)
  biosynthesis, 115
  chromatography, TLC, 109
  distribution
    in fruits containing chloroplast carotenoids, pattern (2), 136–160 *passim*, 190–256 *passim*
  most widespread xanthophyll, 90
  predominating carotenoid in chloroplasts, 122–124
  derivatives
    Chrysanthemaxanthin, (5,8-epoxide of),
    distribution, 123, 137, 176, 196–256 *passim*
    structure, 95
  isolutein, (5,6-epoxide of)
    biosynthesis, 116
    distribution, 123, 176, 190–256 *passim*
    during ripening, 173
    structure, 95
  during ripening, 169, 170, 173, 174
  esters, 178, 194–196
  stereoisomers, 101
  structure, 95
Lycopene
  absorption spectrum, 97
  biosynthesis of, 91, 112, 115
  crystals of, in chromoplasts, 125
  cyclization of, 112, 114, 115
  during ripening, 171, 175, 257
  factors affecting, 183–186
  fruits abundant in, 142, 143, 155, 158, 160, 160, 219
  in others, 217–258 *passim*
  prototype of carotenoids, 88–89
  poly-*cis*-lycopene (prolycopene)
    in fruits containing pattern (7), 135
    in mutants
      orange-fleshed watermelon, 167, 258
      tangerine tomato, 160
    systematic name, 160
    xanthophylls of, 160, 171

Maize leaf chloroplasts, 10

Malus domestica, see Apples
Malvidin, structure, 61
Mandarin (Tangerine; *Citrus reticulata*), *see also* Satsuma mandarin
  carotenoids, 147–148, 226–229
    chromatography, 108, 109
    during ripening, 24, 26, 168, 171, 172, 226–229
    ethylene effects, 44, 54
    on ultrastructure, 45, 46, 49
    patterns, 148, 226–229
    pulp and peel, 227
  chlorophyll content, 23, 24, 226, 227, 228
  Clementine, 129, 148, 165, 171, 226
  Dancy (tangerine), 23, 108, 109, 129, 148, 168, 171, 172, 173, 176, 226, 228–229
  Michal, hybrid of, 129, 148, 171, 227
  Robinson (tangerine), 54, 147, 148, 165
  source of citraurin and citraurinene, 129
Mango (*Mangifera indica*)
  anthocyanins, 230
  carotenoids, 148–149, 231
    during ripening, 231
    patterns, 149, 231
    temperature effects during ripening, 185
Manzanillo olives
  anthocyanins, 81, 234
Maturation, see Ripening
Maturity, index of fruits
  level of anthocyanins, 79
  level of carotenoids, 178–180
Measurement of pigment content, *see* Determination
Medicine, carotenoids in, 121–122
Melon (cantaloupe, muskmelon, *Cucumis melo*, green-fleshed, orange-fleshed), *see also* Watermelon
  green-fleshed
    carotenoids (mesocarp and exocarp), 149, 233
      during ripening, 25, 169, 172, 233
      pattern (2), chloroplast carotenoids, 149, 233
    chlorophyll content (peel), 11
      during ripening, 24, 25
      variation with depth (pulp), 23
    plastid ultrastructure (peel), 11
      changes during ripening (mesocarp and exocarp), 30, 31, 34, 36, 38, 176
    fluorescence emission spectra, 35, 40
  orange-fleshed
    carotenoids, 149, 150, 232
Membranes, achlorophyllous, 45, 175
Metabolism, *see* Biosynthesis; Degradation
Mevalonic acid metabolism, 111–112
Mimulaxanthin, 137, 193
Musa × paradisiaca, see Banana
Muskmelon, *see* Melon

Neochrome (allenic triol), 90
  distribution 123, 136–160 *passim*, 190–256 *passim*
  esters, 194–196
  structure, 96
Neoxanthin (allenic triol) 90
  chloroplast carotenoid constituent, 122
  distribution, 123, 136–160 *passim*, 190–256 *passim*
  esters, 178, 194–196
  esterification, 178
  formation from violaxanthin, 116
  stereochemistry, 116
  structure, 96
Nitrogen, effects of
  on anthocyanin levels, 84
  on carotenoid levels, 185, 186
  on chlorophyll levels, 58
  on regreening, 45

Olea europaea, see Olives
Olive (*Olea europaea*)
  anthocyanins, 76, 234
    changes during ripening, 81
    cytokinins, effect, 85
    ethylene effect, 84
  Manzanillo, 81
Optical activity of carotenoids, 99–101
Oranges (*Citrus sinensis*)
  anthocyanins, 75, 234
  carotenoids, 150–151, 235
    during growth/maturation/ripening, 28, 166, 167, 168, 172, 177, 181

Oranges (*Citrus sinensis*) (*cont.*)
   esters, 177
   patterns (pulp and peel), 151, 235
   plastids, 175
   xanthophyll esterification, 177
chlorophyll, during growth, maturation, 25, 26, 28
   degreening, 44
   ethylene effect, 44, 45
   gibberellin effects, 55
   regreening, 44, 45
   Shamouti, 25, 28, 55, 131, 150, 151, 172, 177
Oxidation of carotenoids, 135
   in vitro, 135
   in vivo, 135, 137, 180
Oxygen effects on colour, 183

P_{700}-chlorophyll *a*-protein complex, 13, 14
Papaya (*Carica papaya*)
   carotenoids, 151, 236
   patterns, 151, 236
Paper chromatography
   anthocyanins, 69–70
   carotenoids, 107
Partition chromatography
   carotenoids, 107
   chlorophylls, 9
Passiflora edulis, see Passion Fruit
Passion fruit (*Passiflora edulis*)
   anthocyanins, 237
   carotenoids, 152, 237
   patterns, 152, 237
Peach (*Prunus persica*)
   anthocyanins, 76, 238
   carotenoids, 152–153, 239
      during ripening, 29, 170, 239
      during storage, 185
      esterification, 178
      patterns, 153, 239
   chlorophylls, changes during growth/maturation, 26, 29
Pear (*Pyrus communis*)
   anthocyanins, 77, 240
   carotenoids, 153, 241
      patterns, 153, 241
   unripe and ripe, 241
      chlorophylls, 24, 241

   during ripening, temperature effect, 57
Peel, skin (exocarp), flavedo
   carotenoid content, 127, 136–161 *passim*, 169–171, 179, 181, 190–256 *passim*
   chlorophyll content, 22, 23, 24, 26, 28, 31, 44, 45, 54, 55
   plastid ultrastructure, 10, 11, 38, 45, 48–53
Poenidin structure, 61
Pepper (*Capsicum* spp.)
   carotenoids, 131–132, 133, 153–154, 242
      after storage, 185
      biosynthesis, 131–132
      during ripening, 172, 178, 185, 242
      esterification, 178
      ethylene effect, 185
      patterns, 154, 242
      plastid ultrastructure, 125, 166, 172, 175
      temperature effect, 185
      unique, 127, 165
      structure, 133
   hot (*C. frutescens*), 131, 153–154
      chlorophyll degradation products, 29, 30
   sweet bell (*C. annuum*), 131, 153–154, 242
   use in carotenoid biosynthesis investigation, 111, 112, 126, 132
Perlargonidin
   absorption spectrum, 64, 65
   structure, 61
Persea americana, see Avocado
Persicachrome, 133
   structure, 134
Persicaxanthin, 133
   absorption spectrum, 97
   structure, 134
Persimmon (*Diospyros kaki*), Japanese persimmon
   carotenoids, 154–155, 243–245
      classification, according lycopone content, 155
      during ripening, 171–172, 179, 181, 244, 245
      maturity index, cryptoxanthin level, 179
      patterns (of distribution), 155, 171–

172, 243–245
    pulp and peel, 245
    chlorophyll, 244
Petunidin structure, 61
Phase separation *see* Solvent partition of cartenoids
pH changes, anthocyanin structural changes associated with, 64–66
Phenylalanine ammonia lyase activity, 72
    anthocyanins accumulation correlated with, 82–83
Phenylalanine ammonia lyase inactivity system, 82, 83
Phenylpropanoid metabolism, 71, 72
Phoenix dactylifera, *see* Date
Phosphoenolpyruvate carboxykinase in developing fruit, 37, 39
Phosphoenolpyruvate carboxylase in developing fruit, 37, 39
Photodensitometry, *see* Densitometry
Photoprotection by carotenoids, 117
Photosynthesis, 12–14
    carotenoid roles, 116–117
    in developing fruit, 35–39, 43
Photosystem I, 13
    β-carotene, main carotenoid, 124
Photosystem II, 13, 13–14
    β-carotene, main carotenoid, 124
    in ripe or unripe fruit, 35
Phytoene metabolism, 112
    accumulation in ripe fruits, 169, 174
    distribution in fruits, 192–258, *passim*
    major pigment, in citrus, tomato, orange flesh watermelon, 216, 217, 257, 258
    structure, 91
Phytofluene
    absorption maximum, 97
    accumulation in ripe fruits, 169, 121, 173, 174, 177
    chromophore, 97
    distribution in fruits, 191–258, *passim*
    major pigment in carambola, citrus, quince, 199, 216–218, 223, 249, 250, 252
    metabolism, 112
    structure, 91
Phytohormones
    delaying ripening, 54–55, 56, 85, 182
    promoting ripening, 42–49, 54, 55–56, 84–85, 182
Phytol metabolism, 17–18
Pilgrim cranberry, 74
Pineapple (*Ananas comosus*)
    carotenoids, 156, 246
    patterns, 156, 246
Plastids, *see also* etioplasts, chloroplasts, chromoplasts, gerontoplasts
    avocado, 30, 32, 33
    ultrastructural changes
        ethylene-induced, 45–49
        in ripe/ripening fruit, 30, 31, 32, 33, 38, 45, 50, 53, 160, 175–177
        in yellowing and regreening, 21, 44, 46–47
    grapefruit, 45, 48–49
    lemon, 44, 46–47
    muskmelon, green-fleshed, 31, 34, 36, 38
    pummelo, 45, 50–53
    tomato, 160
Plastoglobules
    accumulation of xanthophyll esters in, 177
    appearance during ripening, 175, 176
    characteristic features of globulous chromoplasts, 125
    in chloroplasts, 10, 11
    in leaves, 21, 22
    in plastids of fruits, 31, 33–39, 45–53
Plum (*Prunus domestica*)
    anthocyanins, 246
    carotenoids, 156–157, 247
    patterns, 157, 165, 247
    cultivar Saghiv, source of persicaxanthin and persicachrome, 156, 247
    subspecies, 156
Polarity
    anthocyanins, 69
    carotenoids, 98
    chlorophylls, 1
    migration in chromatography, determined by, 107
Pomegranate (*Punica granatum*)
    anthocyanins, 76, 248
    carotenoids, 157, 248
    patterns, 157, 248

Porphobilinogen metabolism, 15, 16
Porphyrin
  overproduction in porphyria, 122
  structure, 1, 2
Potassium fertilization effect on fruit colour, 58, 185
Prolamellar body
  avocado ultrastructure, 30, 32, 33, 176
Prolycopene, see Lycopene
Protochlorophyll conversion to chlorophyll, 30
Protochlorophyllide
  metabolism, 15–16
  structure, 19
Protoporphyrin
  metabolism, 15–16
  structure, 18
Provitamins A (carotenoids)
  β-carotene highest vitamin A activity, 119
  calculation of vitamin A activity, 120
  conversion to vitamin A, 120
  factors influencing, 119
  physiological function of, 118
  splitting enzyme 15,15'-dioxygenase, 119
  structural characteristic necessary for, 118
Prunus spp., see Apricot; Cherries, Peach; Plum
Psidium guajava, see Guava
Pulp (flesh, endocarp, mesocarp, pericarp)
  carotenoid content, 25, 29, 127, 136–161 *passim*, 192–258 *passim*
    during ripening, 169, 171, 172, 174, 176
  chlorophyll content, 22–25, 29–31
  fluorescence emission spectra, 40, 41
  plastid ultrastructure, 31–34, 39
Pummelo (shaddock; *Citrus grandis*)
  white
    carotenoids, 157, 158, 249, 250
      during ripening
        accumulation of phytofluene, 158, 249, 250
        in pulp, 249
        in peel, 169, 250
      patterns, 157, 158, 249, 250
    chlorophyll content, 23, 249, 250

  hybrids of, 158, 165, 166
  plastid ultrastructural changes, 45, 50–53, 177
  red
    carotenoids
      in pulp and septa; albedo and flavedo, 158, 159, 251
    patterns, 158, 159, 251
Punica granatum, see Pomegranate
Purification, see aponification, Chromatography, Crystallization, Removal of Sterols
Pyrrole structure, 2
Pyrus communis, see Pear

Quality parameters, fruit, 6
Quince (*Cydonia oblonga*)
  carotenoids, 159, 252
  patterns, 159, 252
  phytofluene, major pigment, 159, 252

Raspberry (*Rubus idaeus*)
  anthocyanins, 252, 254
    content, 78
    during ripening, 81, 223
    ethylene effect, 84
  patterns, 76, 79, 254
Red currant (*Ribes rubrum*)
  anthocyanins, 205–206
    content, 78
    during maturation/ripening, 81, 206
    patterns, 76, 79, 81, 205–206
  carotenoids, 140, 207
    during maturation/ripening, 207
    patterns, 140, 207
  chlorophyll content, 207
Reduction of carbonyl containing carotenoids, 102, 103, 129, 135
Reflectance, fruit colour measurement, 7–8
Regreening
  in citrus fruits, 44–45, 47
  leaf, 21, 22
Retinol, see Vitamin A
Retinol equivalent, 119, 120, 136–161 *passim*, 163, 188–258 *passim*
Ribes spp., see Black currant; Gooseberry; Red currant

Ripe fruit
  carotenoids in, 136–163 *passim*
  chlorophyll containing, 22, 23, 30–35
Ripening, *see also various fruit*
  changes during
    in anthocyanins, 79–81
    in carotenoids, 25–29 *passim*, 166–181
    in chlorophyll, 22–30
  phytohormones, role in, 42–56, 84–85, 182
Ripening curves, variation of total carotenoid content, 25–29, 167–169
Robinson tangerine, *see* Mandarin
Root temperature effects on colour development, 57, 184
Rosaceae
  anthocyanin patterns in, 78
  carotenoid patterns in, 164
Rubus spp., *see* Blackberry; Raspberry

Saponification of carotenoids, 105
Saponfication test in bananas, 194–196
Satsuma mandarin (*Citrus unshiv*), *see also* Mandarins
  carotenoids, 147
    during ripening, 26, 167, 184
    patterns, 147
    temperature effects, 184
  chlorophyll changes during ripening, 26, 27
  degreening, 44, 54
  plastid changes, ethylene-induced, 45, 49
  temperature effect, 57
Semi-β-carotenone, 131, 132
Separation (of pigments), 8–9, *see* Chromatographic methods
  anthocyanins, 68–70
  carotenoids, 106–110
  chlorophylls, 8–9
Shaddock, *see* Pummelo
Shamoutichrome, shamoutixanthin (Tetrols; unique citrus carotenoids), 131, 132
  absolute configuration, 131
  distribution, 217–249 *passim*
Shamouti orange, 25, 28, 55, 150, 172, 235

Silylation of carotenoid hydroxyl groups, 103
Sinensiachrome, sinensiaxanthin (allylic apocarotenols), 132
  structures, 134
  in fruits, 133, 135
Solvent partition of carotenoids, 106
Solubility of carotenoids, 98
Sour cherry (*Prunus cerasus*), *see* Cherry
Spectra, absorption, *see* Absorption spectra
Spectrofluorometry (low temperatures fluorescence emission spectra measurement), 6
  chlorophylls determination using, 6
  chlorophyll-protein complexes of photosystems I and II, characterization, 14
  in ripe and unripe kiwifruit, avocado, cantaloupe, 35, 40–42
Spectrophotometric, determination of, *see also* Absorption spectra
  anthocyanins, 68
  carotenoids, 105–106
  chlorophyll, 5–6
Star fruit, *see* Carambola
Stereochemistry of carotenoids, 98–101
Sterols, removal, 106
Storage, *see also* Harvest (post)
Strawberry (*Fragaria* × *ananassa*)
  anthocyanins, 78, 255
    during ripening, 24, 27
  carotenoids, 159–160, 256
    during ripening, 24, 27, 168
    patterns, 159–160, 256
  chlorophylls during ripening, 24, 27, 256
Stroma matrix, filling the plastids, 9, 11, 47, 124
Structure
  anthocyanins, 59–63
  carotenoids, 88–96, 128, 130, 132–134
  chlorophyll, 1–2, 3
Sugar, effects on anthocyanin levels, 83–84
Sugar moieties of anthocyanins, 60, 62
  distribution in fruits, 76
  identification, 70–71
  variation in anthocyanin patterns, 77
Sweet cherry (*Prunus avium*), see Cherry

Tangerine, *see* Mandarin
Taxonomy, pigments, markers in
  anthocyanins, 78–79
  carotenoids, 163–166
Temperature effects
  on anthocyanin content, 83
  on carotenoid content, 183–185
  on chlorophyll content, 57
Thin layer chromatography
  anthocyanins, 70
  carotenoids, 107–108
  chlorophylls, 9
Thylakoids, 9–11, *see also* Grana
  carotenoids in, 124
Tomato (*Lycopersicon esculentum*)
  carotenoids, 127, 160, 257
    during ripening, 167, 171, 181, 183, 257
    factors affecting, 54, 56, 58, 182–186
    metabolism, 112, 167, 171, 181, 257
    patterns, 160, 161, 171, 181, 257
  cherry-, 24, 37, 124, 125, 160, 167, 171
  chlorophyll content during ripening, 24, 58, 167
  colour assessment, 7–8
  colour mutants, 160
  plastid ultrastructure, 124, 125, 160, 166
  photosynthesis, 37
  use in biosynthesis investigations, 111, 112, 115, 126, 160

Unique carotenoids (citrus and citrus relatives), 131, 132
Uroporphyrinogen (Urogen)
  metabolism, 15
  structure, 17
Use of carotenoids, 121–122

Vaccinium spp., *see* Blueberry; Cranberry
Violaxanthin (Zeaxanthin diepoxide)
  esters, 178
  fruits containing, 136–160 *passim*, 163, 190–256 *passim*
  isomerization to auroxanthin (epoxide test), 101, 102
  levels, as an index of maturity, 179
  metabolism, 116, 117–118
    during ripening, 169, 171–174, 178, 179
  structure, 194
Violaxanthin cycle, 117–118
Violeoxanthin (9-*cis*-isomer), 180
Vitamin A (retinol)
  calculation of vitamin A value, 120
  conversion of provitamins A (carotenoids) into, 119
  structure, 118, 120
  unit (retinol equivalent), 119
  values of various fruits, 136–161 *passim*, 163, 191–258 *passim*
Vitis vinifera, *see* Grapes

Watermelon (*Citrullus lanatus*)
  carotenoids, 161, 258
    patterns, 161, 163, 178, 258
    lycopene, major pigment in red fleshed, 161
    prolycopene, major pigment in orange fleshed, 161

Xanthophylls (oxygenated derivatives of carotenes)
  during ripening, 177–178
  esterification, 177–178
  metabolism, 115–116, 117–118, 177–178
  in various fruits, 137, 143
  structure, 90, 92–96, 128, 130, 132–134

Yellowing
  during fruit ripening, 166, 169, 173–174, 177
  in lemons, 44, 46
  leaf, chloroplast structure during, 21, 22

Zeaxanthin (dihydroxy-β-carotene)
  biosynthesis, 115–116
  chromatography, 109
  derivatives (epoxides), 94
  in fruits containing pattern (5), 188–256

distribution
  minor pigment in chloroplasts, 122, 123
  in fruits containing pattern (4), 135, 136–159 *passim*, 163, 188–256 *passim*
  during ripening, 171, 172, 174
  in violaxanthin cycle, 118, 122
structure, 90, 94

FOOD SCIENCE AND TECHNOLOGY

A SERIES OF MONOGRAPHS

Maynard A. Amerine, Rose Marie Pangborn, and Edward B. Roessler, *Principles of Sensory Evaluation of Food.* 1965.
Martin Glicksman, *Gum Technology in the Food Industry.* 1970.
L. A. Goldblatt, *Aflatoxin.* 1970.
Maynard A. Joslyn, *Methods in Food Analysis*, second edition. 1970.
A. C. Hulme (ed.), *The Biochemistry of Fruits and Their Products.* Volume 1—1970. Volume 2—1971.
G. Ohloff and A. F. Thomas, *Gustation and Olfaction.* 1971.
C. R. Stumbo, *Thermobacteriology in Food Processing*, second edition. 1973.
Irvin E. Liener (ed.), *Toxic Constituents of Animal Foodstuffs.* 1974.
Aaron M. Altschul (ed.), *New Protein Foods*: Volume 1, Technology, Part A—1974. Volume 2, Technology, Part B—1976. Volume 3, Animal Protein Supplies, Part A—1978. Volume 4, Animal Protein Supplies, Part B—1981. Volume 5, Seed Storage Proteins—1985.
S. A. Goldblith, L. Rey, and W. W. Rothmayr, *Freeze Drying and Advanced Food Technology.* 1975.
R. B. Duckworth (ed.), *Water Relations of Food.* 1975.
Gerald Reed (ed.), *Enzymes in Food Processing*, second edition. 1975.
A. G. Ward and A. Courts (eds.), *The Science and Technology of Gelatin.* 1976.
John A. Troller and J. H. B. Christian, *Water Activity and Food.* 1978.
A. E. Bender, *Food Processing and Nutrition.* 1978.
D. R. Osborne and P. Voogt, *The Analysis of Nutrients in Foods.* 1978.
Marcel Loncin and R. L. Merson, *Food Engineering: Principles and Selected Applications.* 1979.
Hans Riemann and Frank L. Bryan (eds.), *Food-Borne Infections and Intoxications*, second edition. 1979.
N. A. Michael Eskin, *Plant Pigments, Flavors and Textures: The Chemistry and Biochemistry of Selected Compounds.* 1979.
J. G. Vaughan (ed.), *Food Microscopy.* 1979.
J. R. A. Pollock (ed.), *Brewing Science*, Volume 1—1979. Volume 2—1980, Volume 3—1987.
Irvin E. Liener (ed.), *Toxic Constituents of Plant Foodstuffs*, second edition. 1980.
J. Christopher Bauernfeind (ed.), *Carotenoids as Colorants and Vitamin A Precursors: Technological and Nutritional Applications.* 1981.

Pericles Markakis (ed.), *Anthocyanins as Food Colors*. 1982.
Vernal S. Packard, *Human Milk and Infant Formula*. 1982.
Geoge F. Stewart and Maynard A. Amerine, *Introduction to Food Science and Technology*, second edition. 1982.
Malcolm C. Bourne, *Food Texture and Viscosity: Concept and Measurement*, 1982.
R. Macrae (ed.), *HPLC in Food Analysis*. 1982.
Héctor A. Iglesias and Jorge Chirife, *Handbook of Food Isotherms: Water Sorption Parameters for Food and Food Components*. 1982.
John A. Troller, *Sanitation in Food Processing*. 1983.
Colin Dennis (ed.), *Post-Harvest Pathology of Fruits and Vegetables*. 1983.
P. J. Barnes (ed.), *Lipids in Cereal Technology*. 1983.
George Charalambous (ed.), *Analysis of Foods and Beverages: Modern Techniques*. 1984.
David Pimentel and Carl W. Hall, *Food and Energy Resources*. 1984.
Joe M. Regenstein and Carrie E. Regenstein, *Food Protein Chemistry: An Introduction for Food Scientists*. 1984.
R. Paul Singh and Dennis R. Heldman, *Introduction to Food Engineering*. 1984.
Maximo C. Gacula, Jr., and Jagbir Singh, *Statistical Methods in Food and Consumer Research*. 1984.
S. M. Herschdoerfer (ed.), *Quality Control in the Food Industry*, second edition. Volume 1—1984. Volume 2—1985. Volume 3—1986. Volume 4—1987.
Y. Pomeranz, *Functional Properties of Food Components*. 1985.
Herbert Stone and Joel L. Sidel, *Sensory Evaluation Practices*. 1985.
Fergus M. Clydesdale and Kathryn L. Wiemer (eds.), *Iron Fortification of Foods*. 1985.
John I. Pitt and Ailsa D. Hocking, *Fungi and Food Spoilage*. 1985.
Robert V. Decareau, *Microwaves in the Food Processing Industry*. 1985.
F. E. Cunningham and N. A. Cox (eds.), *Microbiology of Poultry Meat Products*. 1986.
Walter M. Urbain, *Food Irradiation*. 1986.
Peter J. Bechtel, *Muscle as Food*. 1986.
H. W.-S. Chan, *Autoxidation of Unsaturated Lipids*. 1987.
J. Gross, *Pigments in Fruits*. 1987.
P. Krogh, *Mycotoxins in Food*. 1987.
J. M. V. Blanshard and P. Lillford, *Food Structure and Behaviour*. 1987.